Camaro
RESTORATION HANDBOOK

Ground-Up or Sectional Restoration Tips & Techniques
for 1967 to 1981 Camaros. All Models Included.

by Tom Currao & Ron Sessions

HPBooks

HPBooks are published by
The Berkley Publishing Group
A division of Penguin Putnam Inc.
375 Hudson Street, New York, New York 10014.

Printed in the United States of America

24   23   22   21   20   19   18   17

The Penguin Putnam Inc. World Wide Web site address is
http://www.penguinputnam.com

CAMARO is a registered trademark used with permission from General Motors Corporation. Although the cooperation of the Chevrolet Division of General Motors Corporation is gratefully acknowledged, this book is a wholly independent production of the publisher.

Cover photo by Scott Dahlquist. Interior photos by Tom Currao and Ron Sessions unless otherwise noted.

NOTICE: The information in this book is true and complete to the best of our knowledge. All recommendations on parts and procedures are made without guarantees on the part of the authors or the publisher. Because the quality and accuracy of parts, procedures and methods are beyond our control the authors and publisher disclaim all liability incurred in connection with the use of this information. Parts and procedures may not be legal for sale or use in some states, particularly those which may affect emission and smog-control devices. Check local laws.

Library of Congress Cataloging-in-Publication Data

Currao, Tom.
       Camaro restoration handbook : ground up or sectional restoration
tips & techniques for 1967–1981 Camaros, all models included / by
Tom Currao and Ron Sessions.
          p.   cm.
       ISBN 0-89586-375-8
       1. Camaro automobile—Conservation and restoration.   I. Sessions,
Ron   II. Title.
TL215.C33C87   1990
629.28'722—dc20
                              90-4303
                         CIP

TABLE OF CONTENTS

About the Authors

TOM CURRAO

Co-author Tom Currao began tinkering with cars at an early age, his first project being his father's '64 Chevelle. After taking it apart and putting it back together several times, Tom moved on to other projects, and was a frequent fence-monger at the local drags. By the time he was 22, he was instructing others on how to disassemble, diagnose, repair and assemble cars by writing technical service manuals for General Motors. Ten years and 14 writing awards later, he continues to research and write technical training programs for Chevrolet service technicians.

Tom purchased his first Camaro 15 years ago, and has owned 10 since. His current prizes are a '67 RS convertible, one that comes to life in the following pages, and a 1970 Z/28. He is also a member of the United States Camaro Club and an associate member of the Society of Automotive Engineers. He resides in Fraser, Michigan.

Personal Acknowledgements—I would like to thank the following: My family for their understanding and assistance, especially my father who helped fuel my automotive ambition; to Bob Ranucci, Rudy Stagl, Mike Nizza, Karl Kohl, Russ Corley, Brian McKinnie, Jim Campbell, Ken Radzikowski and Rick Vanlerberghe who helped with every facet of the project; to Mr. A. Kohl for installing the convertible top; to Eric Cluck for getting a "load" off my chest; to Henry Longhi for helping me obtain quarter panels; to Chuck Gallup at Mike Savoie Chevrolet for his enthusiasm and help in procuring parts; to Gloria Smith for showing me the true meaning of love; and to Lea Anne for providing that love.

In addition, my project would not have been possible without the support of the following contributors, many of whom provided parts and expertise at little or no cost. They are: Camaro Connection, Ssnake-Oyl Products, Classic Camaro, C.A.R.S. and the Eastwood Company.

Co-author Tom Currao wheels his freshly restored 1967 convertible out into the noon-day sun.

RON SESSIONS

After graduating from Temple University, Ron Sessions landed a job with the Chilton Book Company in Radnor, Pennsylvania writing and editing workshop manuals. Five years later, Detroit beckoned Ron to write and produce product training films and videos for Chevrolet. It was during the production of a film on the 1979 Camaro that Ron first met co-author Tom Currao.

From Detroit, Ron migrated to the sunnier climate of Tucson, Arizona to write and edit how-to manuals for HPBooks. During his tenure there, Ron wrote his first two books, *Turbo Hydra-matic 350 Handbook* and *How To Work With and Modify the Turbo Hydra-matic 400 Transmission*. It was about this time that Ron joined forces with Tom to undertake a dual restoration project that forms the basis for this book. Ron purchased a 1973 Type LT Z/28 Camaro, joined the local Camaro club and embarked on a lengthy restoration process that appears in the following pages.

After a stint as the Feature Editor for *Road & Track Magazine*, Ron was promoted to Editor of *Road & Track Specials*, a publishing group responsible for *Exotic Cars Quarterly* magazine, R&T buyer's guides, annuals and other specialty publications. He resides in Mission Viejo, California.

Personal Acknowledgements—The writing of this book required more than just sitting down in front of a computer and describing how to take apart, recondition and assemble a Camaro. In reality, it involved hundreds of phone calls, scores of field trips and countless hours and late nights spent in the shop.

To that end, a special thanks goes out to the following companies without whose help this book would not have been possible. Their addresses and phone numbers appear throughout the book. They are: Jeff Leonard and Bob Brennan of Classic Camaro Parts & Accessories, Huntington Beach CA, for opening up their prodigious mail-order catalog of new and reproduction parts for our disposal; Don Bevers of Bevers Welding, Tucson

AZ, for putting a roof back over our project 1973 Camaro; John Anderson and Gary Nori of Z&Z Auto, Orange CA, for digging out countless used parts for the camera and helping with numerous photo sequences; and last but not least, Jim Altieri of Arizona GM Classics, Tucson AZ, for going the extra mile with expert paint and bodywork. Not to mention, (but we will anyway) numerous others who contributed their expertise and products for photography: The Eastwood Company, Malvern PA; Kanter Auto Products, Boonton NJ; Beverly Hills Motoring Accessories, Hollywood CA; Meguiar's, Newport Beach CA; Camaro Country, Marshall MI; H.C. Fasteners, Alvarado TX; Dagley's Camaro & Firebird Wrecking Yard, Phoenix AZ; Bob Toth and the Goodyear Tire Co., Akron OH; Ssnake-Oyl Products, Dallas TX; Stencil & Stripes Unlimited, Park Ridge IL; Mike Riley and The Carb Shop, Costa Mesa CA; Dick Guldstrand and Guldstrand Engineering, Culver City CA; Intercoach Products, Lewistown PA; Tom Hoxie and Kay Ward of Chevrolet Public Relations; and Jay Jett, seat-cover wiz, wherever you are.

And of course, a special thanks to Tom Currao, whose enthusiasm for Camaros involved me with these beloved Chevrolets in the first place.

Researching the long-lived second-generation Camaros fell to co-author Ron Sessions, posing here in front of his 1973 Type LT Z/28. Photo by Dean Siracusa.

HOW TO USE THIS BOOK

If you were to read *Camaro Restoration Handbook* front to back, the general progression would be figuring out what kind of Camaro you want, buying it, getting yourself organized, tearing it down, fixing or replacing everything that needs attention, putting it together and finally caring for and enjoying it.

However, each chapter has been written to stand pretty much on its own. You may want to keep your car running and roadworthy as you perform a new paint job, reupholster the seats or install a new wiring harness. Not all Camaros are in need of a subframe-off, ground-up restoration—a fact recognized by the authors.

During the five years or so of nights and weekends it took to put this book together (as well as two Camaros), we've tried to cover procedures and tips that supplement any manuals you may already have. In fact, elsewhere in this book, we tell you how to go about ordering a factory service manual for the particular model year of your Camaro. Also, the *Camaro Restoration Handbook* is not a photo essay of what is and isn't correct for any given model year. For a litany of what molding is correct on what Camaro and so on, see our listing of recommended reading on page 25.

This book is, however, a greasy fingernail, how-to-do-it book for all first- and second-generation Camaros. We sincerely hope that the *Camaro Restoration Handbook* becomes an indispensable tool in your quest to bring back to full glory one of Chevrolet's finest—the 1967-81 Camaro.

1 A Brief History

As his Ford counterpart, Lee Iacocca, did with the first Mustang, then-Chevrolet General Manager Elliot "Pete" Estes basked in the favorable publicity surrounding the introduction of the 1967 Camaro. Estes would later move on to become President of General Motors. Photo courtesy Chevrolet.

The Camaro was introduced to the public on September 29, 1966 as Chevrolet's entry into the ponycar market pioneered by the Ford Mustang. Offered in coupe and convertible models, with 69 available factory-installed options and 12 available dealer-installed options, the Camaro was designed to be many things to many people.

First and foremost, the Camaro was to be a *driver's* car, which was reflected in its handling, ride, high-performance engine availability and styling. It was intentionally designed to do what a Corvette did

for less money. And judging by its popularity, it accomplished this goal exceedingly well.

Camaro's long-hood, short-deck styling was intended to give a dynamic feeling of motion, even when the car was standing still. The chassis was also a first in that it used a separate front subframe attached to a unit-body via bolts and computer-tuned double-biscuit rubber mounts. This design helps minimize road noise and vibration into the passenger compartment from the drivetrain and front suspension.

The subframe is bolted to the body and front sheet metal at six points. Four bolts and rubber mounts are used to attach the rear of the subframe to the body. Two bolts and rubber mounts attach the front of the subframe to the radiator support, to which the front sheet metal is attached. The engine, transmission, front suspension, brakes and steering gear and linkage are secured to the subframe. With minor revisions for steering linkage and front-suspension geometry for the second-generation Camaros which began production in 1970, this basic design was used through the 1981 model year.

The rear portion of Camaro was initially designed to be very similar to the compact Chevy II sedan, using a solid Salisbury axle and splayed monoplate leaf springs. Computer analysis indicated that placing

the rear shock absorbers on the outboard side of the rear leaf springs and mounting near the vertical would allow the wheels to more closely follow irregular road surfaces and improve cornering. To combat wheel hop with the monoleaf springs, a rear-axle traction bar was installed on the passenger-side rear spring of all 1967 Camaros with high-performance engines.

Engine choices were what made a Camaro a Camaro. At its introduction, Camaro was offered with a choice of either 230 or 250 cu-in. 6-cylinder engines. For those who wanted a Camaro with more authority, two 327s and a 295-hp 350 cu-in. small-block V8 were offered. In May of 1967, a 396 cu-in. big-block V8 with 325 hp became available in the SS model. And this was only the beginning of a series of big-blocks that could be installed in

Camaros over the next several years. Camaros with these engines required a driver with a very disciplined right foot or a good lawyer to keep their record clear of speeding tickets.

Braking action on the first Camaros was provided by 9 1/2-in. drums, front and rear. Later models offered ventilated front disc brakes as optional or standard equipment depending upon the model and year. Rear disc brakes adapted from the Corvette were offered as a factory-installed option on 1969 Camaros on a very limited basis. This system was offered in order to meet the homologation requirements of the Sports Car Club of America, who sanctioned the Trans-Am races Camaro competed in.

But we're getting a little ahead of ourselves here. Follow along as we chart the evolution of the Camaro on a year-by-year basis.

1967

Fresh out of the gate, Camaro was available in coupe and convertible body configurations with seating for four passengers. Standard fare were front bucket seats and a rear bench, but a front center console (with or without auxiliary gauges), a front bench seat and a fold-down rear seatback were available as options.

The SS (short for Super Sport, although Camaros were never called Super Sports) package included a special hood and ornamentation, paint stripes, safety-wired fuel-filler cap, SS identification, performance suspension, tires and wheels.

Another upgrade was the Rally Sport (RS) package, which included hideaway headlamps operated by electric motors, front valance-mounted parking lamps, rear valance-mounted backup lamps, safety-wired fuel-filler cap, RS identification (unless combined with the SS package), a special black-out grille and other specific trim. Camaros could be ordered with both SS and RS packages, the resulting car having hideaway headlamps, black-out grille and valance-mounted parking and backup lamps, but otherwise SS trim, identification and running gear were dominant. Some 100 white-on-blue convertibles with this SS/RS trim combination and bearing *Indianapolis 500 Pace Car* decals were sold to the public.

The Z/28 was a mid-year introduction, with production not getting underway until December 29, 1966. Central to the first Z/28 was its 302-cu-in. V8, derived by installing a short-stroke crankshaft from a 283-cu-in. V8 in a 327 V8 block. This

Looking diminutive in its base form with small hub caps and minimal chrome, 1967 Camaro convertibles are nevertheless highly sought after. This was the only year Camaros had front vent windows. Photo courtesy Chevrolet.

For 1967, Camaro SS engines included an all-new 350 cu-in. V8 and several versions of the big-block 396 cu-in. V8. This SS features Rally Sport hideaway headlamps. Photo courtesy Chevrolet.

enabled Chevrolet to homologate the Z/28 for the SCCA's Trans-Am sedan racing series, which at that time had a 305 cu-in. (5.0-liter) displacement limit. Available on coupes only, it also included a close-ratio Muncie 4-speed transmission, power-assisted front disc brakes, quick-ratio manual steering, 15x6-in. Rally wheels and red-stripe nylon-cord tires, a heavy-duty cooling system, 3.73:1 final drive, sport suspension and special "runway" hood and deck lid stripes. Also, Z/28s could be ordered with a special plenum

air intake and tubular steel exhaust headers. No external badging identified the cars as Z/28s. The RS package could be combined with any Z/28.

Other than the Z/28's standard 302, engine choices included the 230 and 250 cu-in. inline-6s, two 327 cu-in. V8s (one 2-bbl, one 4-bbl), the 350 cu-in. 4-bbl V8 and two 396 big-block V8s. Available transmissions included the synchromesh 3-speed manual, M20 (wide-ratio) and M21 (close-ratio) 4-speed manuals, Powerglide 2-speed automatic and the

M40 3-speed Turbo Hydra-matic 400 automatic.

Unique to 1967 models are front vent windows. Also, 1967 Camaros are the only models without side marker lamps.

A total of 220,906 1967 Camaros were produced.

1968

Camaro styling was largely a carryover for 1968, with notable changes being the addition of side marker lamps and the deletion of the front vent windows. Cloth and vinyl seat trim were available for the first time in Camaros (in a houndstooth pattern). The optional console-mounted auxiliary gauges changed to a two-tier, stacked arrangement and the automatic transmission floor-mounted shifter changed to a "stirrup" design. This was the last year you could order a Camaro with a front bench seat.

The RS package continued as before, but its hideaway headlamps were now vacuum-operated instead of by electric motors. The Camaro SS package added a 350-hp 396 V8 to its specification sheet and all SS models had a black-painted rear panel. The Z/28 finally got exterior identification at the leading edge of the front fenders: a "302" insignia early in the model year and a full-blown Z/28 badge later that year. Some 1968 Z/28s had dealer-installed rear disc brakes adapted from the Corvette and a dual-quad, cross-ram manifold was sold over-the-counter for Z/28 installation. Again, the RS could be combined with the SS or Z/28 and the Z/28 package could not be ordered on convertibles.

Underneath, staggered rear shock absorbers were used to help minimize rear-axle hop. And all Camaros equipped with 350 cu-in. and 396 cu-in. V8s switched to multi-leaf rear springs.

Chevrolet put a short-stroke crank from the 283 V8 into a 327 V8 cylinder block and created a high-winding, Trans-Am-winning 302 cu-in. V8 that began the Z/28 legend. Photo courtesy Chevrolet.

Detail refinements for 1968 included deletion of front vent windows and addition of side marker lights. Z/28 finally got front fender badging in mid-1968. Photo courtesy Chevrolet.

Dual-quad, cross-ram intake manifold set-up with big Holley 4150s was sold over-the-counter for 1968-69 Z/28s, but was not installed on production cars.

Camaro convertible production ended after 1969. See-through louvers covering RS hideaway headlamps were a 1969 exclusive. Photo courtesy Chevrolet.

The Z/28 that gets many an enthusiast's heart pumping is this 1969 model. Cowl-induction hood proved a popular option. Photo courtesy Chevrolet.

Performance goodies of 427 cu-in. ZL-1 engine included aluminum block and heads. 1969 marked the high point of Camaro engine performance. Photo courtesy Chevrolet.

Previous engine and transmission choices continued with a few additions. The mid-level performance 350-hp 396 V8 was new as was a heavy-duty, close-ratio M22 "Rock Crusher" 4-speed manual (available only on the Z/28 and 375-hp 396) and a Torque Drive manually shifted 2 speed automatic (available on 6-cylinder Camaros only).

Chevrolet produced 235,147 1968 Camaros.

1969

Of all first- and second-generation Camaros, the 1969 model stands alone as the most unique of the bunch. Except for the hood, roof and deck lid, no sheet metal carried over from 1968. Neither did the instrument panel, which was completely new for 1969 and would change again in 1970.

Why did Chevrolet go to all the trouble and expense to freshen the Camaro with a deeply recessed grille and scalloped wheel openings for 1969 when an all-new replacement was due one year later? Truth be known, Chevrolet was locked in a knock-down, drag-out battle for the number-one sales position with Ford in the late 1960s and a three-year-old Camaro needed help if it was to gain any ground on archrival Mustang, which was all-new inside and out for 1969.

Other elements set the 1969 Camaro apart from all the others. A few hundred 1969 Camaros were factory-equipped with a 427 cu-in. V8, either the all-aluminum ZL-1 or iron-block L-72. While the 427 V8 option never appeared on dealer order forms, these could be special-ordered under codes COPO 9560 and COPO 9561. The resulting COPO rat-motor Camaros (COPO standing for Central Office Production Order) are valuable collector items today. Along with the COPO Camaros was an electrically operated, cowl-induction hood, with a ram air set-up, which could also be ordered on SS and Z/28 models. Also 4-wheel disc brakes (adapted from the Corvette) could be factory-ordered as options on the Z/28 and SS. Headlamp washers made their first and only appearance in 1969, standard on the RS and optional on all other models.

Various under-the-skin improvements made their debut in 1969. Single-piston, floating-caliper front disc brakes replaced the corrosion-prone four-piston design of 1967-68. Also new was a steering column mounted ignition lock, variable-ratio power steering and a medium-duty, 3-speed Turbo Hydra-matic 350 transmission option for all applications except Z/28 (4-speed manual mandatory) and SS396 (which got the heavy-duty Turbo Hydramatic 400 when an automatic transmission was specified). Under the hood, the 230 cu-in. inline-6 and venerable 327 V8 made their last appearances and a new low-compression 307 cu-in. 2-bbl V8 was introduced. And the gas filler neck was moved to a concealed location behind the license plate and beneath the rear bumper.

The ever popular Rally Sport option continued to use vacuum-operated hide-away headlamps, but for 1969, new louvered "see-through" headlamp doors were used to permit night driving in the event the doors became stuck in the closed position. For the second time in three years, Camaro was named as the Official Pace Car for the 1969 Indianapolis 500, and Chevrolet commemorated the event by producing 3675 Pace Car replicas. Sold under RPO code Z11, all were orange-on-white SS/RS convertibles. Model year 1969 also marked the last time a convertible would be offered in the Camaro line until 1986.

A total of 243,085 1969 Camaros were produced during its extended model year, which ran 18 months—well into the 1970 calendar year.

1970

An all-new, second-generation Camaro was introduced on February 26, 1970 to rave reviews from *Road & Track, Car and Driver* and *Sports Car Graphic* magazines. It would prove to be one of the longest-running and most profitable production runs in GM history, the basic design evolving over 12 consecutive model years until being superseded in 1982 by the third-generation Camaro. Because of the mid-February introduction, many have described these first second-generation Camaros as 1970-1/2 models. Nevertheless, GM certified these as 1970 models, and the vehicle identification number verifies it. For state motor vehicle, insurance and other official purposes, these are known as 1970 models.

In true Detroit parlance, the 1970 Camaro was longer, lower and wider than its predecessor, 2.0 in., 1.1 in. and 0.4 in. respectively. Although the new car shared the first-generation's 108-in. wheelbase, the entire passenger cabin was swept back, accentuating the long hood, short deck ponycar theme. The second-generation Camaro had an international styling flair, with hints of Ferrari sprinkled about and a Jaguar influence in the front fenders and

hood. The European influence was evident in the new Camaro's aggressive road stance, now 1.7-in. wider up front and 0.5-in. wider at the rear. Variable-ratio power steering was available for the first time and front disc brakes became standard equipment.

Yet another performance 427 V8, this iron-block COPO 9561 engine appeared in the torrid Yenko SC of 1969.

Completely redesigned stem-to-stern for 1970, second-generation Camaros went on to enjoy a long-lived 12-year model run. On the Z/28, displacement increased to 350 cu. in. and automatic transmission was available for the first time.

For 1969, SS396 Camaros sported simulated velocity stacks on the hood and a blackout rear end panel.

Longer, lower, wider second-generation Camaro eliminated rear quarter windows. Small, one-piece rear spoiler was standard on 1969-71 Z/28s.

Available in a coupe body style only, various improvements were incorporated. A strengthened front subframe aided structural integrity and a new double-panel roof with acoustic interliner helped minimize road noise. To reduce cost and manufacturing complexity, the rear quarter windows were eliminated. The doors, now 8 in. longer than before, ran all the way to the C-pillars and featured steel side impact beams for passenger crash protection. A new, larger, more swept-back windshield helped increase glass area by 10%. Hideaway windshield wipers (hidden below the hood line when not in use) became optional equipment (standard on RS and SS models). And the radio antenna was incorporated into the center of the windshield.

Inside, a completely new interior included a deeply recessed, wraparound instrument cluster and soft instrument-panel upper section for crash protection. A new, 150-mph speedometer announced the second-generation Camaro's road worthiness. The rear view mirror moved from the roof to a mount bonded to the windshield. Low-back bucket seats with adjustable headrests were standard, but would be discontinued for 1971.

The Rally Sport option (RPO Z22) continued for second-generation Camaros, but without the stylish yet troublesome hideaway headlamps. Distinguishing the 1970 RS from other Camaros was its unique front-end appearance with split bumperettes, protruding snout with Endura grille surround and large, European-style parking lamps in the "catwalk" area of the header panel between the grille and headlamps.

This trim and elegant front end was used through 1973 and represents a high mark of Camaro styling evolution. In addition to the front-end styling, Rally Sports received upgraded trim, blackout grille, RS identification (unless combined with SS or Z/28 packages which then became dominant) and 14x7-in. Rally wheels.

The SS option (RPO Z27) continued as well, in both SS350 and SS396 variations. Included with the SS package was the 300-hp 350-cu.-in. V8, plus power brakes, F70x14 bias-belted wide ovals on 14x7-in. Rally wheels, a blackout grille and rear panel, upgraded trim and SS350 identification. Opt for either the 350-hp or 375-hp big-block 396 V8 (actually displacing 402 cu in.), and you'd get the F-41 special performance suspension along with SS396 identification and a blackout tail panel. Also, Positraction was a required option when the limited-production 375-hp big-block was specified. By the way, the 402 cu-in. displacement was achieved by overboring the 396 block 0.032 in. For marketing reasons, however, the SS396 designation was continued from the 1967-69 years. The 402 was not the killer big-block of the late 1960s, having switched from Holley to Rochester carburetion and a 2-bolt main block for 1970.

The Camaro's big performance gun for 1970 was the Z/28. Chevrolet took advantage of a rule change for the Trans-Am series which permitted destroked engines and made the 350 cu-in. V8 standard in the Z/28. Boasting 360 hp @ 5800 rpm, solid lifters, a Holley double-pumper sitting atop an aluminum high-rise intake manifold and 11:1 compression exhaling though dual exhausts, the Z/28 powerplant was the Camaro equivalent of Corvette's heralded LT-1. Other go-fast Z/28 goodies included F-41 performance suspension, power brakes, Positraction, heavy-duty radiator, choice of M20 4-speed manual or M40 Turbo Hydra-matic 400 transmission (a first in the Z/28) and raised-letter F60x15 wide-ovals on new 5-spoke 15x7-in. steel wheels. Also part of the package was special hood insulation, full instrumentation, rear bumper guards, one-piece rear spoiler, blackout grille, Z/28 identification and the familiar black or white "runway" stripes running the length of the hood and deck lid. As in previous years, the Z/28 could be combined with

Through 1972, SS396 model continued, but sales were poor. Due to a 0.030-in. overbore, 1970-72 SS396 Camaros actually displaced 402 cu. in. Photo courtesy Chevrolet.

Handsome front-end treatment of 1970-73 Rally Sports incorporated flexible Endura grille surround, minimalist front bumperettes and Euro-style round parking lamps. This 1971 RS is combined with a Z/28 and shows chin spoiler and three-piece high rear spoiler that became available that year. Photo courtesy Chevrolet.

the Rally Sport, the resulting car having the split-bumper RS front end with Z/28 mechanicals and identification.

Another outgrowth of Trans-Am racing, a larger, 3-piece rear spoiler with greater downforce, became available mid-year. Using the center section from the Firebird T/A spoiler, but with Camaro-specific end caps, the 3-piece spoiler could be special-ordered for the Z/28 as COPO 9796. It would eventually replace the one-piece spoiler on the Z/28 and appear on hundreds of thousands of Camaros through 1981 under RPO D80.

Chevrolet produced 124,901 Camaros in the abbreviated 1970 model year.

1971

After the major changes of 1970, the 1971 Camaro was basically a carryover with a few refinements. Externally, Camaro received new emblems front and rear, and the side-marker lamps now flashed in unison with the turn signals. The D80 spoiler option was the new 3-piece fiberglass design and included a black ABS chin spoiler as well, though the Z/28 continued with the smaller one-piece rear spoiler as standard equipment.

Inside, the 1971 Camaro got new high-back front bucket seats adapted from the Chevy Vega. These highly bolstered, full-foam buckets would continue, basically unaltered, through 1981. Optional on the driver's seat was a 2-position seatback recliner. For occupant crash protection, protruding knobs and switches were padded for 1971. And there was a new 2-spoke steering wheel.

Much to the dismay of enthusiasts, performance ratings of Camaro's engines plummeted an average of 30%. Part of this loss was real, part imagined. An industry-wide trend away from leaded gasoline forced automakers to lower compression ratios to the 8.5:1 to 9.0:1 ranges resulting in a real power loss of about 10%. Simultaneously, automakers also changed the way they measured output from SAE gross to SAE net specification. This was accomplished by adjusting output calculations to an 85 degree Fahrenheit correction factor rather than the previous (denser) 60 degrees Fahrenheit. The effect was a more realistic rating system, reflective of an engine with all accessories mounted in an as-installed condition.

A total of 114,630 Camaros were produced for 1971. Camaro production was discontinued at Van Nuys, CA.

1972

Refinement was once again the Camaro watchword for 1972. The most obvious change was a coarser, egg-crate grille. Inside, revised door trim panels with built-in map pockets and change holders made their debut and a 4-spoke soft-vinyl sport steering wheel was optional. The speedometer was scaled back to a more realistic 130-mph maximum. This was also the last year for the Camaro SS, in both 350 and 396 (402 cu-in.) versions.

The Camaro and its Pontiac stablemate, Firebird, almost didn't survive the 1972 model year. Sales of ponycars (Mustang, Cougar, Barracuda, Challenger, Javelin, Camaro and Firebird) had taken a nosedive beneath a tidal wave of safety and environmental criticisms, high insurance rates, declining performance levels due to tightening emission controls, and a mild economic recession that hit youthful buyers hardest. Camaro/Firebird production had ended at Van Nuys, California the year before. GM brass were frowning on the Camaro/Firebird balance sheet when a grinding 117-day UAW strike paralyzed the sole F-car assembly plant in Norwood, Ohio. When the strike finally ended late in the summer of 1972, over a thousand cars had to be sent to the crusher because they didn't meet stricter 1973 bumper-protection and emission-control regulations. Only an eleventh-hour reprieve kept GM from pulling the plug on the Camaro that year.

A dismal 68,651 Camaros were produced in 1972.

1973

Model year 1973 turned out to be something of a pleasant surprise. At a time when unsightly, battering-ram front bumpers were festooning the noses of many cars in response to Federal 2-1/2-mph front-bumper regulations, Chevrolet engineers devised an ingenious bracing and spacing design that kept Camaro's svelte and shapely bumpers intact for one more year. They even managed to save the Endura-nosed, split-bumper RS for 1973 with hidden-from-sight reinforcements that met crash standards.

Inside, the good news continued with the 4-spoke Sport steering wheel now standard. Available was a new soft-vinyl-padded console, adapted from the Firebird, which replaced the previous hard-plastic unit. When an automatic transmission was ordered with the console, a new single-stalk shifter replaced the stirrup design used from 1968-72. Late in the model year, optional power windows became available, the switches mounted atop the new console. Full-foam rear seat cushions were introduced in 1973. Also, early in the model year, a molded hardboard headliner replaced the previous cut and sew cloth headliner.

Responding to a changing (vanishing?) performance-car market, Chevrolet introduced an all-new luxury-touring Camaro model, the Type LT. Taking a cue from the sales success of the Monte Carlo, Chevrolet launched the Type LT with a full complement of standard equipment. Included were previous extra-cost goodies

For 1973, GM engineers avoided battering-ram bumpers on the Camaro by reinforcing the existing units. Photo courtesy Chevrolet.

such as a 145-hp 350 cu-in. V8, variable-ratio power steering, extra sound insulation, 14x7-in. Rally wheels, hideaway wipers, dual sport mirrors, full instrumentation, deluxe seat and door trim, woodgrain accents, bright exterior accents, black-accented rocker panels and Type LT identification. The Type LT model could be combined with *both* RS and Z/28 option packages, the resulting car carrying Type LT interior and exterior trim and identification, but with RS split-bumper nose treatment and Z/28 chassis and drivetrain pieces (plus the D80 spoiler package and 5-spoke wheels). These cars were true sleepers because the Z/28 badges and runway stripes were deleted.

The Z/28 was something of a good news/bad news proposition. The good news was that it was still available (unlike the dropped SS350 and SS396), and could be ordered with air conditioning for the first time. The bad news was that power was down to 245-hp SAE net. For 1973, the Z switched to hydraulic lifters and Quadrajet carburetion on a low-rise, cast-iron intake manifold. These changes plus the introduction of exhaust-gas recirculation (EGR) on all 1973 engines sapped another 10 hp from the previous year's output.

The Powerglide 2-speed automatic was dropped for 1973, but a new standard coolant recovery system and optional inflatable spare tire made their debuts.

Sales picked up a few notches, with Chevrolet producing 96,751 Camaros that year.

1974

For 1974, the Camaro finally succumbed to the battering-ram bumpers to meet Federal crash standards. But rather than just tack the squarish bumpers and their bulky leaf-spring energy-absorbing zones onto Camaro's curvaceous body, stylists crafted a new front and rear appearance. Up front, a long, sloping nose constructed of sheet molded compound (SMC) tied the hoodline and bumper together. The headlamps, now deeply recessed with "sugar-scoop" bezels and Rally Sport-style parking lamps, flanked a steeply raked two-piece grille. At the rear, the previous concave tail treatment was replaced with a flat panel and triangular, wraparound taillamps, necessitating a new rear quarter-panel design.

Inside, new three-point front seat/shoulder belts with roof-mounted inertia

Aluminum bumpers on huge leaf springs, a sloping nose and wraparound taillamps made their debut in 1974. Type LT model, introduced the previous year, could still be combined with Z/28 to make a stealth musclecar. Chevy temporarily discontinued Z/28 after 1974 model year.

retractors replaced the clumsy, old separate seat and shoulder belt design. Also, 1974 marked the introduction of the dreaded federally mandated seat-belt interlock system. In order for the engine to start, the driver and front-seat passenger had to have their seat belts fastened. Seat-cushion and seat-belt latch sensors wired into the starter circuit would prevent starter-solenoid engagement in the event the driver and/or front-seat passenger failed to buckle up. Trouble was, the system sometimes prevented starter operation even when both parties did what they were supposed to do. To handle just such problems, Chevrolet put a bypass switch in the engine compartment which, when pressed, would permit one start regardless of whether the front-seat occupants were buckled or not. The system proved so unpopular that Congress repealed the law requiring the interlock and by mid-1974, Camaros were rolling off the Norwood assembly line without the device.

With its svelte bumperettes and Endura nose legislated into oblivion, the Camaro RS was canceled for 1974.

Not so the Z/28, which enjoyed one more year in the sun, 4-bolt mains, forged crank, baffled, cast-aluminum valve covers, dual-snorkel air cleaner and all. The 245-hp Z/28 engine was the first Camaro to be equipped with breakerless high-energy ignition (HEI). And just to make sure everyone knew you were driving a Z/28, huge block-letter Z/28 decals covered the hood and deck lid where the distinctive runway stripes once lived.

The Type LT was back with more luxury touches: extra sound insulation in the doors, firewall, rear deck and underfloor, plus a color-coordinated lower instrument panel. And as in 1973, the Type LT could be combined with Z/28 mechanicals to make a luxury touring sleeper.

Elsewhere on the Camaro landscape, the 307 cu-in. V8 was dropped. Front disc-brake wear sensors and lower ball-joint wear indicators became standard equipment, as did a new, larger 21-gal. fuel tank adapted from the Nova. And radial-ply tires and AM/FM stereo radios became optional for the first time.

A rousing 151,008 Camaros were born in 1974.

1975

A long-standing second-generation Camaro weak spot was remedied in 1975. The rear three-quarter vision blind spot caused by the large C-pillars was minimized by a new roof design that incorporated a wraparound backlite. But this refinement was overshadowed by disturbing developments beneath the sheet metal.

The catalytic converter made its grand entrance in 1975 and for the most part, caused the exit of the Z/28 that year. Use of the platinum and palladium-packed muffler-like devices allowed automakers to meet stricter 1975 federal exhaust-emissions regulations. But they also caused a huge airflow restriction that limited output of high-winding performance engines such as the Z/28's. So rather than offer an emasculated stripe and tire Z/28 caricature, Chevy canceled the evergreen performance favorite. Catalytic converters were used across the board on other Camaro engines along with the maintenance-saving HEI system. With memories of long gas lines from the 1973-74 Arab Oil Embargo still fresh in many people's minds, gas-saving radial tires became standard equipment on all Camaros.

The Rally Sport returned for 1975 as a two-tone paint and trim option with dual-sport mirrors and Rally wheels. With the lower body color of your choice separated by a tri-color stripe, the hood, header panel, grille, headlamp bezels, plus the tops of the front fenders, doors and front portion of the roof received a matte-finish blackout treatment. Power door locks were offered for the first time on second-generation Camaros. Also, two new suspension options made their debuts. An FE8 radial-tuned suspension featured front and rear anti-roll bars and shock absorbers selected to work with Camaro's newly standard radial tires. Plus, a Z86 Gymkhana suspension with 15x7-in. wheels, 60-series radials, fast-ratio steering gear and many of the canceled Z/28's underpinnings was offered too.

Camaro production in model year 1975 totaled 145,770 units.

1976

Despite the fact that there was little new to herald the coming of the 1976 Camaro, a booming economy and strong Camaro demand caused Chevrolet to add Van Nuys, CA once again as a second F-car assembly plant, joining the Norwood, Ohio facility. A new 305 cu-in. 2-bbl V8 made its debut as the smallest available V8. Power brakes were now standard with all V8s. And cruise control, PEl polycast wheels (cast-aluminum with decorative polyurethane center section) and a new "halo" partial vinyl-roof covering were released as options. The FE3 radial-tuned suspension was dropped and F41 sport suspension replaced the previous Z86 Gymkhana suspension option.

Detail refinements to the Type LT included a "leather-look" instrument cluster, "hockey stick" combined armrests and door pulls and a brushed aluminum panel between the taillamps.

With Norwood going full-steam and Van Nuys picking up the pace, Chevrolet produced 182,981 1976 Camaros, the best year yet since the second-generation cars hit the streets back in 1970.

1977

Once again, Camaro styling for 1977 was a carryover from the previous year. Yet sales climbed to new heights, with Camaro finally ousting archrival Mustang from the number-one ponycar spot after a 10-year chase.

Enthusiasts heralded the mid-year return of the Z/28 and voted with their checkbooks, the 14,349 sales of the vaunted performance model no doubt helping to tip the Mustang/Camaro sales race in Camaro's favor. This was a different Z/28 than before, down on power but with excellent handling and bold graphics. The Z/28's 350 cu-in. V8 managed 185 hp breathing through a catalytic converter and single exhaust that split into two tailpipes and resonators with 40% less back pressure than a single-outlet system. Other familiar Z/28 elements returned: M21 close-ratio 4-speed transmission, revised F41 sport suspension with higher-rate front springs and a thicker front anti-roll bar, GR70-15 radials on color-keyed 5-spoke steel wheels, D80 spoilers, Z/28 identification decals, dual sport mirrors, blackout grille and full instrumentation. Chevrolet quickened the Z's power-steering ratio from 14.3:1 to 13.0:1, but dropped the variable ratio. Plus the Z/28 now sported color-keyed bumpers and spoilers, front and rear.

Elsewhere on the Camaro front, hideaway windshield wipers became standard fare. And the Rally Sport was now available in a variety of exterior two-tone paint schemes other than matte-black upper color and body-color lower only.

Total 1977 Camaro production hit 218,854.

1978

The long-awaited integrated soft-fascia bumper systems, inspired by GM styling chief Bill Mitchell's 1973 ZL-1 Berlinetta show car, made their debut for 1978. The new bumpers absolutely transformed the Camaro from a tacked-on battering-ram look to an almost bumperless appearance. These soft-fascia bumpers literally appeared to be made for the then eight-year-old second-generation design from the beginning. Actually, the new bumpers were to have made their debut on 1976 models, but engineering problems and other delays cropped up.

What appeared on the outside to be a simple, elegant, body-color urethane skin mated to the Camaro's sheet metal was in fact a complex structure that was at once flexible for minor impacts (up to 5 mph) and rigid for passenger protection above that speed. The urethane skin, itself able to spring back from minor dents and dings, stretched over a honeycombed, energy-absorbing fiberglass skeleton, designed to provide support and give shape to the skin while collapsing in a front-end collision. This skeleton, in turn, bolted to a steel basher bar which tied into the front subframe and radiator support. A similar albeit less hefty system comprised the rear bumper as well.

Overall, the front-end styling retained cues from the 1970-73 Rally Sport and 1974-77 models. A horizontally split grille was flanked by single, round headlamps in "sugar scoop" bezels and RS-style parking lamps, now rectangular rather than oval. At the rear, the wedge-shaped, wraparound taillamp theme continued, but with the license plate now moved to the center

Glorious return of Z/28 occurred in 1977-1/2 with body-color bumpers and 5-spoke wheels, body graphics, F-41 suspension and low-restriction exhaust. Wraparound backlite debuted in 1975.

Soft fascia front and rear bumpers gave an integrated look to 1978-81 models. Hood scoop and fender louvers on this 1978 Z/28 were non-functional. Ever-popular T-roof became a factory option in late 1978. Photo courtesy Chevrolet.

Camaro Rally Sport resurfaced in 1975-80 with this unique two-tone paint treatment. Shown is 1979 model with T-roof. Photo courtesy Chevrolet.

Chevy dropped the Type LT in 1979 in favor of this luxury-oriented Berlinetta. Composite cast-aluminum and polyurethane PEI Polycast wheels were part of the package. Photo courtesy Chevrolet.

of the soft "bumper" and the gas filler hidden behind a hinged door in the plastic trim panel.

Structurally, Camaro received a reinforced front subframe and rear spring perches to handle the additional weight of these bumper systems. And a factory-built T-roof option (RPO CC1) with twin removable smoke-gray glass panels became available for the first time.

The smash hit of 1977-1/2, the Z/28, continued going great guns for 1978 with 54,907 Z's produced. Aside from a down-rated engine for buyers in California and high-altitude areas (175 hp versus 185 hp in the 49 states), all of the go-fast hardware from the previous year returned. The Z/28's exterior appearance did get jazzed a bit, though, with a fake NACA duct hood scoop, simulated front fender louvers, blackout grille and tail panel and an integrated front spoiler (part of the urethane soft fascia). Inside, a simulated rope-wrapped, thick-section Z/28-only steering wheel helped drivers keep a grip on things.

Elsewhere in the lineup, the RS package and Type LT model continued as before. An all-time high of 272,633 1978 Camaros were produced, and several thousand were exported to Europe.

1979

The 1979 model year started out with Camaro sales ablaze, but ended with a whimper. Political upheaval and a moratorium on oil shipments from Iran snowballed into the second oil crisis in five years, with odd/even gas rationing and long gas lines chilling the market for gas-guzzling performance cars by mid-year.

All Camaros received a new, squared-off vinyl-covered upper instrument panel, dash pad and gauge cluster faceplate, although the actual instruments remained the same. A new electric-grid rear window defroster replaced the previous blower type. And a power mast-type radio antenna and CB (citizen's band, in case you forgot) radio were available for the first time as options.

The Z/28 suffered a slight performance decline due to retuning to meet emissions regulations. Power dropped to 175 hp in the 49 states and 170 hp in California and high-altitude areas. Outside, the Z/28 got a new, more aggressive, body-color chin spoiler that wrapped up into the front wheel wells as well as tri-color Z/28 decals and stripes running from the front spoiler back along the lower portion of the front fenders and doors.

As its luxury flagship, Chevrolet replaced the Type LT with a new Berlinetta model for 1979. The Berlinetta's suspension was retuned for a "boulevard" ride and the PE1 Polycast wheels were made standard. But otherwise, the upgraded trim, convenience items and extra soundproofing of the Type LT were carried over to the Berlinetta.

Despite the economic problems, more 1979 Camaros were produced than any other year (282,582) and a record 84,877 Z/28s were built that year.

1980

In an effort to improve Camaro's EPA fuel-economy rating, Chevrolet made a new, small-bore, 120-hp 267 cu-in. V8 available in 49 states. To save weight (and thereby improve fuel efficiency), Chevy also gave the venerable 250 cu-in. inline-6 the old heave-ho in favor of a pair of compact, 90-degree V6s. In California, the base engine became the even-firing, 110-hp Buick-built 231 cu-in. V6, while all other customers got an odd-firing, 115-hp Chevy-built 229 cu-in. V6 (essentially 3/4ths of a 305 V8).

All Camaros, except the Z/28, got a new fine-mesh grille for 1980. The Berlinetta now came with wire wheel covers instead of the Polycast wheels. The two-tone paint treatment on the Rally Sport changed a bit for 1980, the upper color on the nose and hood now only as wide as the grille. In keeping with the emphasis on better fuel economy, the Camaro's speedometer now only read to 85 mph. And vinyl roofs were no longer available, swept away in the quest for better aerodynamics.

The good news was that Chevrolet engineers were able to squeeze a few more horses out of the Z/28's 350 cu-in. V8 for 1980 with a functional, rear-facing cold-air hood and functional engine-compartment vents in the front fenders. At wide-open throttle, a solenoid opened a trap door in the high-pressure area at the rear of the hood to admit dense, ambient air to the carburetor air cleaner. The bad news was that the 190-hp 350 V8 was limited to 49-states applications, California buyers forced to settle for a 165-hp 305 cu-in. V8. A heavy-duty radiator once again became part of the Z/28 package. Outside, a new horizontal-bar grille and rear wheel-opening skirts distinguished a 1980 Z/28 from the previous model.

On the heels of the Iranian oil crisis and

For 1980, Chevy replaced the long-running inline-6 with a pair of 90-degree V6 engines adapted from V8s: one Buick and one Chevrolet. Previous Berlinetta Polycast wheels were dropped in favor of more traditional wire wheel covers. Photo courtesy Chevrolet.

1981 Z/28 marked end of long-running second-generation Camaro. In 1980-81, Z/28 boasted functional cold-air hood, fender louvers horizontal-bar grille, full front air dam and rear wheel-opening skirts. Photo courtesy Chevrolet.

the beginning of an economic recession in the U.S., Camaro production was nearly cut in half, sinking to 152,021 cars in the 1980 model year.

1981

This would be the last year for the second-generation Camaro, the end of an incredible 12-year model run. And it seemed that many Camaro enthusiasts had adopted a wait-and-see attitude, postponing purchase decisions until the all-new, third-generation Camaro would be unveiled for 1982.

Exterior appearance was a carryover from 1980. Chevrolet made power brakes and the space-saver spare standard. New options included halogen headlamps and for the Berlinetta, locking wire wheel covers. The Rally Sport was dropped.

A Computer Command Control emissions control system made its debut this year with electronic control of the carburetor fuel mixture and torque converter lockup feature. Due to emissions considerations, the Z/28's 350 cu-in. V8 was only available with automatic transmission and a 305 cu-in. V8 was required when manual transmission was selected. Output was down for the Z/28, 175 hp with the 350 V8 and 165 hp with the 305 V8.

For the 1981 model year, 126,139 Camaros were produced, 43,272 of them Z/28s.

IDENTIFICATION

Like being able to recite your social-security number without peeking, knowing your Camaro's serial number or vehicle identification number is important for registration, insurance and other purposes. But what really separates the more valuable, collectible Camaros from the street rods and daily beaters is the additional documentation that verifies special equipment and rare options or a low-volume model. In today's restoration-conscious market, a performance-optioned Camaro with matching numbers (i.e. original engine) may fetch two times the price of a similar-but-non-original model. Originality counts, even more so when you consider how many fake first-generation Z/28s and Indy Pace Cars have been pawned off as the real thing.

So learn all you can about how to identify your Camaro. Knowing exactly what year and model you have will be invaluable when it comes time to order parts for the restoration, whether they be original equipment or reproduction.

In 1967, VIN plate was riveted to driver's door hinge pillar. Cracking this car's code, 1 means it's a Chevrolet, 2437 indicates a V8 Coupe, 7 gives the 1967 model year, L says the car was built at Van Nuys, CA, and 126711 is its consecutive serial number.

VEHICLE IDENTIFICATION NUMBER

First and foremost, find your Camaro's VIN plate, jot down the number and compare it to the appropriate chart nearby. On 1967 models, the reverse-stamped, aluminum VIN plate is riveted to the lower section of the driver's door hinge pillar. Starting in 1968, the VIN plate was moved to the top of the driver's side of the instrument panel, visible from outside the car.

From 1967 to 1980, the VIN is a 13-digit alphanumeric code describing car division (Chevrolet), body type and model, engine type, year of manufacture, assembly plant and a 6-digit consecutive serial number. Beginning in 1981, GM stretched the VIN code to 17 digits, adding such things as a code for the type of passenger restraint system (non-passive seat belts) used. For details, see the nearby VIN code charts for your model-year Camaro.

The VIN code is also stamped onto a pad on the cylinder block (right front on all V8s and V6s, behind the distributor on inline sixes) and onto a pad on the transmission case. So if the Camaro you are looking at is being touted as a numbers-match car, check these components to see if the VIN code is indeed correct.

From 1968 on, VIN plate was riveted to driver's side of instrument panel, visible from outside. Deciphering this plate, I indicates Chevrolet Division, FQ87 is a Sport Coupe, H indicates a 350 cu-in. 2-bbl V8 under the hood, 4 says it's a 1974 model, N gives its Norwood, OH place of birth, and 117373 is its consecutive serial number.

Body Series/Style—As mentioned earlier, GM used a 13-digit VIN code from 1967 to 1980. But along the way, they changed the way they differentiated between models. From 1967 to 1969, 12337 was a 6-cylinder Sport Coupe, 12437 a V8 Sport Coupe, 12367 a 6-cylinder convertible and 12467 a V8 convertible. For 1970, the convertible was dropped, but the new body got an 87 designation, so through 1972 a 6-cylinder Sport Coupe got a 12387 and a V8 sport Coupe a 12487 designation.

In 1973, GM switched to an alphanumeric body series/style designation and began a new letter designation for engine size, not just number of cylinders as in 1967-72. See the nearby chart for the engine-size breakdown. Also in 1973, Chevrolet introduced the Camaro Type LT and made it a bonafide model, FS87, while the base Sport Coupe became FQ87. When the Berlinetta replaced the Type LT in 1979, it inherited the FS87 designation. Also, for reasons only known to Chevrolet, the designation for the Sport Coupe was changed in 1980 to FP87. Got that? See the nearby VIN code charts for a more complete breakdown.

BODY NUMBER PLATE

The second major identifier is your Camaro's body number plate, or color and trim tag as it is known by enthusiasts. This reversed-stamped metal tag is affixed to the driver's side on the firewall in the engine compartment. On 1967-69 Camaros, the tag is riveted to the vertical face of the firewall, adjacent to the windshield-wiper motor and brake master cylinder. On 1970-81 models, the tag is affixed with Phillips-head sheet-metal hex screws to the horizontal face of the firewall near the left-rear hood bumper.

On 1967-69 Camaros, body plate is riveted to the vertical face of the firewall in the engine compartment, adjacent to the wiper motor. From the looks of this tag, we've got a 1969 V8 Sport Coupe, gold with a black vinyl roof and black custom interior. It was assembled at Van Nuys, CA on the first week of March, 1969.

ONES TO WATCH

While many Camaros are appreciating in value, the models listed below show the greatest appreciation potential. Production totals for the models in question are included for reference.

1967 Convertible	25,141
1967 Z-28	602
1967 Indy Pace Car	*100
1967 SS396	5,141
1968 Convertible	20,440
1968 Z-28	7,199
1968 SS396	18,199
1969 Convertible	17,573
1969 Z-28	19,014
1969 Indy Pace Car	3,675
1969 SS396	13,970
1969 ZL-1 (COPO 9560)	69
1969 Yenko SC	201
1969 L-72 (COPO 9561)	*50
1970 Z-28	8,733
1970 SS396	2,464
1971 Z-28	4,862
1971 SS396	1,533
1972 Z-28	2,575
1972 SS396	970
1973 Z-28	11,574
1974 Z-28	13,802
1977 Z-28	14,349

* Unofficial estimate

The trim tag is much more descriptive than the serial number plate, giving the restorer clues as to the car's personality. Included on the trim tag is the following: model year, division (in this case, Chevrolet), body series/style (coupe or convertible), assembly plant (Norwood, OH or Van Nuys, CA), sequential unit number (an in-plant processing number that has *nothing* to do with the VIN), interior trim code (color, material, see chart in appendix), exterior trim code (some models only) modular seat code (highback bucket, lowback bucket or bench front seat on second generation models only), lower body color (or full body color, see chart in appendix), upper body color (RS or vinyltop only, see chart in appendix) and the time build code. In the late 1970s, the trim tag also detailed vinyl roof options (full or halo) and accent colors, where applicable. Some tags listed significant options, such a RPO Z/28 on 1970-74 Norwood, OH-built cars so equipped. Remember that the

Beginning in 1970, body plate is screwed to the top of the firewall on the driver's side. This plate indicates a 1974 Sport Coupe, yellow with black vinyl upholstery and high-back bucket seats. It rolled off the assembly line at Norwood, OH on the first week of August, 1974. Oh, it's a Z/28 to boot!

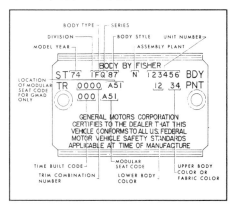

Typical body number plate breakdown—1974 shown. Courtesy Fisher Body.

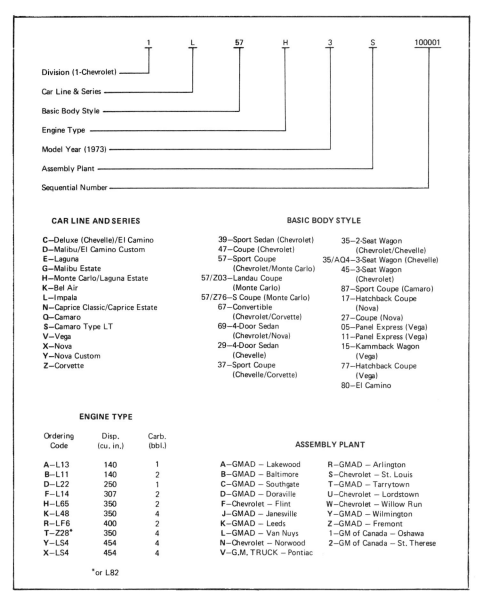

VIN code breakdown—1973-80 models. Courtesy Chevrolet.

CAR LINE AND SERIES

C—Deluxe (Chevelle)/El Camino
D—Malibu/El Camino Custom
E—Laguna
G—Malibu Estate
H—Monte Carlo/Laguna Estate
K—Bel Air
L—Impala
N—Caprice Classic/Caprice Estate
Q—Camaro
S—Camaro Type LT
V—Vega
X—Nova
Y—Nova Custom
Z—Corvette

BASIC BODY STYLE

39—Sport Sedan (Chevrolet)
47—Coupe (Chevrolet)
57—Sport Coupe (Chevrolet/Monte Carlo)
57/Z03—Landau Coupe (Monte Carlo)
57/Z76—S Coupe (Monte Carlo)
67—Convertible (Chevrolet/Corvette)
69—4-Door Sedan (Chevrolet/Nova)
29—4-Door Sedan (Chevelle)
37—Sport Coupe (Chevelle/Corvette)

35—2-Seat Wagon (Chevrolet/Chevelle)
35/AQ4—3-Seat Wagon (Chevelle)
45—3-Seat Wagon (Chevrolet)
87—Sport Coupe (Camaro)
17—Hatchback Coupe (Nova)
27—Coupe (Nova)
05—Panel Express (Vega)
11—Panel Express (Vega)
15—Kammback Wagon (Vega)
77—Hatchback Coupe (Vega)
80—El Camino

ENGINE TYPE

Ordering Code	Disp. (cu. in.)	Carb. (bbl.)
A—L13	140	1
B—L11	140	2
D—L22	250	1
F—L14	307	2
H—L65	350	2
K—L48	350	4
R—LF6	400	2
T—Z28*	350	4
Y—LS4	454	4
X—LS4	454	4

*or L82

ASSEMBLY PLANT

A—GMAD — Lakewood
B—GMAD — Baltimore
C—GMAD — Southgate
D—GMAD — Doraville
F—Chevrolet — Flint
J—GMAD — Janesville
K—GMAD — Leeds
L—GMAD — Van Nuys
N—Chevrolet — Norwood
V—G.M. TRUCK — Pontiac

R—GMAD — Arlington
S—Chevrolet — St. Louis
T—GMAD — Tarrytown
U—Chevrolet — Lordstown
W—Chevrolet — Willow Run
Y—GMAD — Wilmington
Z—GMAD — Fremont
1—GM of Canada — Oshawa
2—GM of Canada — St. Therese

VIN ENGINE CODES

Engine	1973	1974	1975	1976	1977	1978	1979	1980	1981
229 V6	—	—	—	—	—	—	—	K	K
231 L6	—	—	—	—	—	—	—	A	A
250 L6	D	D	D	D	D	D	D	—	—
267 V8	—	—	—	—	—	—	—	J	J
305 V8	—	—	—	Q	U	U	G	H	H
307 V8	F	—	—	—	—	—	—	—	—
350 V8, 2bbl	H	H	H	—	—	—	—	—	—
350 V8, 4bbl	K	K	K	L	L	L	L	L	L
350 Z/28, V8	T	T	—	—	—	—	—	—	—

6-digit sequential unit number on the trim tag will *not* match the last six digits on your Camaro's serial number. These are two entirely different numbers and the only thing they have in common is the number of digits.

Body Series/Style—See description under *Vehicle Identification Number.*

Time Build Code—This three-digit alphanumeric code tells you when your Camaro's body started rolling down the Fisher Body, Chevrolet or GMAD assembly line. The first two numbers describe the month of production, 01 being January and 12 being December. The last letter indicates the week of production, A being the first week, B the second week and so on. So if your Camaro had a time build code of 05D, that would mean it began life on the fifth month, fourth week; in other words, the last week of May. The time build code does not indicate which day a car was built, so weeding out Monday or Friday cars (notorious for their questionable build quality) is impossible.

Assembly Plant—Another interesting nugget of knowledge you can glean from the trim tag is the actual place of your Camaro's birth. It's right there on the top row of the tag, between the body series/style code and sequential unit number. Spot an **N** or **NOR** there and your Camaro was assembled in Norwood, OH. An **L,**

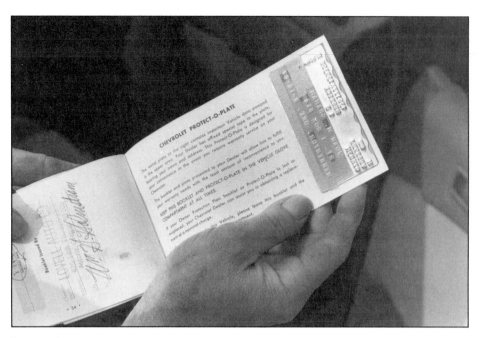

Protect-0-Plate came with the warranty book stashed in the glovebox of all 1967-72 Camaros. Think of it as your Camaro's G.I. dogtags.

Certification sticker on rear edge of driver's door on 1970-up models gives VIN, month of manufacture and GVWR.

LOS or **VN** in that spot indicates a Van Nuys, CA (also known as the Los Angeles plant) assembled car.

VEHICLE CERTIFICATION STICKER

This federally required sticker began appearing on the rear face of the driver's door of Camaros in 1970. It's a paper sticker with a clear vinyl overlay that basically states that on the month and year of manufacture, your Camaro met all applicable federal safety and emissions regulations. The sticker also includes the car's gross vehicle weight rating (total, front axle and rear axle) and repeats the vehicle identification number. On an original car that hasn't been repainted and still has its original driver's door, the VIN should match that on the VIN tag visible atop the instrument panel.

PROTECT-0-PLATE

Originally intended as an identifier to speed warranty service and replacement parts, the Chevrolet Protect-0-Plate was furnished in the glovebox of every 1967-72 Camaro when delivered to the customer. Chevrolet stopped the practice before the 1973 model year. The reversed-stamped metal plate is like a credit card that gives a complete description of the car's standard and optional equipment. Imprinted on the plate is engine, transmission and rear-axle type, exterior color, month of production, vehicle identifica-

tion number, the aforementioned optional equipment and other useful data. Needless to say, the Protect-0-Plate is an invaluable aid to the Camaro restorer, especially for documenting rare or special interest Camaros.

Two different Protect-0-Plate formats were used in Camaros, the first on 1967-68 models and the second from 1969 to 1972.

BUILD SHEET

Who would have ever guessed that a single 8-1/2 x 11-in. sheet of paper left behind by an assembly-line worker would be so sought after. Yet no other document describes the Camaro you see before you in more detail, right down to the color of the speedometer gear. At one time during the assembly process, every Camaro had one of these sheets accompany it down the line. If you find one in your Camaro, it will be easier to confirm its authenticity.

The build sheet takes the basic car and its selected options and tells the assembly-line workers what parts go in or on the car. For example, the nearby sample build sheet (pulled from a wrecking yard 1979 Z/28, IQ87L9L548317) shows an **NX** code under item 84, instrument cluster. Code **NX** instructs the assembly-line worker to install the RPO U14 performance gauge cluster and clock in the car. I pulled the cluster from this very same car and voila, atop the cluster housing is a 1x3-in. sticker marked **NX.**

Other useful information on the build

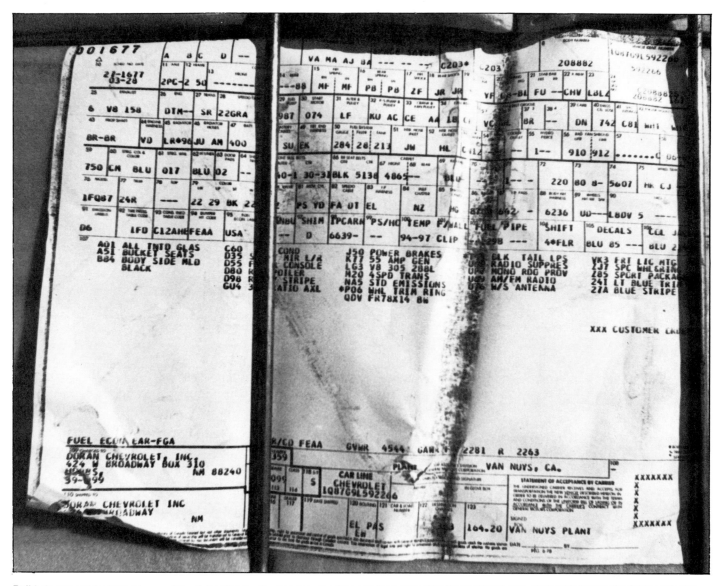

Build sheets can be found in a variety of locations—if you get lucky. The sheet on this 1979 Sport Coupe was stashed between the foam and seat frame of the rear seat back.

sheet includes a description of all the options the car was ordered with and their RPO numbers, the car's VIN, exterior paint code, interior trim code and the sequential body number (which should match that on the trim tag).

The bad news is that not every Camaro has a build sheet. These were sometimes discarded as the completed cars rolled off the assembly line. So your best chance of finding one is if it was hidden from view all these years. The sample build sheet was pulled from between the seat frame and foam backing of the seatback on a 1979 Z/28 destined for the crusher. Other loca-

tions build sheets have been known to be lurking are between the fuel tank and trunk floor, and under the driver's seat cushion. Go ahead and have a look. If you find one, it was time well spent.

LET'S GET STARTED!

Hopefully, we've piqued your interest to a fever pitch and you're ready to scour the countryside for a project car (if you don't have one already, that is). But before you get crazy, you should realize that the road to a completely restored, mint-condition vintage Camaro is fraught with

frustration, hard work, skinned knuckles, drained bank accounts and a sinking feeling that it will never get done. After restoring both a 1967 convertible and a 1973 Type LT Z/28, we are a testament to what it takes. The question is, was it worth it? You bet it was. The sweet melodious rumble of a V8, coupled with a pristine and shining 20-year old car that you've transformed from a planter to one that looks like it just rolled off the assembly line with your own hands, offer sensations that are difficult to describe. As you restore your Camaro, with the help of this book, you'll soon know what we mean. Good luck.

Selecting a Project Car

Finding that 1967 convertible alongside the road may get your juices flowing. But make sure you look before you leap and determine how much time and money it will take to get that dream Camaro into pristine condition. This one has extensive rust rot.

So you've got the bug for a Camaro you saw parked along the road on the way to work. It's just like the Camaro you've always longed for, a Rally Sport cloaked in red with black racing stripes. Your mind starts wondering what it would be like to drive it. "Wonder if it has a 4-speed or an automatic? Never really liked automatics. Like to row my own gears. Hmmm. Maybe it has a console and gauges like the car I saw at the auto show? If I'm lucky, it'll have a nice LT-1 or maybe even a big-block!" You just can't stop thinking about it. "Wish I knew how much they wanted for it." It seemed to be in pretty decent shape as you drove past, but you want to take a closer look before you make any commitments. That's a good idea.

When combing the market for the Camaro of your dreams, bring your good sense with you and leave your emotions at home. If you don't, you may get burned. You might even want to bring a friend along who can play devil's advocate, helping you to decide between the car you think you see and the car that you may end up buying. Many people see a car and fall in love with it if it's pretty on the outside. Unfortunately, beauty may be only skin deep.

Buying a used musclecar like a Camaro takes some research and thought before doling out your hard-earned cash. Many articles and books have been published about Camaros and you'd be doing yourself a big favor by reading as many of them as possible before taking a look at that "mint" Camaro.

SHOW CAR OR DRIVER?

But first, ask yourself one very important question. "What do I intend to do with this Camaro once I've plunked down the green? Do I want to build a show car? A rare model with investment potential?

Or an everyday or occasional driver?" The answer to these questions will color everything you do to your Camaro during the restoration. For that matter, it should be an important factor in the type of Camaro you buy and how much you pay for it.

THE DRIVER

If you're looking for a sound Camaro to drive and enjoy, your basic purchase considerations are no different than they would be for any other used car. You want a Camaro in solid mechanical and structural condition that has stopped depreciating and can be maintained with minimum fuss and investment. That means a strong, emissions-legal engine and drivetrain, tight suspension and steering, a firm brake pedal, no electrical gremlins, a straight body with no major crash damage and minimal or zero rust. If this particular Camaro has aftermarket wheels, a few chrome goodies under the hood and an aftermarket stereo, so what? Originality isn't as important here as good overall shape is.

We'd recommend that you stay away from high-mileage Camaros, although you should pay more attention to the condition of the brake and clutch pedals, front carpeting, driver's seat and weatherstripping than what the odometer shows. Aside from the illegal activity of rolling an odometer back, some cars are run with the speedometer cable disconnected to keep the miles off and some cars have "broken" speedometers replaced, then the owner "forgets" to tell the buyer that the speedometer is a lot newer than the car. The aforementioned "soft" items in the interior of a car will show quite a bit of wear on a high-mileage car and sellers seldom go to the trouble of replacing them. If you spot worn-through pedals and driver-footwell carpeting, broken-down and torn front-seat upholstery and torn driver's-door weatherstripping on a

Camaro with 40,000 miles showing on the odometer, be wary. It's probably more like 140,000 or 240,000 miles. Your guess is as good as ours.

Cleanliness—It's certainly more pleasant to buy a clean car, one that's enjoyed obvious care and service over the years. Perhaps the carpet is worn, but isn't littered with coffee stains and fast food wrappers. Maybe there's a little tear in the upholstery, but the owner stitched or patched it to keep it from spreading. Sure there's an assortment of the inevitable parking-lot dings, but someone went to the trouble of touching up the paint and installing bodyside moldings. Under the hood, you might find several year's worth of oil mist and road grime, but service stickers show regular maintenance and the oil on the dipstick is nearly clear. If the Camaro you're inspecting is giving off these kinds of cues, it probably deserves a new home—yours.

It is possible, though, for a Camaro to be *too* clean. By this, we mean over-detailed, or with fresh paint or undercoating that's there to hide some major faults. Usually, you'll find this type of Camaro on a used-car lot from the other side of the tracks. Maybe they've used so much ArmorAll on the interior that you're barely able to keep the steering wheel from slipping out of your hands. Or perhaps they steam-cleaned the engine compartment, then went hog wild with a can of Krylon and sprayed everything that didn't move—belts, hoses, spark plugs and all. Keep your distance from these Camaros.

Reality—Sure, a pedigree would be nice, but what you're really after with a driver is a Camaro you can depend on. A nice Sport Coupe with a medium-performance small-block V8, automatic transmission and Deluxe Interior that may not be cosmetically perfect isn't a bad way to go. Perhaps one

you can drive to the supermarket or movies and leave it parked in the lot without deploying a SWAT team nearby. You wouldn't do that with a ZL-1.

THE SHOW CAR

You can also view your purchase as an investment because many Camaros offer the combined advantages of being suitable for occasional driving yet appreciating at the same time. This is true of practically all first-generation cars and some second-generation cars as well. The Pace Car Replicas, ZL-1s and COPO big-block cars, Z/28s, Rally Sports and convertibles are the most coveted of the first-generation cars. Among the second-generation cars, 1970 through 1977 Z/28s, 1970-72 SS396s and 1970-73 Rally Sports are highly desirable.

Value—What makes a car valuable? It's that old rule of supply and demand again. Some Camaros are worth more than others

A clean-running, legally registered Camaro with minimal rust can make a decent daily driver. Early second-generation Rally Sports offer fantastic looks without the fussiness and insurability hassles of a high-performance drivetrain.

Yenko SC and other COPO big-block Camaros offer excellent investment opportunities and are an absolute blast to drive. Just make sure it's the real thing before you pay top dollar at auction.

In lieu of a complete service record, lube stickers can give clues as to a Camaro's past use or abuse.

Likewise, 1969 Indy Pace Cars are as good as gold. Look for the ZIO code on the trim tag.

17

because of their low-volume production, such as the aforementioned high-performance models and convertibles. Others are worth a bundle because they're equipped with lots of options, such as a bench-front seat, fold-down rear seat, tilt steering wheel, fiber-optics light monitor, 8-track cassette player, 4-wheel disc brakes or inflatable spare tire. The less popular the option originally was, the more desirable it is now.

Or then there's the combinations of trim and drivetrain pieces that work particularly well. One such car would be the luxury-oriented 1973 Type LT optioned with Z/28 drivetrain and suspension pieces and the split-bumper Rally Sport front end. Numerous first-generation SS/RS and Z/28/RS combinations look equally "right."

Still other qualifiers that tend to increase the value of a car is "first of a series" or "last of a series." Here, any 1967 or 1969 Camaro would be a candidate, as would the 1970 models, particularly Z/28s with the LT-1 engine. Likewise the 1973 models are significant as the last well- proportioned Camaros before the onset of the battering-ram bumpers. The 1974 Z/28 is special as the last of the pre- catalytic-converter cars, as is the 1977 Z/28 as evidence of Chevrolet rediscovering the performance formula.

What a car's value really boils down to is how much the buyer—you—are willing to pay for it. Beauty, or in this case the opportunity to reap a big, fat, juicy capital gain, is most certainly in the eye of the beholder. A rusted-out hulk of a 1969 Camaro pace car or ZL-1 is probably worth investing in, even though it'll take lots of time and money to restore it. All of which raises the next big factor—how do you determine the car's authenticity?

The authors of this book do not claim to have all the answers on how to determine originality. Read all you can about serial numbers and trim tags (page 12) in this book, gather together all of the magazine articles you can find and go to the library and read up on Camaro history. If you're going to pop big bucks for a rusty hulk that is supposed to be a very valuable Camaro, go in with your eyes open and your check-book welded shut. There are probably more 1969 Pace Cars and Z/28s on the road today than Chevrolet built—so beware of fakes.

And remember, you'll need enough money not only to buy the car but to restore it as well. Once you've taken the car apart and run out of money, its value will be decreased unless you find a buyer for it with a very strong imagination.

ORIGINAL VS. MODIFIED

If you plan to show the car, *originality is everything!* Unlike the driver, where hood pins, an aftermarket air cleaner or a stereo with speaker holes in the doors won't matter much as long as the car runs right, a show car must be restored to the same condition as it left the assembly line. Anytime a car is missing any of its original parts or had holes cut in its flanks to install aftermarket equipment, it will cost you money to find and buy the original parts and fix any of the sheet-metal hackery. Frankly put, and no matter how nicely done, modified cars are worth less than original ones. Period.

FINDING A CAR

Now, if you've got the bug to buy a Camaro but don't know where to look, don't despair. With nearly 700,000 first-generation Camaros built and almost 2

million second-generation Camaros originally registered, you've got a good chance of finding your car.

Start with the local newspapers. Watch the listings under Chevrolet, Classic Cars or Hot Rods for a few weeks and see what asking prices sellers are listing. Go ahead and look at a few cars before you're ready to buy. Then when you are ready to open your checkbook, you'll have a better idea of what's available. If there's an illustrated automotive classified newspaper in your area, check it out as well. At least, you'll have a fuzzy black and white photo to go by.

Enlist the help of your peers and friends. If you've joined a Camaro club, make a big announcement at the next meeting that you'd like to buy a car. Tell the guys down at the auto parts store and gas station. Hell, tell the butcher, the baker and the candlestick maker. They may have seen one alongside the road in a different part of town. If you frequent a parts store that has a bulletin board, check it out as well.

If you're really serious about buying a rare Camaro to restore, look in some of the nationally distributed magazines. Scan *Hemmings Motor News, Cars & Parts, Super Chevy, Road & Track, Autoweek. Chevy/Corvette Buyer's Guide.* just to name a few. It could be that if you live in the northern or eastern parts of the U.S., this is the only way you'll find a clean, rust-free Sunbelt Camaro. But before you buy a one-way ticket to Phoenix, AZ, and hop on a plane, be sure to have the Camaro checked out by a certified appraiser.

THE WALKAROUND

Now that you've found a Camaro for sale that sounds promising, it's time to check it out. But first, a few unbendable rules to adhere to:

Rule #1: Never inspect a car at night.
Rule #2: Be suspicious of fresh paint or undercoating.
Rule #3: Make sure the person selling you the car has clear title or proof of ownership.
Rule #4: Make no exceptions to the first three rules.

Exterior—Begin your inspection with a quick tour of the Camaro's exterior. Crouch down alongside each front fender and sight down the side of the car. If the doors and quarter panels look wavy, chances are this Camaro suffered some major collision damage. Crazed, cracking or crowfooted paint, probably the result of

Do your homework and research the market before popping for the first Camaro that catches your eye.

hastily conceived body and paint work, is another indicator of a quickie rust-out or collision repair. With your knuckles, rap the sheet metal along the lower edges of the fenders, doors and rear quarter panels. Healthy sheet metal will emit a hollow sound, while rusty metal will crunch and plastic filler will sound flat and non-metallic. Better yet, take along a refrigerator magnet that won't scratch the paint and stick it to those same areas you just rapped. The magnet will fall off areas filled with plastic filler and stick to the good metal. However, it may stick to areas where plastic filler was used correctly; that is 1/16 to 1/8-in. thick. So be aware. For a more detailed list of locations to check for rust on Camaros, see *Rust Never Sleeps* which follows.

Does this Camaro have fresh paint? If so, be suspicious of it being offered for sale so soon after painting. Look for dead giveaways of a quickie paint job, such as overspray on the fuel tank, weatherstripping, engine compartment, springs, exhaust and door stickers.

How does the car sit? If it leans to one side, the springs may be sagged or worse. Push down on each corner of the car and release. If it bobs up and down more than once, new shock absorbers are in the offing. And if you hear a groan when pushing down on the front end, suspension bushings may be shot.

Begin your walkaround by sighting down the side of the car, looking for wavy bodywork.

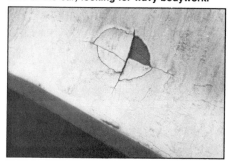

Check for plastic body filler with a magnet. If applied too thick, plastic filler can chip out as shown.

If the car has a vinyl roof, check it for cracked fabric that would let moisture in. If you spot any bubbles under the vinyl, push on the roof with your fingers. If you hear a crunch, the roof is rusted under the vinyl requiring expensive repair, perhaps replacement of the entire steel roof. This rust is predominately located around the edges of the vinyl roof and rear window, since the edges are most vunerable to water leakage.

Also check for exterior modifications that will be expensive to remedy. Items such as hood pins or an aftermarket hood scoop can only be corrected by replacing the hood. Radiused rear fenders or fender flares may require new rear quarter panels and outer wheel houses to return the car to original form. And run, don't walk, away from a Camaro with an aftermarket sunroof installed; once that big hole is cut in the roof, the only remedy is welding in a new one and that's a big, expensive job.

Don't worry so much about damaged or missing emblems, chrome trim, lights, bumpers or grille—these can be obtained from GM or Camaro aftermarket parts and accessories outlets. Just deduct their value from the cost of the car.

Oh yes, if the Camaro you're inspecting is a 1967-69 Rally Sport, make sure the hideaway headlamps are present and accounted for and functioning properly. Buying all these pieces from scratch will set you back hundreds, perhaps over a thousand dollars.

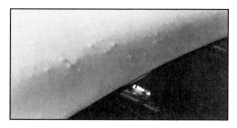

Bubbles under a roof indicate rust. It's very expensive to fix properly.

On a restoration, the only way to deal with aftermarket hood pins is to replace the hood.

Likewise, an aftermarket sunroof means bringing this 1973 Type LT back to stock will entail welding on a new roof.

Sagging door indicates the need for new door hinge bushings.

Interior—Open the driver's door. If it sags as it comes off the striker, the door's in need of new hinge pin bushings. It's a given that the Camaro you're looking at probably shows some wear on the soft interior trim parts such as the carpeting, seats, dash pad and door trim panels. No big deal here as all carpeting and most of the other trim pieces are available from the aftermarket, especially for first-generation Camaros. Take a whiff, though, and if the interior smells musty, check the carpeting for mildew that signals a leaking windshield or worse yet, a rusty floor. If you have any doubts, lift the rug and check underneath it. Camaros usually rust in the footwell areas. These areas can be replaced, but they take time and money. Also, a considerable amount of rust in this area may indicate the rest of the car is quite rusty and should be passed by.

Fire up the engine and check the gauges, lights, wipers, heater blower and radio. Do they work? If not, investigate why and deduct more points. Take a flashlight and have a look under the dash. If it's a rat's nest of spliced wires, connectors and electrical tape, you could be buying into a dash wiring harness job, which is about as much fun as standing on your head and spitting wooden nickels—for hours upon end!

If the interior smells musty, check for wet carpeting and rust holes in the floor. This rear seat cushion anchor has been rendered useless and the car may be unsafe.

Cracked steering wheel and gauge face plate, loose-fitting carpeting and sagging lower instrument panel indicate sloppy care on this 1970 RS. Several gauges were inoperative too.

Drooping headliner and torn seats with frame exposed through decomposed foam suggest a complete interior restoration is in order.

Holes cut in lower door panels for aftermarket speakers make interior restoration all the more difficult and expensive. Some interior trim pieces are no longer available, so used parts will have to be located.

Is the steering wheel cracked? If the wheel rim is shod with an aftermarket laced cover, ask the owner for a look underneath. Many 1967-72 Camaros will have cracked steering wheels and these are not inexpensive to replace.

Camaros that spent their formative years in the Sunbelt may be virtually rust-free, but as sure as the sun beats down 300 days a year, all of the molded plastic parts will eventually burn black and turn to dust. Some of these, such as rear sail panels, upper door panels, rear quarter trim panels and the like, are nearly impossible to find used and may not be available from GM or aftermarket sources. If you buy a Sunbelt car, plan on replacing lots of ABS plastic parts, all interior upholstery, plus all rubber weatherstripping, the dash pad and rear package shelf, paint and decals—in short, anyplace the sun shines.

Watch out for ham-fisted modifications, too. A dash cut out for an aftermarket radio, gauges or A/C system can be difficult to repair. Also, be on the lookout for doors, kick panels, rear shelf or rear quarter-trim panels hacked up for speakers. Some of these trim pieces will be difficult to come by.

Trunk—Out of sight, out of mind, they say. Like sweeping dirt under a rug, many sellers figure that what's under the deck lid is safely hidden from view. And that's exactly why you should take a peek. Does the trunk interior color match the rest of the car? If not, it's been changed from original. Are all the pieces to the jack there and does the spare hold air? How about the rear wiring harness; was it cut up to install a trailer harness? Lift up the trunk mat (or what's left of it) and check the trunk floor for rust. A serious rust-out here could involve the fuel-tank mounts and the leaf spring shackles—big, expensive repairs.

Check the inside of the rear quarter panels and rear end panel for signs of a collision repair. Small drill holes with hardened plastic filler oozing through tell you the repair was done the slide-hammer and Bondo way, not what you want as a foundation for a restoration.

Check trunk for condition of spare, jack, wiring and trunk floor. Here, evidence of recent quickie interior spray paint job was discovered.

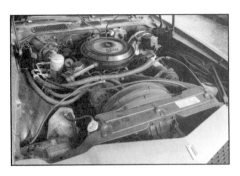

If you're lucky, engine will have all original emission-control equipment, such as air pump, EGR valve, heated air stove, vacuum control hoses and so on. If not, get out your wallet!

Engine Compartment—Pop the hood and have a gander. Is the engine original? Does its VIN match that on the dash or door jamb? If not, is it at least the correct type of engine for this year and model of Camaro? Big points penalties if they don't match up (see page 12 in Chapter 1 for details on reading serial numbers and trim tags).

Is all of the original emission-control system hardware in place? How about the air cleaner and intake and exhaust manifolds? If you're planning to use this Camaro for shows, these pieces must be correct and can be a bear to find in salvage yards. On a driver, aftermarket speed equipment may be OK, as long as the car meets any applicable state emission-control inspection or tailpipe checks.

How does the engine compartment look? If you've seen cleaner coal bins, chances are the rest of the car didn't receive much TLC or maintenance either. Pull the dipstick. If you spot white goo at the bottom of the dipstick—or the oil on the dipstick is a light, milky brown which feels like cream—a cylinder head or the block may be cracked, allowing coolant into the crankcase. This may also indicate a blown head gasket. Rust on the inside of the oil filler cap would tend to verify this. And unless the seller was running graphite motor oil, black, gritty oil on the dipstick says that it's been a long time since the last oil change. Add the cost of an engine rebuild into your restoration plans if you buy this car.

Unscrew the radiator cap and have a look inside. If you spot rust here, there's more of it inside the engine which can eventually clog water jackets and cause overheating. And while you're in the neighborhood, look for cracked and leaky radiator and heater hoses.

Undercarriage—Here's where you separate the honest sellers from the rest. Ar-range with a service facility to put the Camaro up on a lift. Any serious rust-outs or crash damage should be readily apparent. Be wary of fresh undercoating on the floorpan. Ask yourself, what could it be hiding? We once came upon a Camaro that had been cut in half, a weld running from rocker panel to rocker panel behind the front seats. It appeared to be a professional job, but you have to ask yourself, was it done right? Unless this is a one-of-a-kind Pace Car or ZL-1, we'd go on to the next Camaro. The chances of incurring structural or front-to-rear tracking problems would be too great.

How does the front end look? Do the tires show uneven tread wear or cupping? A front-end job may set you back hundreds of dollars. Are the shocks leaking? Are the grease fittings caked over with a buildup of old grease and dirt? Are the rubber boots cracked? Are the power-steering hoses leaking? All of these things will add to your restoration costs.

How about traction bars or a trailer hitch? If you find one of these, the Camaro has seen tough duty as a quarter-mile pavement burner or a workhorse. Either activity is tough on drivetrain parts.

Engine, transmission and rear-axle leaks can be fixed, but they're indicative of the car's overall condition. So is the exhaust system. If it's hung with coat hangers and patched with tape, beware of other repair and maintenance shortcuts elsewhere. Also check the fuel and brake lines. Leaks here could prove fatal.

This Camaro is really two cars welded together behind the front seats. Fresh undercoating hid most of weld, except at rocker panels (arrow).

Traction bars or helper springs are evidence of a hard life.

RUST NEVER SLEEPS

Rust is often referred to as *body cancer,* and aptly so. Chances are, if you find serious rust in one location, you'll find more of it somewhere else on the car. So if you find a rust-out when the car's on the lift, better have a more detailed search and destroy before making any offers. All the more so if the car is from the Rustbelt.

Camaros have their unique areas where rust usually starts. And while you're looking for rust, keep an eye out for body filler. It's not that you should avoid a car that has body filler on it, but keep in mind that the presence of body filler usually indicates extensive rust or possible accident damage. And this should be reflected in your offer for the car if you want to give it a new home.

Front-End Sheet Metal—Begin the inspection by checking the exterior sheet metal. Start at the front of the car and look at the valance panel. Since it's practically the first piece of sheet metal in the line of fire for stone damage, check it for chipped paint and rust.

Then move toward the rear of the car and look at the area of the front fender that lies directly behind and below the front tire. A small reinforcing panel is welded to the fender at this point to strengthen its lower attachment. Unfortunately, it also provides a trap for dirt. And once water gets in, the rusting process proceeds unnoticed until the exterior of the fender exhibits paint bubbles and sheet metal repair is in order. So check for this type of damage and keep it in mind if and when bargaining time comes along.

Doors—Moving to the doors, sight down the sheet metal and look for tell-tale waves of previous damage topped by poor repair. Also open the door and look at the door frame for evidence of rust. It's not unusual to find rust where the door skin wraps around the frame at the bottom of the door. Rust can also exist in the area of the lower door hinge. Doors with these types of damage may require a new skin and repair of the door frame to bring the door into presentable shape, so keep this in mind before handing over the big green.

Rear Quarter Panels & Tail Leaks—The quarter panel is another favorite area of the rust bug. With salt being used to clear the roads in many northern states, cars from this area are more prone to have rusted fender lips. And if the car has rusted fender lips, be sure to check the outer half of the wheel house for rust damage as well. If no signs of rust are apparent, gently tap on the

fender lip with your knuckles. A car with no or minimal filler will have a light ring to it but a car with doctored lips will produce a dull thud.

If the car you're looking at has this type of damage and you want to bring it back to original condition, keep in mind that quarter panels for first-generation coupes are still available but add to your restoration costs. And convertible quarter panels are practically non-existent. Coupe quarters have to be adapted to this application.

And while you're checking the quarter panel, check the quarter-panel extension directly behind the lower rear portion of the quarter. If the quarter panel is rusty, chances are the extension panel will be too.

Most common rust spot at front fenders is at the lower rear. Luckily, fenders unbolt and are easily replaced without welding.

This 1974 Camaro has nary a rust-free panel. Note rust atop windshield and backlite, at door bottoms as well as surface rust. Flush and dry rocker panels circulate air, so cancer of the rocker panels is rare.

Plan on welding on a new rear quarter panel and outer wheelhouse (as we did, page 92) if you buy this 1969.

The tail panel houses the taillights and ties the quarter panels together. This piece usually isn't prone to rust but it's one of the first things a tail-gater will rearrange if he hits the car. So check its exterior and interior surfaces for signs of accident damage such as dents or bowing.

Windshield Leaks—From outside the car, take a look at the area of the dashboard where it meets the windshield. The windshield and backlite reveal moldings sometimes chafe against the paint and the drive nails used to install the reveal-molding clips can cause a break in the paint. With the paint chipped or worn through, any condensation that collects here can cause rusting. Rust here can also be caused by an improperly sealed windshield which can be leaking at any point on its circumference. Minor rust in this area can be repaired by removing the windshield and replacing the rusted out areas with sheet metal. Major rust can necessitate replacement of the entire dashboard. And if the rust has perforated a sizable hole in the dash, chances are good that the cowl has an advanced case of rust also.

Convertible Considerations—Convertibles have more potential for rust because of the folding top. If the top wasn't or isn't properly sealed to the windows and windshield header, water can collect in the footwell areas or the trunk and cause rust. So be sure to check these areas. Also unlatch the top and look at the mating surfaces of the windshield header and top. Rust in these areas is not unusual. Severely rotted header panels can be replaced. However,

the top is a different story. A replacement part isn't available for the portion of the top that fits against the header. So if this area is rusted, it'll have to be repaired by welding.

TEST DRIVE

OK, So maybe the Camaro you're looking at isn't perfect, but after decades of use, it is still in one piece and an excellent restoration candidate. Let's go for a drive!

Start the engine and listen for unusual noises: the rattle of worn connecting-rod or main bearings, the slap of a piston with excess clearance, clattering valve lifters or a chattering clutch throw-out bearing. Pop the hood, remove the radiator cap (make sure it's cold first!) and look at the coolant. Bubbles may indicate a blown head gasket.

Back around to the rear and see (and smell) what's coming out of the tailpipe(s). White smoke on initial startup may just be water vapor from condensation inside the exhaust system that'll burn off after a few minutes (except in really cold or humid weather). But if white smoke continues, it could be a blown head gasket or leaking automatic transmission vacuum modulator. A vacuum test will settle that question. Black smoke could be a sticking choke or misadjusted carburetor. No big deal. Blue smoke, however, is burned oil and will require compression or leak-down testing to isolate the cause. If the blue smoke disappears soon

after startup, the cause is likely worn valve seals.

Still, it's a good idea for a friend to follow behind you on the test drive. He can check for tailpipe smoke under varying conditions and see if the car tracks straight. Ever see a car that "crabbed" down the road as a result of a bent frame? We have.

For goodness sake, check the brakes for a firm pedal before venturing out. Take side roads so you can evaluate the Camaro without traffic breathing down your neck. And once you've determined that it works or not, turn off the radio and listen for unusual noises from the deep.

Steering is also important. Does the car pull to one side with your hands on or off the wheel? Is there any lag in the steering? Stop and turn the steering wheel to full-right lock, then spin it to full-left lock. Does the power steering chatter and groan? When you hit a bump (and don't be afraid to seek a few out for the sake of science), do you hear any unholy clunks or groans from the suspension?

Nail the brakes. Does the car stop straight and without juddering? If the Camaro's an automatic, are the upshifts and downshifts positive, without slippage? On a stick-shift car, are the synchros in good shape? Is the clutch slipping or grabbing only at the end of its travel?

If this Camaro meets all of your expectations, warts and all, then start thinking about making the seller an offer. At the very least, you now have a better idea of what your restoration project will entail.

Before venturing out on that test drive, make a few underhood checks and listen for unusual noises. Watch for blue smoke emitting from the exhaust too, which would indicate that the car is an oil-burner.

It may not look like much now, but squint a bit and imagine it's a sunny afternoon in May of 1969. . .

Now that you have a Camaro to restore and you've researched its serial number and trim-tag numbers to determine what equipment it should have, it's time to get serious about its restoration. Perhaps your Camaro is looking a bit ragged around the edges. But squint a bit as you take another gander at it sitting in your driveway and imagine shiny paint, gleaming chrome, an inviting interior and that heavenly Bow Tie rumble from under the hood. Your Camaro's keys are jingling in your pocket as you anticipate that brisk autumn morning or warm summer night you fire it up and head out for your favorite winding stretch of two-lane road.

But that's not the only thing you're anticipating. If it was, you could have gone to an auction or perused the musclecar classifieds, dipped into the ol' checkbook and driven off in a pristine Camaro

Key to a successful restoration is getting organized and staying that way. The Camaro stashed in a barn will never see the road again if you don't develop and stick to a plan.

someone else already restored. Or maybe you'd take this Camaro to a restoration shop and say, "see ya six months and $10,000 later." No, the reason you bought a Camaro that needs tender loving care and the purpose behind buying this book is that you want to get involved. You want the satisfaction of bringing it back to life yourself. And even if you'll be farming out some of the work to shops and subcontractors, you'll know exactly what went into the restoration. Sweat equity has its rewards!

THE WORK AREA

It may sound trite, but don't take your Camaro apart in just any old place. Once it is up on jackstands, major pieces come off and it becomes an immovable object, you'll find out just how friendly the neighbors or the local police are. They may not share your vision of a dream Camaro weeks or months down the road. Driveway or curbside projects invite disaster: messy fluid leaks with stains that don't come out, lost tools, insufficient electrical power, nosy neighbors, kid's bikes, dogs and cats, traffic, weather....need we say more?

Obviously, the best place to undertake a major restoration project is within the cozy confines of your own garage. Just don't make it too cozy. At the very least, you need working space approximating that of a two-car garage—with your Camaro in the very middle. Think of your Camaro as a pool table, with enough room for Minnesota Fats to swing his cue around the entire perimeter of the table. Leave

Owning a Camaro is a lot more fun if you join a club and do some bench racing at car shows, rallies, tech sessions and so on. Here, United States Camaro Club members assemble for the 1989 West Coast Nationals. Photo by Bob Brennan.

You'll be buying a lot of new, used and reproduction parts during the course of your Camaro's restoration. So get to know who sells what. Here, Bob Brennan of Classic Camaro Parts & Accessories, Huntington Beach, CA (714/894-0651) fills the parts needs of a couple of Camaro enthusiasts.

Find a wrecking yard that will work with you during the restoration. We visited Dagley's in Phoenix, AZ, which specializes in used Camaro/Firebird parts.

Don't overlook swap meet bargains, either. They're fun and can be great exercise. Here, Jim Altieri of Arizona GM Classics, Tucson AZ (602/746-9300) helps a customer find the part he needs.

Leave plenty of working space around your Camaro. Always support the car on jackstands and make sure it's secure before venturing underneath.

space for the doors to open without hitting anything. You'll need room to get a floor jack under the front, sides and back. Planning to pull the engine? Leave room to wheel in that cherry picker. Front-end sheet metal coming off? Leave plenty of real estate to wield those unwieldy fenders and hood, and to store them out of the way where they won't get tripped over and stepped on once they're off the car. Devote another area to storing seats, soft trim, chrome trim, glass and small boxes and cans for small parts such as screws, clips, nuts, bolts and other fasteners. Figure it this way; Your Camaro will take up a lot more space disassembled into hundreds or thousands of pieces than it did as one big piece.

More considerations. Will you be painting the car in the garage? If so, turn off the water heater and furnace pilot lights and be prepared to hose down the floor wall-to-wall to keep dust down. Is the overhead lighting sufficient for you to see what you're doing without dragging a hot drop light, ready to melt a hole in the vinyl door trim or give your elbow one hell of a burn, over, under, around and through your Camaro? Maybe it's time to pop down to the hardware store and install some fluorescent overhead lighting. For that matter, do you have enough 3-prong electrical outlets to run all of your power equipment? And make sure whoever shares the garage with you that they know your Camaro isn't a convenient place to store old newspapers, pots and pans or

whatever. Just because it doesn't move doesn't mean it's a dumpster.

CLUBS & PEER SUPPORT

Whether you now own a Camaro you want to restore or are just thinking about buying one, it's always a good idea to band together with other individuals or join a club with similar interests. You won't be the first person who has restored a Camaro and you can certainly benefit from the experiences of others. They can warn you of common pitfalls, recommend shops and parts houses to buy from and those to stay away from, provide technical advice and moral support.

You can share in the fun of Camaro ownership. Many clubs with local chapters sponsor picnics, car shows, cruise nights, rallies, technical seminars, trips to race tracks or other recreational activities that are fun, often informative and are opportunities for you to show off your car. Plus club membership often entitles you to discounts at various local shops and parts houses which can save quite a chunk of change over the duration of a restoration. Some Camaro clubs, such as the **United States Camaro Club, Dayton, OH (513/426-6494),** have a full-time staff and publish a four-color bi-monthly national newsletter *(Camaro Corral)* chock full of technical how-to articles, club-member car profiles and with a comprehensive Camaro-only want-ad section. The want-ad section alone is worth the price of club membership.

CONTACT SHOPS & SUPPLIERS

Venture down to the local newsstand and pick up copies of magazines that cater to the hobby. Just to name a few, *Hemmings Motor News, Cars & Parts, Muscle Car Review. Hot Rod, Car Craft, Super Chevy, Chevy/Corvette Buyer's Guide.* and *All Chevy* are packed with mail-order companies that specialize in Camaro and musclecar parts and accessories. Send away for as many catalogs as you can. Most can be had for a few dollars and will credit that amount towards your first parts purchase. The catalogs will make excellent outhouse reading and whet your appetite to get started on that restoration. You'll be amazed at the comprehensive model-year coverage and variety that many of these companies offer.

Next, pick up the local Yellow Pages and check out the listings under Auto Body Repair; Auto Body Supplies; Auto Parts, New; Auto Parts, Used and Rebuilt; and Automotive Restoration, to name a few. Let your fingers do the walking and contact a few for further investigation. Pay them a visit and ask the owner/manager about past experience with Camaros or Firebirds. If an auto wrecking yard will let you walk the premises, check if they have a car like the one you're restoring that could function as a donor for needed pieces. You can learn a lot about what's missing or incorrect on your car by perusing similar models.

Establish a relationship with the companies you'll be doing a lot of business with. If a parts house manager knows he'll be selling you $1000 worth of merchandise over the next several months, he may be more willing to go the extra mile finding that hard-to-get piece. Likewise, if you're going to send your Camaro to a bump and paint shop for the bodywork, talk to the manager and explain what a special project this is for you. Maybe some of that enthusiasm will wear off and the shop will take extra care in making your Camaro's paint job the very best they can do.

ESTABLISH A BUDGET

But first, make a realistic appraisal of what parts of the restoration you can do yourself and what jobs you'll have to farm out. Doing a job yourself will save money and probably give you lots of satisfaction. But make sure you have the time to put your Camaro back together without taking shortcuts after scattering it to smithereens. And remember, your time is worth something. After you've perused through a few of the aftermarket Camaro parts and accessories catalogs, met your local friendly Chevrolet parts counterperson, toured a few wrecking yards to see what used parts availability and prices are like and talked to a few shops (upholstery, welding, body and paint, front-end alignment, radiator, air-conditioning, radio repair, to name a few) to determine what they charge for their services, you'll have a better idea of what you *could* spend, given an unlimited budget.

But the question is, how much *should* you spend? Well, that depends on a few factors. Factor number one is how much your Camaro will be worth restored when you (sob!) go to sell it. Most 1967-81 Camaros have stopped depreciating and many are rising in value at a rapid rate for the same reasons of supply and demand that govern most other economic forces. Old Camaros aren't being made anymore. Every time some drunken cowboy plows one into a light pole or rust eats the car in two, the remaining 1967-81 Camaros become just a bit more dear. That, of course, presupposes a continuing demand which should continue, barring any untoward economic catastrophe, government crusade against sporty performance cars or oil crisis. Your guess is as good as ours.

We'd be real frugal about spending much of the family fortune restoring a 1976 Sport Coupe to 100-point condition. On the other hand, we're not sure we'd take a second mortgage out on the house, but we'd not be shy about lavishing thousands, maybe tens of thousands on a 1969 ZL-1 or other COPO 427 car, Baldwin/Motion or Yenko special, 1967 or 1969 Pace Car or 1967-69 Z/28. For that matter, we'd find the extra bucks to pump into a worthy 1967-73 Rally Sport, 1970-74 Z/28, any convertible or big-block car, and most any first-generation car provided it was solid, rust-free, original and all there. For example, a rust hole in the quarter panel might get treated differently depending on the car. Getting back to our 1976 Sport Coupe example, perhaps welding in a low-cost patch panel and covering it with a little plastic body filler would be appropriate to a car worth maybe $2000-$3000 at best. On the other hand, that 1967 Z/28 would deserve nothing less than a new quarter panel devoid of any plastic body filler!

The trick is to establish a budget and then *stick to it*! Leave yourself enough time to do the project, too, because generally any time you have to do things in a hurry, the job doesn't get done right or you have to pay more to get it done. And after you've tallied up the total for getting all the special NOS (new-old-stock) and reproduction parts and contracted for all the special plating, assembly and other services, don't forget to add in the cost of the more mundane parts. Four tires (five if you need a spare to match), a battery, shock absorbers, brakes, exhaust system, wiper blades, air and oil filters, plugs, points, cap, rotor and ignition wires, belts and hoses, radiator cap, battery cables. . . . the list goes on and on. . . can easily add hundreds of dollars to the total.

LITERARY SOURCES

In addition to this book, the authors recommend that you read up on Camaros

Get a set of factory manuals for the model year of your Camaro.

as much as possible. This includes the aforementioned periodicals plus other reference material, such as factory manuals, and historical works.

At the top of the list are *The Fisher Body Service Manual, Passenger Car Chassis Service Manual, and the Unit Repair (Overhaul) Manual*. These manuals are available new from **Helm Inc., P.O. Box 07130, Detroit, MI 48207, (313/865-5000)**, used from various used book stores, automotive swap meets or used literature dealers such as those advertising in *Hemmings Motor News., Cars & Parts* and other magazines of the old car hobby. Also, reprinted manuals are available from some of those literature dealers and from many of the mail-order Camaro parts and accessories specialists, such as **Classic Camaro Parts & Accessories, Huntington Beach, CA (714/894-0651)**.

While you're shopping around, keep an eye out for a copy of the *Chevrolet Chassis & Body Parts and Illustration Catalogs*. These phone-book size treasure troves are available used from some literature dealers or new from GM's Warehousing and Distribution Division in Flint, MI, and are the same source your local Chevy dealer uses to look up and order parts for you—the Real McCoys.

An Au-Ve-Co body hardware catalog is an invaluable aid when searching for those hard-to-find fasteners. Ask the Au-Ve-Co distributor near you for one. H.C. Fasteners in Alvarado, TX (817/783-8519) provided the catalog shown here.

Another "must-have" is the *Passenger Car Assembly Manual*. It was originally developed for the Chevrolet or GM Assembly Division assembly plants as a guide to assemble the cars. One exists for each car line for each model year. GM does not and never did make these available to the public, but reproductions are available from literature dealers and Camaro parts specialists. The books offer fantastic detail on assembling every nut and bolt of your Camaro.

Year by Year Identification—The *Camaro White Book* by Michael Antonick, a pocket-sized compendium of facts on year-to-year changes, standard and optional equipment, color codes, production numbers and engine codes, is highly recommended. This highly accurate book covers 1967-85 models, but is lightly illustrated.

Camaro, Chevy's Classy Chassis by Ray Miller, a large-format, hardbound photographic treatise on Camaros from 1967-81. This book gives incredibly detailed information on 1967-69 Camaros, devoting some 200 pages to photo details on these models, standard and optional feature. There's a bit less detail (only 100 or so pages) and level of accuracy available for 1970-81 models, however.

Historical—An excellent book on history is *Camaro, From Challenger to Champion,* by Gary Witzenberg and published by *Automobile Quarterly,* 216 pp, plus a breathtaking color portfolio.

The Camaro Book From A to Z/28 by Michael Lamm is another excellent choice. It's published by HPBooks; 144 pp., 12 in color.

Both of the above historical titles present the reader with a wealth of detail on the design, development and evolution of first- and second-generation Camaros. These well-illustrated books follow the Camaro from its inception and first pencil sketches to full-scale clay models, engineering mules and early prototypes. Both devote considerable space to the Z/28 and other special performance models. Both feature a section on Camaro's exploits in racing over the years. And they're packed with informative charts and interviews with the movers and shakers behind the Camaro.

SAFETY TIPS

From time to time and whenever appropriate throughout this book, we stress safety when working on your Camaro. While all of these warnings are common sense, some bear repeating. You wouldn't want to have a pristine Camaro sitting in your driveway and be unable to drive it because of a needless accident.

Whenever you're working under the car, *always* make sure it is firmly supported. If you don't have access to a lift or a pit, jack up your Camaro and place sturdy jack stands underneath. At the rear, the outboard ends of the rear axle housing, the reinforced box members in front of the rear spring eyes or the rear spring perches themselves are good support locations. In front, place jack stands under the front subframe, as close to the wheel centerline as possible. If only one end is jacked up, make sure the wheels are securely chocked to prevent the car from rolling off. If your Camaro is going to be up in the air for a prolonged time, it may be a good idea to install two additional jack stands under the middle of the car (at the rear of the subframe) to keep things from sagging.

When working around a running engine, keep your hands off the front of the engine, clear of the drive belts and pulleys. Never wear a necktie, scarf, a necklace or bracelets or any loose-fitting article of clothing around your neck. Remember Isadore Duncan? Keep your mitts off the hot exhaust manifolds, too. And by all means, leave enough ventilation to prevent a build-up of carbon monoxide fumes. Also, keep a fire extinguisher handy.

There are a lot of things you shouldn't breathe when working on a car. One is brake dust. If your Camaro has old brake linings, they may contain asbestos which was used until the mid-1980s as a friction-material additive. Asbestos is a known carcinogen, which means it can cause lung cancer. Also hazardous to the lungs are a host of chemicals used to clean, prep and paint cars. Wear a respirator, not a dust mask, when working with acid-based paint strippers and whenever you paint. Period. Also, be sure to use a charcoal-activated respirator, and work in a well-ventilated area whenever you work with catalyzed primer or paint. Particulates generated by sandblasting can get deep into the lungs. For that matter, if you can see a cloud of dust as you sand a car, you should be wearing some sort of air filter.

Likewise, many things are bad for your eyes. Goggles should be worn whenever you're working with a grinder, polishing wheel, air chisel, sandblaster or any operation that may create particles that could penetrate the eye. Wear goggles and gloves when servicing or replacing any air conditioning system component.

Wear rubber gloves when working with paint stripper or other types of corrosive chemicals.

Never use gasoline to clean parts. It's highly toxic and can be absorbed into the blood system if you have any open cuts or wounds. And in a closed area, gasoline vapors are highly explosive.

Don't even go near a welder without undergoing some training such as that available at an evening adult vocational training class. Aside from the obvious fire danger, electric welders must be properly grounded or you could receive a mega-jolt, and you must wear a welder's hood or special dark-tinted goggles to avoid permanent damage to your eyes.

Whenever possible, have a friend help you when doing the heavy work; like installing a transmission or engine, mounting a subframe or rear axle, removing or installing a door. If something heavy were to fall on you, someone would be there to lift it off or call for help. If you are working alone, make sure someone knows where you are and can check up on you. A small, imported car once fell off its jack stands onto one of the co-authors and he was pinned there until a friend stopped by. That friend may have saved his life. But you may not be so lucky, so don't take any chances—it could happen to you.

Chevy took this publicity photo upon the introduction of the first Camaro to show under-the-skin details. Don't laugh. On a subframe-off, ground-up restoration, your Camaro will be in more pieces than this. Photo courtesy Chevrolet.

It's time to gather up your tools and your courage and begin taking your Camaro apart. We'll presume for now that at the very least, you want to prepare your car for body repair, painting and interior refurbishing, as well as reconditioning or replacing all the little trim parts that make such a difference in the finished product as you go along. Why remove all emblems and every piece of chrome? These pieces are terribly difficult to mask off properly and tend to collect old wax, road grime and car wash bristles that prevent new paint from adhering properly. Besides, you'll want to inspect behind these trim pieces for rust that often starts around the fastener mounting holes.

We'll also figure that in addition to a cosmetic restoration, you'll be refurbishing some mechanical items on your Camaro. For these items, we refer you to the chapters on *Driveline Reconditioning, Electrical System & Wiring Harness, Brakes, and Suspension & Steering.*

This chapter, *Disassembly*, is divided into five major areas: *Exterior Trim, Interior Trim, Engine Compartment, Trunk* and *Undercarriage.*

Organizing the disassembly process in this manner gives you the option of skipping over areas that are not applicable. Perhaps your Camaro's interior is like-new and all your car needs is a coat of fresh paint. Then, you'd bypass the interior disassembly section and go right to the areas that need the most attention.

Get Organized—You've probably heard the story about the guy who took his engine apart, rebuilt it and when it wouldn't start, realized that he had a few parts left over. It's no joke. Your Camaro is

constructed out of thousands of parts that will scatter to the four corners of the earth during the restoration process if you don't have a system to keep them organized.

The best organization method, we think, is to keep fasteners with the parts they fasten to. This is particularly helpful when a part is retained with many similar-looking but different specification fasteners, such as screws with the same head but different-length shanks. The only drawback to this method is when the fasteners are cleaned or polished or when the part is painted or reconditioned, the fasteners must be temporarily removed, and this is when they can get permanently lost. Still, it's easier to keep track of say 30 or 40 screws, clips, grommets, nuts or whatever for an hour, a day or a week than it is the thousands of parts in your Camaro over the duration of its restoration.

One method that works reasonably well is to keep fasteners and other small parts in labeled baby food jars, coffee cans or small boxes. Be sure to use a reasonably wide-tipped permanent, indelible marker so the label can be read six months or six years down the road. One can or jar could be marked "interior trim screws" and be the repository for the hardware for seats, door trim panels, dash panel, seat belts, radio and so on, although it's best to keep the screws with the part they belong to if at all possible.

The screws can be further separated by sandwiching them between two layers of wide, good-quality masking tape and labeled, for example, "headliner front garnish molding screws" and so on. Just don't make the same mistake one of the co-authors did by

27

On 1967-73 Camaros, some bumper bolts are exposed and easy to get to; others are hidden from view. See the nearby exploded views for details.

If equipped with vertical bumper guards, these must come off first.

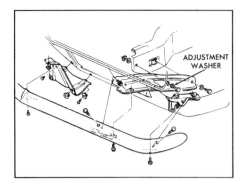

Front bumper details—1967-68 models, 1969 similar. Courtesy Chevrolet.

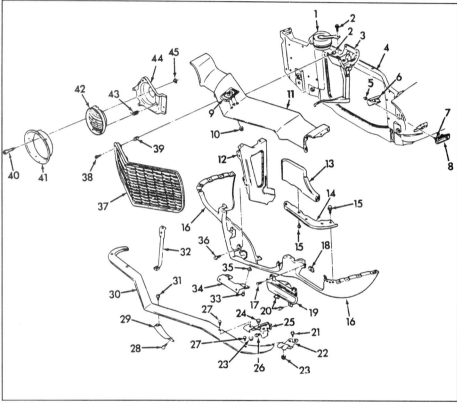

1970-72 CAMARO GRILLE AND BUMPER (EXC. R.S.)

1. Horn, Low Note
2. Screw
3. Catch Asm, Hood Lock
4. Support, Rad
5. Nut
6. Retainer, Side Marker Lamp
7. Lamp (#194)
8. Lamp Asm, Frt Fdr Side Marker
9. Emblem Unit, Hdr Pnl
10. Nut (3/16")
11. Panel Asm, Complete
12. Support Asm, Hd Lk Cat
13. Part Of Header Asm
14. Panel, Frt Bpr Filler
15. Screw (#10-16 x 1/2")
16. Panel, Frt Lower

17. Screw (#10-16 x 3/4")
18. Nut (#10-12)
19. Lamp Asm, Parking
20. Bumper, Frt Lic Plate
21. Screw (1/4"-14 x 5/8")
22. Bracket, Frt Face Bar Otr
23. Nut (3/8"-16)
24. Spacer, Frt Bpr Bracket
25. Bracket, Frt Face Bar Inr
26. Screw
27. Bolt Asm
28. Screw, (3/8"-16 x 1")
29. Guard Asm, Front
30. Bar, Front
31. Screw (3/8"-16 1 1/8")
32. Brace, Face Bar To Rad Supt

33. Screw (1/4"-14 x 3/4")
34. Bracket, Front
35. Nut, (1/4"-14)
36. Screw, (5/16"-18 x 3/4")
37. Grille, Rad
38. Screw (#8-18)
39. Nut, Radiator Grille
40. Screw, Hdlp Adj
41. Bezel, Hdlp
42. Sealed Beam, Hdlp
43. Retaining Ring Screw, (#8-12 x 11/32)
44. Mounting Ring
45. Nut, Hdlp Adj

Front bumper and grille details—1970-72 models except RS. Courtesy Chevrolet.

writing in non-permanent felt-tip pen and then leaving the taped screws out in direct sunlight where the ink faded and the tape dried out, allowing the screws to fall in a heap with other screws to the bottom of a confusing pile. Larger parts can be identified with paper tags tied on with string. These are available at many office-supply stores.

Keep Old Parts For Reference—We simply can't emphasize this enough. Never throw away old parts until you have the new ones and they've been successfully installed. Always compare new or especially reproduction parts against the old ones to see if the item is correct. Parts counterpersons and mail-order fulfillment people are only human and do make mistakes. Sometimes an old part will give clues as to how the new part should be installed. This could be a depression or scratches from a fastener, dirt smudges or paint overspray.

Even when you are absolutely sure you have the correct replacement part, don't go running for that dumpster just yet. Find out if a certain part is hard to come by, no longer available or in extreme demand and therefore valuable. Maybe you want to keep that old part around as a guide to have a new part made from or a similar part adapted to fit. If it's an assembly, perhaps some of the components can be reused (a screw, a gasket, a molding that's not supplied with the replacement part). And just maybe, your old discard in fair to poor condition is just what someone else is looking for because their's is in worse shape or is missing altogether. Possibly, you can trade it for something you do need or want. After all, isn't that what swap meets and flea markets are all about?

Surplus small parts that don't take up much storage room are a good hedge against future needs. Who knows, maybe you'll buy another Camaro to restore a year or two from now and wish you had saved those old parts.

Enough preaching already. Down your last cup of coffee or hot chocolate, put down the sports section of the paper and head out to the garage. The Camaro of your dreams awaits you.

EXTERIOR TRIM

FRONT SPOILER

On 1967-77 models so equipped, remove the front spoiler. On 1967-69 models, unbolt the three support struts and retaining nuts and bolts from the front valance panel and fender extensions. On 1970-77 models, there are just two support struts—one at each outboard end with nuts and bolts through the front valance. The spoiler is a 3-piece unit on 1974-77 models. Getting the spoiler out of the way gives better access to the front bumper.

FRONT LICENSE BRACKET

Odds on this item being reusable vary between slim and none. However, they are available from aftermarket suppliers. Unbolt the license plate first. Then, depending on model year, remove the screws which hold the bracket to the valance panel and/or the front bumper. The license bracket is built into the urethane cover on 1978-81 models.

FRONT BUMPER

Before you do anything, find the bumper-to-bumper bracket bolts on the rear face of the bumper and give them a liberal dousing with penetrating fluid. It'll also kill any spiders that may be lurking there as well. Tap the bolts lightly with a hammer, then let the penetrant soak in for 10-15 min.

On 1967-73 models, reach up behind the bumper and remove the nuts for the bumper carriage bolts. On first-generation Camaros with front bumper guards, remove these first; the Federally mandated bumper guards on all 1973 non-RS Camaros cannot be removed until the bumper is. The 1970-73 RS obviously has two bumperettes. Keep in mind that the outboard brackets for all 1967-73 Camaros are sandwiched in between the fenders and valance panel or fender extensions.

With 1974-77 Camaros, you have two options: remove just the aluminum basher bar or take out the entire bumper assembly. To remove the aluminum bar alone, remove the two bolts retaining each outboard end to the brackets and four nuts

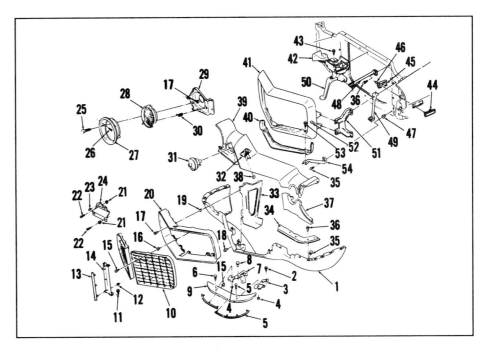

1973 CAMARO FRONT GRILLE "RALLY SPORT"

1. Valance Panel	20. Filler, Grille	38. Nut (1/8")
2. Screw (3/8"-16 x 1)	21. Attachment Unit, Lic. Plate	39. Header of Panel
3. Bracket	22. Attachment Unit, Lic. Plate	40. Reinforcement
4. Nut (3/8"-16)	23. Bumper	41. Face Bar, Center
5. Cushion	24. Bracket	42. Horn ASM, Vibrator
6. Bolt	25. Screw (1/4"-28 x 1 5/8")	(High Note)
7. Bracket	26. Screw (#8-18 x 3/4")	43. Screw (5/16"-18 x 7/8")
8. Spacer	27. Bexel	44. Lamp ASM, Side Marker
9. Face Bar	28. Headlamp	45. Retainer, Side Marker
10. Grille	29. Housing	46. Nut (1/2")
11. Screw (3/8"-16 x 1)	30. Spring	47. Nut (5/16"-18)
12. Nut (#10-24)	31. Lamp ASM	48. Brace
13. Cushion	32. Emblem, Header Pnl	49. Brace
14. Bar, Vertical Behind Grille	33. Support ASM, Hook Lk Catch	50. Catch ASM, Hood Lock
15. Screw (#8-18 x 3/4")	34. Panel	51. Bracket
16. Nut (#8-18)	35. Nut (5/16"-18)	52. Screw (3/8"-16 x 1 1/8")
17. Screw (1/4"-14 x 1")	36. Screw (5/16"-18 x 1")	53. Screw (3/8"-16)
18. Screw (5/16"-18 x 13/16")	37. Header ASM	
19. Nut (1/4-14)	(Part of Item #39)	

Front bumper and grille details—1973 RS, 1970-72 RS similar. Courtesy Chevrolet.

retaining the center of the bar to the brackets, then lift off the bumper bar. If you want to remove the whole shooting match, unbolt the valance panel. Then get a floor jack to support the bumper assembly (over 100 lbs!) and a friend to keep it balanced on there while you crawl underneath. Unbolt the bumper assembly (leaf springs, supports, braces and all) from the subframe and radiator support and carefully lower it to the ground.

On 1978-81 models, the bumper is integrated into the soft fascia nose piece and really need not be removed for painting. If you need to remove the nose piece to repair it or to gain access to remove the front end sheet metal, refer to Chapter 7, *Front End*

Sheet Metal & Subframe, page 76 for the removal procedure.

Remove all nuts with a hex socket to avoid rounding off the flats. If the nuts are rusted in place, chances are the carriage bolt will break and you'll be home free. However, if your toast always lands jelly side down, the bolt won't break but the square portion of the bolt will spin in the bumper.

If this happens, find the exact center of the bolt head and mark it with a center punch. Then drill into the mark 1/2-in. deep with a 1/8-in. drill bit. This allows you to make sure you're in the center of the bolt head and acts as a pilot for the second drilling operation. Provided everything's

On 1970-81 models, grille screws are accessible from the front. But before pulling on that pricey plastic piece, look for any hidden screws.

alright, use a 3/8-in. drill bit to drill a 1/2-in.-deep hole using the first hole as a pilot.

When you've finished drilling, insert a 5/16-in. or 3/8-in. punch into the bolt and hammer it out. Just don't get over-enthusiastic and nail your car with the hammer.

GRILLE

Generally speaking, getting this big item out of the way gives you elbow room to wrench on other front-end trim pieces. On all first-generation Camaros, grille removal is facilitated by first removing the bumper. On 1967-68 models, however, you must first remove the headlamp bezels and disconnect the parking-lamp leads on non-RS cars. Some fasteners are aluminum rivets which must be drilled out, then replaced later with new rivets or screws and nuts. Other fasteners are nuts and bolts, so you'll have to get a wrench down inside the header panel and valance panel to keep the nuts from spinning as you loosen the bolts.

Keep in mind that some of the nuts used on 1967-69 models to retain the grilles are not clipnuts. Therefore they'll spin if you try to turn the bolt or screw without using a wrench on the nut. On first-generation Rally Sports, you'll also have to contend with the headlight doors and linkage, headlight motors or actuators, wiring and hoses.

On 1970-81 Camaros, removing the grille is a simple matter of finding the hex screws recessed inside the front face of the grille (remember, two grilles—one above and one below the bumper—on 1974-81 models), removing them and lifting off the grille(s). On 1970-73 Rally Sport models, also unbolt the urethane grille surround from the header panel, valance panel and hood catch support and remove the surround. NOTE: You may later want to reinstall the surround before the car is painted to ensure a good paint match.

1967 Models—On models without the RS option, the grille is attached to the header panel with screws accessible from inside the panel. The bottom of the grille is attached to the valance panel with one screw in the center and 8 rivets. The rivets are accessible for drilling through the valance panel vents. 1967 RS models use nuts to retain the grille. They are accessible from the inside.

1968 Models—On 1968 non-R/S cars, nuts retain the top of the grille. Screws are used at the bottom of the grille and can be reached through the valance panel vent openings. R/S models use the same arrangement. However, the lower bolts and nuts are only accessible from behind the valance panel.

1969 Models—For 1969, GM changed the design of the grille so that its retaining screws could be reached from the front. As an added benefit, both non-R/S and R/S models have grilles that are held in the same manner. The screws that hold the upper portion of the grille can be found at the grille's leading edge. And the lower section of the grille uses screws that are right behind the front bumper.

1970-81 Models—All second-generation Camaros feature front-loaded grilles retained by screws that thread into snap-in plastic nuts.

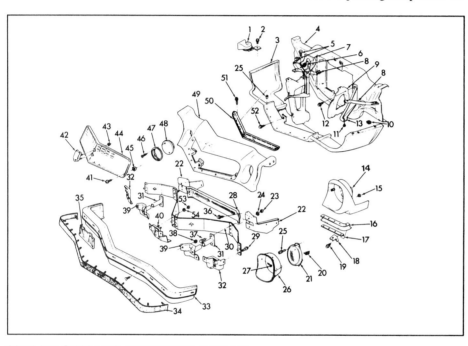

1974-77 CAMARO RADIATOR GRILLE

1. Horn Unit, High Note/Low Note
2. Screw, Hex (5/16"-18)
3. Panel, Front Bumper, Valance
4. Support Asm, Radiator
5. Screw, Hex (5/16"-18)
6. Nut, Hex Flg (5/16"-18)
7. Catch Asm, Hood Latch
8. Brace Asm, Front-LH, RH
9. Support Asm, Rad. Grille/Hdr. Pnl.
10. Nut, "U" (5/16"-18)
11. Bumper, Special
12. Screw, Hex (5/16"-18)
13. Support, Frt Bpr Jacking-LH
14. Housing, Headlamp-LH/RH
15. Nut, Spl Hdlp Adj
16. Panel, Frt Bar Flr-RH/LH
17. Retainer, Frt Bpr Flr Inr
18. Retainer, Frt Bpr Fil Pnl Otr-LH
19. Screw Asm

20. Spring, Headlamp Adj
21. Headlamp Capsule (7")
22. Bracket Asm, Frt Brp Spring-LH
23. Nut (7/16"-14)
24. Washer (15/32")
25. Screw (1/4"-28)
26. Bezel-LH/RH
27. Screw (1/4"-28 x 1 21/32")
28. Crossmember
29. Bolt
30. Spring Asm, Frt Bpr Face Bar
31. Tube Asm
32. Bracket Asm, Outer-LH/RH
33. Bar Asm (w/guards)
34. Strip Asm
35. Bracket, License Plate
36. Bolt Asm (3/8"-16)
37. Bolt

38. Nut (3/8"-16)
39. Guide Asm-LH/RH
40. Bracket Asm, Center
41. Screw (8-18 x 3/4")
42. Emblem Unit (1974-75)
43. Nut
44. Grille, Rad Upr (1974)
45. Nut, Special Anchor
46. Screw (8-18 x 1")
47. Ornament, Park & Sig Lamp
48. Lamp Asm, Park & Sig
49. Panel Asm
50. Grille, Lwr
51. Screw (8-18 x 3/4")
52. Screw (5/16"-18 x 13/16")
53. Nuts (7/16"-14)
54. Washer (15/16")

Front bumper and grille details—1974-77 models. Courtesy Chevrolet.

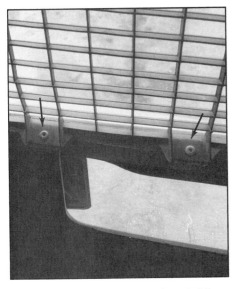

If original grille is still in place, rivets holding grille to valance panel (arrows) must be drilled out on 1967-68 Camaros.

When you've removed all the screws, gently pull on the grille to make sure that it's free. If it doesn't move, chances are there's another screw or bolt holding it in position. Don't just yank on the grille or you'll probably break it.

HEADLAMPS

All Except 1967-69 Rally Sport—If you haven't already done so, remove the Phillips-head screws retaining the head-lamp bezels and lift off the bezels. Next, with a long, flat-blade screwdriver, re-move the three screws for the headlamp retaining ring, and while holding the head-lamp, lift off the ring and disconnect the headlamp from its connector plug. Do the same for the other headlamp.

If further disassembly is required, re-move each headlamp adjusting bucket by disengaging its tension spring and sliding the bucket free of its two adjusting/aiming

screws. To remove the headlamp brackets, it is first necessary to remove the battery and evaporative canister to gain access to the mounting bolts on the back side of the radiator support. Remove these bolts and from the front, those retaining the brackets to the fenders, then lift out the brackets. With the headlamps and brackets re-moved, you can unbolt the outboard fen-der brackets on 1967-73 cars.

Except on 1967-69 RS cars, after removing the headlamp bezels, remove the retaining ring, pull out the sealed beam and disconnect its plug.

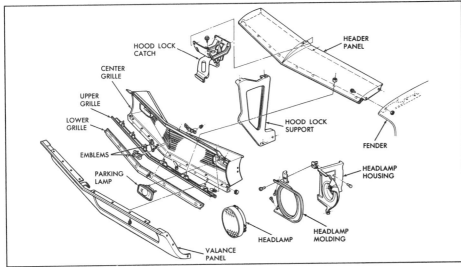

Details of 1968 grille. 1967 is similar. Courtesy Chevrolet.

Disengage headlamp tension spring and slide bucket free of its aiming screws.

On all 1969 models, remove all screws from the perimeter of the grille surround and headlamp bezels and lift off the grille.

On 1967-77 models, headlamp brackets must first be unbolted to remove the front fenders, header and valance panels.

On RS cars, headlamp doors can be unbolted from their pivots for painting and detailing.

To remove headlamp brackets of 1968-69 RS models, unbolt vacuum-operated linkage and lift out headlamp assembly complete with brackets. On 1967 RS models, electric motors for doors are on reverse side of brackets. Two bracket bolts are accessible from the front; two from inside engine compartment.

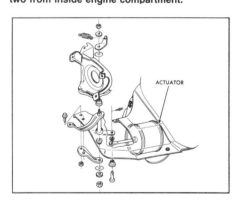

RS headlamp linkage details—1968-69 models. Courtesy Chevrolet.

Access to side marker nuts is from inside fenders. Twist bulb socket 1/4 turn to disconnect lamp assembly from harness.

1967-69 Rally Sport—To just remove the headlamps, you can turn the ignition on (and start the engine on 1968-69 models), turn on the headlamps (which should open the headlamp doors) then shut off the ignition before turning off the headlamps (this should leave the doors in the open position). Or you can open the doors on 1968-69 models by pressing the bypass switch on the vacuum reserve tank, or manually pull the headlamp doors open, taking care not to strip the plastic gears on the electric motors on 1967 models. With the doors open, merely unscrew the retaining rings and disconnect the headlamps from their 3-way plugs as you would for a non-RS car.

To remove the headlamp doors for painting and detailing, remove the nut and bolt for the electric motor or vacuum actuator arm at the lower outboard edge of each door, then the two nuts and bolts at the door's inner pivot. Take care not to lose any of the washers and bushings at these pivot points and take note of their orientation. With the doors off, the headlamp bracket assemblies can be removed complete with their adjusting buckets from the fenders and radiator support. To gain access to the bolts for the bracket assemblies, it's necessary to remove the battery and (on 1968-69 models) vacuum reserve tank.

PARKING LAMPS

For simple bulb replacement, remove the Phillips-head screws for the parking-lamp lens, remove the lens and rotate the burned-out bulb 45 degrees counterclockwise and remove it.

Except on 1970-73 non-RS models, all parking lamp assemblies are removed after unbolting their stud nuts from the rear and disconnecting their leads at the quick disconnect near the radiator support. On 1978-81 Camaros, you'll also have to remove the parking-lamp bezels, which are retained with Phillips-head screws.

On 1970-73 non-RS models, remove two Phillips-head screws and pull off the lens. If the lens is stuck to its gasket, carefully pry with a flat-blade screwdriver on one end with pulling up to break the seal. With the lens removed, you gain access to the two hex screws retaining the parking lamp assembly to the valance panel. Remove these hex screws, disconnect the parking-lamp lead at the harness near the radiator support and angle the assembly out through the valance panel opening.

SIDE MARKER LAMPS

These Federally mandated items are used on the front fenders of all 1968-81 Camaros and on the rear quarter panels of 1968-73 models. Wraparound taillamps eliminated the need for rear side markers beginning with 1974 models.

To remove front side markers, open the hood and remove the battery if you haven't already done so. From inside the fenders, unplug the bulb (at the end of its lead) by rotating 1/4 turn. Remove the two shaft nuts, lamp assembly retainer and remove the lamp (and bezel on 1968-69 models) from the outside. Rear side marker lamp removal is similar, except access is gained through the trunk.

HEADER PANEL EMBLEMS

With your Camaro's nose minus most of its trim, access to any badges, emblems and insignias you might want to remove is good. Using a box-end wrench or 1/4-in.-drive socket, remove the stud nuts for any header emblems and lift off the emblems. Also remove the engine-size emblem or trim-level emblem from the front fenders of first-generation Camaros in the same manner. However, any fender emblems that are secured with nuts and positioned between the rear of the wheel opening and the rear of the fender can only be removed with the fender off the car.

Before removing emblems from car, check first to see if they are retained with stud nuts, generally located underneath the panel. After removing both the nuts and the emblem, screw the nuts back on the emblem so they don't get lost.

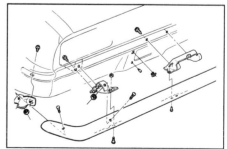

Rear bumper details—1967-68 models, 1969 similar. Courtesy Chevrolet.

REAR BUMPER

On 1967-73 Camaros, the rear bumper is retained to the body with eight bolts: two bolts inside each frame reinforcement, one at each corner (both accessible from inside the trunk) and two bolts running vertically into the center bumper bracket (accessible from outside). Before removing these bolts on 1967-69 models, first disconnect the license-plate lamp lead at the harness (it will come out with the bumper). Also, unbolt any rear bumper guards from the rear valance before removing the bumper.

On 1974-77 models, the aluminum basher bar can be unbolted from its leaf-spring support struts by undoing the nuts on the rear face of the bar. Or you can unbolt the bracing from inside the trunk (as on 1967-73) and underneath (adjacent to the fuel tank) and, after supporting the bumper with a floor jack, lower it to the ground.

On 1978-81 models, you may want to leave the soft fascia bumper in place for painting. But if it must come off for repairs, support it with a jack and from underneath, remove the Phillips-head screws retaining the corners of the soft urethane cover to the lower rear quarter panels and the four stud nuts retaining the bumper support through the rear end panel (access to these is partially blocked by the fuel tank and rear spring shackles). Then, from inside the trunk, remove the two small stud nuts retaining each corner of the bumper support to the rear quarters and four stout stud nuts through the rear end panel. Lower the assembly to the ground.

REAR SPOILER

Two types of fiberglass rear spoilers have been used on Camaros: a one-piece on 1967-70 models and a three-piece (adapted from the Firebird) on 1970-81 models.

Open the deck lid and unbolt the eight stud nuts retaining the spoiler to the deck lid. With three-piece spoilers, also remove the two nuts for each end piece inside the rear quarter panels. Unlike the front-end sheet metal, the rear spoiler should be painted off the car to assure good paint coverage at its leading and trailing edges.

REAR LICENSE BRACKET

On 1967-68 Camaros as well as 1978-81 models, remove the rear license plate from its bracket, then unbolt the bracket from the body.

A flip-down license-plate bracket which conceals the fuel-filler pipe is used from

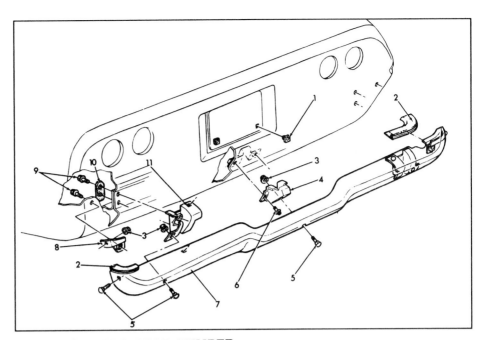

1970-72 CAMARO REAR BUMPER

1. Nut	4. Bracket	8. Bracket, Outer-LH/Bracket, Outer-RH
2. Filler-LH Filler-RH	5. Bolt (3/8"-16 x 15/16")	9. Bolt (3/8"-16 x 1 1/8")
3. Nut (3/8"-16)	6. Bolt (3/8"-16 x 1 1/8")	10. Seal
	7. Bar	11. Bracket, Inner-LH/Bracket, Inner-RH

Rear bumper details—1970-72 models, 1973 similar. Courtesy Chevrolet.

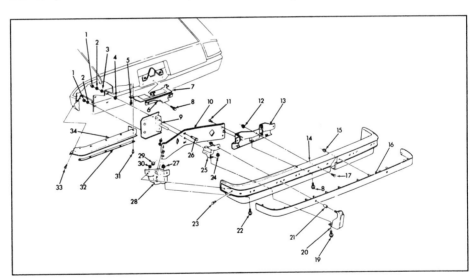

1974-77 CAMARO REAR BUMPER

1. Nut (7/16"-14)	14. Bar, Rear Bumper Face	26. Bolt (7/16"-14 x 3")
2. Bolt (7/16"-14 x 3)	15. Bolt (7/16"-14 x 1 1/2")	27. Nut, Support Bracket (3/8"-16)
3. Washer	16. Strip RR Bumper	28. Bracket Rear Bumper
4. Nut (7/14"-20)	17. Bolt (3/8"-16 x 1")	29. Washer (15/32" x 15/16")
5. Screw (8-18 x 5/8")	18. Screw (8-18 x 5/8")	30. Washer (15/32")
6. Retainer, Rear Bumper	19. Bolt (3/8"-16 x 1 1/4")	31. Screw (8-18 x 5/8")
7. Filler, Rear Bumper	20. Guard Asm, Rear	32. Retainer, RR Bumper Face Bar To Body
8. Screw, Hex (8-18 x 5/8")	21. Spacer	33. Retainer, RR Bumper Filler Panel
9. Spacer, Rear Bumper	22. Screw (3/8'-16 x 1 1/8")	34. Filler, RR Bumper Face Bar To Body
10. Spring Asm, Rear Bumper	23. Bolt (3/8"-16 x 3/4")	
11. Bolts, Hex	24. Nut, Support Bracket (3/8"-16)	
12. Nut (3/8"-16)	25. Bracket Rear Bumper	
13. Bracket Rear Bumper		

Rear bumper details—1974-77 models. Courtesy Chevrolet.

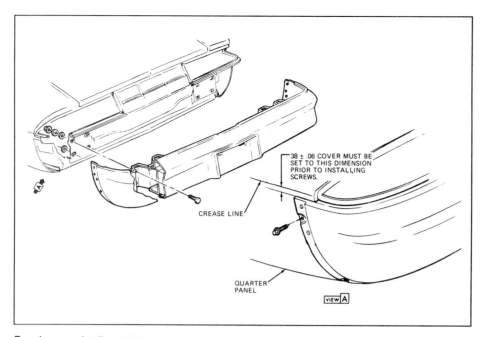

38 ± .06 COVER MUST BE
SET TO THIS DIMENSION
PRIOR TO INSTALLING
SCREWS.

CREASE LINE

QUARTER
PANEL

VIEW A

Rear bumper details—1978-81 models. Courtesy Chevrolet.

Remove eight nuts holding rear spoiler to deck lid. On 1971-81 models with 3-piece spoiler, remove two nuts for each end cap inside rear quarter panels.

All Camaro taillamp assemblies are retained from inside trunk.

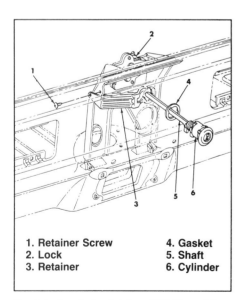

1. Retainer Screw   4. Gasket
2. Lock   5. Shaft
3. Retainer   6. Cylinder

Trunk lock cylinder details—1967-69 models.
Courtesy Fisher Body.

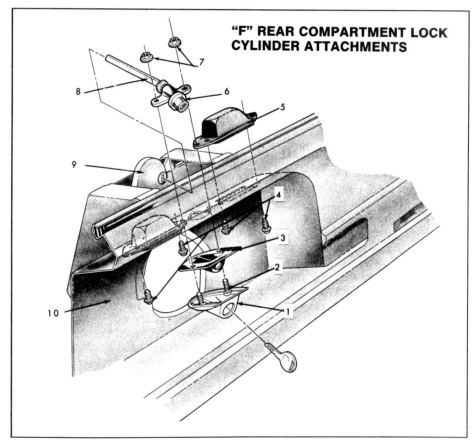

"F" REAR COMPARTMENT LOCK CYLINDER ATTACHMENTS

1. Lock Cylinder Housing
2. Lock Cylinder Housing Attaching Bolts
3. Lock Cylinder Housing Gasket
4. Lamp Assembly Attaching Bolts
5. Lamp Assembly Rear License
6. Lock Cylinder Assembly
7. Lock Cylinder Housing Nuts
8. Lock Cylinder Shaft
9. Rear Compartment Lock Assembly
10. Rear End Panel

Trunk lock cylinder and license plate lamp details for 1970-73 models. 1974-81 lock cylinders are similar. Courtesy Fisher Body.

1969 through 1977. If the rear plate is already off, the two bracket hex screws can be accessed through holes in the bracket. If not, flip down the bracket and remove the screws. NOTE: Do not lose the rubber bumpers for the license plate bracket.

TAILLAMPS

All Camaro taillamp assemblies are accessed from inside the trunk. For simple bulb replacement, unplug the applicable socket and insert a new bulb. To remove 1967-77 taillamp assemblies, remove their stud nuts with a deep-well socket. Plastic wing nuts retain the assemblies on 1978-81 Camaros. Unplug the bulb sockets and remove the lamp housings and lenses as a unit.

TRUNK LOCK CYLINDER

All Camaro trunk lock cylinders are removed from inside the trunk. Remove the lock cylinder retainer (one screw on 1967-69 models and two stud nuts on 1970-81 models), angle the lock cylinder and pull its actuating rod clear of the lock assembly. On 1970-73 models, remove the lock-cylinder bezel and its gasket.

LICENSE PLATE LAMPS

On 1967-69 Camaros, the license plate lamp comes out with the rear bumper.

On 1970-73 models, two small lamps straddle the trunk lock cylinder. From underneath, remove two Phillips-head screws for each lamp and push the lamp assemblies up and into the trunk. Disconnect the bulb sockets from the lamps and remove the lamp assemblies.

On 1974-77 models, remove two screws from inside the license plate cavity, then push the lamp up into the trunk. Disconnect the bulb from its socket.

TRUNK & REAR PANEL EMBLEMS

The rear of your Camaro is nearly devoid of all trim now. Remove any emblems from the deck lid or rear-end panel. From inside the trunk, check to see what kind of retainer the emblems have. Some will have hex-type stud nuts; others, nylon barrel-type nuts. Use a wrench to remove the stud nuts, then push the emblem out from the inside. Emblems with barrel nuts can be removed by protecting the paint immediately around the emblem with masking tape, then gently prying the emblem out with a thin, flat-blade screwdriver from the outside.

MAST-TYPE RADIO ANTENNA

All original-equipment radio installations in 1970-81 Camaros utilized an antenna molded into the windshield. But on first-generation cars, a conventional mast-type antenna (and many will argue that it's more effective, if not as stylish) was used. The standard location is atop the rear of the right front fender, but a rear antenna option mounts it on the right rear quarter panel.

Regardless of location, the antenna must be removed for painting and bodywork. With an open-end or adjustable wrench, remove the bezel nut and slide the upper gasket/washer and lower washer off the mast. If the antenna lead is threaded at the base of the mast, unscrew it and remove the mast. Otherwise, drop the mast down inside the fender for now.

REAR QUARTER PANEL & ROOF EMBLEMS

On 1969 Camaros, the die-cast scallop moldings in front of the rear wheel wells are affixed with stud nuts. Access to these is gained after removing the interior rear-quarter trim panels, and on convertibles, after removing the quarter window inspection panel. Or, if you don't mind snaking your hand through a small, blind hole, you can reach the nuts retaining the moldings after removing the Astro Ventiltion air-extractor grilles in the door jambs.

On 1973-81 Camaros, the model identifier badge at the lower part of the C-pillar is retained with barrel nuts. Pry these badges off carefully with a thin, flat-blade screwdriver.

SIDE MIRRORS

Two types of outside mirrors are used on Camaros: one is manually adjustable and the other is remotely controlled by a joystick located at the upper front section of the door trim panel.

Both types of mirrors are attached to the door in the same manner. A screw at the bottom trailing edge of the mirror secures the mirror to the mounting bracket. The mounting bracket and mirror gasket are attached to the door with two screws or rivets. To remove a riveted bracket, drill out the rivets, then reinstall after painting with new rivets. Even body-color fiberglass sport mirrors should be removed and painted off the car.

Remote-control mirrors have a hole in the door directly under the mirror mounting bracket for routing of the control cable. The control cable is attached to the door by

a clip and two screws. This cable uses three wires with crimped ends inside of a protective plastic sheath to control the position of the mirror.

Sometimes these crimped ends pop out of place or the wires corrode preventing adjustment of the mirror. The ends can be slipped into position with a pair of needle-nose pliers. If the wires are corroded, try spraying some WD-40 on each of the wires and working the control lever. This usually frees up the wires. If the wires are so badly corroded they won't move, you'll have to purchase a replacement mirror.

Emblems installed with barrel nuts can be carefully pried off. But look first to make sure!

Regardless of type, remove small set screw at rear of side mirror(s) and lift mirror off base.

Unbolt mirror base from door. On some 1970-81 models, base is riveted to door and must be drilled out to remove and replace mirror gasket.

HOOD & FENDER REVEAL MOLDINGS

These are thin chrome strips running along the upper rear edge of the hood and fenders along the beltline. They were available on early second-generation Camaros as part of the Style Trim group or Rally Sport option. The strips are a press fit onto the panels and usually can be gently pried off with a flat-blade screwdriver. But first, check the underside of the moldings for small Phillips-head screws.

Bright roof drip moldings on 1967-69 models are crimped on. To remove, pry delicately with a small putty knife. Brute force will bend and ruin the moldings.

If you don't plan to remove the front sheet metal, the front fenders will have to be un-bolted at the bottom rear and wheelhouse and pulled out to get the stud nuts for the fender emblems.

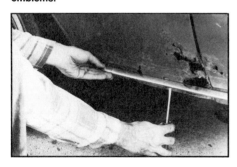

This rocker panel molding is retained with a screw at either end and clips to a rail along its middle. Remove the screws, then lever underneath the molding with a screwdriver to remove it.

BRIGHT ROOF DRIP MOLDINGS

On 1967-69 models, these stainless-steel moldings crimp onto the outside of the drip rails. Removal is a painstaking process of uncrimping the ends with a small, flat-blade screwdriver.

On 1970-72 models, these dress-up items were stainless-steel and screwed to the underside of the roof drip rails, beneath the door window (and quarter window on 1967-69 models) weatherstrip channels with Phillips-head screws. Beginning in 1973, the rails became slip-on chrome-look Vinyl pieces which unfortunately were not very durable. Then, in 1978, GM switched to press-on aluminum drip moldings that were only marginally better. The late-style plastic or aluminum drip-rail moldings pull right off. The 1970-72 style requires removal of the door weatherstrip and weatherstrip channel beforehand.

FENDER EMBLEMS

The "Camaro" script and (on 1970-74 models) engine size emblems at the rear of the front fenders are the next items to remove. On 1967-74 models, these were attached with stud nuts, requiring that you remove the fender lower attaching bolts and wheelhouse-to-fender attaching bolts and pull the fender out at the bottom to gain access to the nuts. Wear an old long-sleeve shirt to avoid scratching your arm on a jagged piece of sheet metal.

On 1975-77 Camaros, these emblems are attached with barrel nuts and can be pried off with a flat-blade screwdriver. Remember to protect the paint around the emblem with masking tape.

Beginning in 1978, the rearmost front fender emblems are adhesive-backed plastic. Remove these with a heat gun (or hair dryer) and a putty knife.

ROCKER PANEL MOLDINGS

These fall basically into two categories: one-piece moldings running the full length of each rocker panel and three-piece moldings running along the bottom of each door, fender and quarter panel between the wheel wells.

On 1967-69 models, the one-piece rocker-panel molding is retained by two small Phillips-head screws at each end. The screws are visible from the underside of the ends. Remove these screws, then gently pry the molding out of its 4-ft.-long hat-section channel. Should you have to remove the channel, it is screwed to the rocker panel.

On 1970-77 models, the three-piece rocker-panel molding is retained with a grocery list of fasteners. On the front fenders, the fasteners are stud nuts, accessed by loosening the bottom fender attachments, pulling it out and unscrewing the nuts. On the doors, there are five or six slide-on nylon clips and a Phillips-head screw at the rear of each door. To remove this molding, open the door and remove the Phillips-head screw. The door molding can then be slid forward and outward off its nylon clips or gently pried off with a screwdriver. The nylon clips can then be slid out from under their drive nails or posts and removed. On the rear-quarter panel, the molding is retained with three of these slide-on nylon clips. Gently pry each molding off the clips.

On 1970-81 models, a one-piece rocker-panel molding is retained with the slide-on nylon clips mentioned above and can be pried off in a similar manner.

BODY SIDE MOLDINGS

Because of the dramatic curvature of the doors on second-generation Camaros, they are particularly prone to receiving parking-lot dings and dents. For this reason, many 1970-81 Camaros were fitted with original, dealer-installed or aftermarket body-side moldings designed to offer protection in this area. Problem is, they tend to clutter the lines of an otherwise clean design. In the course of a restoration, you may want to remove these moldings and vow never to leave your Camaro in a parking lot again. One of the co-authors did on his Camaro type LT.

As with the three-piece rocker-panel moldings, GM-original body-side moldings mount to the front fenders with stud nuts, to the doors with one stud nut and slide-on nylon clips and to the rear quarters with more slide-on clips—and they are removed in a similar manner. Note that the stud nut for the door body side molding is on the inside of its leading edge (near the door hinges). If you plan on doing away with these moldings, you'll have to grind off the weld studs or drive nails for the slide-on clips and braze up the holes for the stud nuts.

WHEEL-LIP MOLDINGS

It's unlikely that these moldings, used only on first-generation Camaros, will be reusable after receiving direct hits over two or more decades of flying stones, sand

and salt spray. But they should be removed to get a good look at the wheel well pinch flanges they cover.

Removal is simple enough. Unscrew the Phillips-head screws and pull off the moldings. If the screw heads are gone, try cutting a new slot in them with a hacksaw. Or, using a 9/64-in. drill bit, drill out the screw heads.

DOOR HANDLE & LOCK CYLINDER

To gain access to these items, you'll first have to remove the interior door trim panel. Each handle is attached with two hex screws and the lock cylinder with a big C-clip. See pages 113 and 114 in the *Door Reconditioning* chapter for removal of the handle and lock cylinder.

VINYL TOP REVEAL MOLDINGS

More slide-on nylon clips fastened to weld studs or drive nails here. Merely pry off one end of the molding, slide it out of the clips, then remove the clips from under the drive nails. If you're deep-sixing the vinyl top (one of the co-authors did), you'll have to grind off the weld studs on the roof, rear quarters and tulip panel, then braze up the holes.

On the other hand, if you plan to keep the vinyl top and it's in absolutely stellar shape (which would be a rare Camaro indeed), it's probably a good idea to leave the vinyl top reveal moldings in place as they help secure the edges of the fabric. Merely mask off the top and moldings during painting.

WINDSHIELD WIPERS

On second-generation Camaros, open the hood. Place a block of wood on the cowl to act as a fulcrum point and to protect first-generation Camaros from paint scratches. Depress the small tab on the underside of each wiper-arm pivot and with a large, flat-blade screwdriver under the pivot head, lever off each arm.

Second-generation cars with Hide-A-Way wipers have an extra articulated link on the left-side arm. It connects to its pivot with a small linkage clip. Lever the clip off, disconnect the link and keep the clip with the arm.

WINDSHIELD & BACKLITE REVEAL MOLDINGS

Here's one operation that requires a special tool—a windshield reveal-molding-clip removal tool (see nearby photo). You can buy one at any good auto-body paint and supply store, or mail-order from the **H.C. Fastener Company, Alvarado TX 76009 (817/783-8519).** This tool looks like an offset putty knife with a hooked end.

Door lock cylinder is retained with a big C-clip, accessed after removing door trim panel.

On 3-piece rocker molding, remove this screw hidden behind the door pinch-weld flange.

Wiper arms can be tricky to remove. On first-generation Camaros, you can use a pry bar and wood block to lever the arms off. Use care not to damage the cowl.

Some molding sections can be popped off their nylon clips with a small flat-blade screwdriver.

Vinyl-top reveal moldings can be pried off with a tack puller.

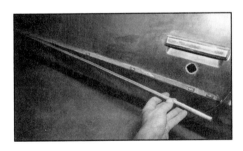
Most 1970-81 Camaros had body side moldings to protect their flanks from parking-lot dings. After removing hidden screw at front of doors near hinges, these can be pried off their clips.

Continue working around the periphery of the window. The molding will pop up when it's loose.

To remove all windshield and backlite reveal moldings except the lower windshield molding on 1967-69 models, you'll need this tool. Insert the hooked end between the glass and molding, engage each clip (about 15 per window) as shown and lever the clip away from the roof slightly to relax its grip on the molding while pulling up on the molding.

On 1967-69 Camaros, lower windshield trim is secured by brackets that are screwed to the cowl.

Header trim screws are usually rusted in place. If you can't screw them out, drill them out. Header and A-pillar trim for convertibles will likely have to be pried loose after removing the screws because of trim sticking to body caulking.

If replacing the windshield on a 1970-81 Camaro, use a 7/32-in. Allen wrench to remove the set screw for the rearview mirror.

Also on 1970-81 models, don't forget to disconnect the radio antenna lead to the windshield.

However, this tool cannot be used to remove the lower windshield moldings on 1967-69 Camaros. These cars use special clips that are screwed to the cowl to retain the molding. Remove the cowl panel to access these screws.

To remove all other reveal moldings, insert the tool under the molding, and above the glass, then slide it along perpendicular to the molding until the hook engages one of the numerous reveal molding clips. Disengage the molding from each clip by pulling the tool away from the roof (which relaxes the clips grip on the molding) while gently lifting up on the molding. Works every time!

After removing the sections of stainless-steel molding from both windshield and backlite, remove the clips. Most clips slide onto drive nails and can be popped off with a screwdriver. Some are screwed in. The area behind these reveal moldings is a notorious area for rust on second- generation Camaros, so inspect it carefully.

REAR WHEEL SKIRTS

On 1980-81 Z/28s, Chevrolet added a ground-effect urethane skirt to the leading edge of each rear wheel well. It merely screws to the wheel opening pinch flange with Phillips-head screws.

QUARTER WINDOWS

On 1967-69 Camaros, you may want to remove the rear quarter windows. Access to their regulators, stops and tracks is gained through the interior rear-quarter trim panels, in a similar fashion to the side glass in the doors. See page 43 on removing the trim panels.

WINDSHIELD & BACKLITE

Over the years, your Camaro's windshield sees a lot of abuse. Whether it's a stone chip, crack, windshield-wiper scratch, sand pit or discoloration due to long exposure to the sun, chances are your Camaro needs a new one. However, replacement is certainly not a necessity just to paint your car, so don't remove the windshield and backlite just for the fun of it. It's no fun, really.

All Camaro windshields and backlites are glued in with butyl-rubber adhesive. This rubber seal has to be cut in order to remove the glass. You'll need a windshield-seal cutting tool, a bicycle cable core or piano wire, electrical tape, two short lengths of PVC pipe, an ice pick and needle-nose pliers to do this job. And windshields are heavy and cumbersome,

so get a friend to help with removal. They make a big mess when they're dropped.

The wiper arms, 1970-81 windshield-antenna lead and inside rear-view mirror, all reveal-molding clips and the windshield mounting pads must first be removed. Also helpful is to remove the inside sun visors and anything else that might get in the way.

To use the windshield-seal cutting tool, force it through the rubber seal between the glass and roof. While exerting downward pressure, pull it toward you along the seal. Do this around the entire perimeter of the glass.

To use the cable core or piano wire method, puncture the seal with an ice pick, then thread a 3-ft. length of cable through the hole so it is half inside and half outside the car. Make some handles by wrapping the ends of the cable several revolutions around some 3/4-in. PVC lawn sprinkling pipe and secure them in place with electrical tape. Then pull the ends of the cable tight and draw it towards you as you work it around the perimeter of the seal, cutting it in the process. It's easier to cut the seal when it has been in the sun to warm it.

Be careful. The windshield is very heavy and needs to be supported as the last of the seal is cut through. Repeat the procedure to remove the backlite, if you're replacing it as well.

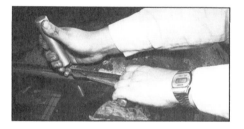

All Camaro windshields and backlites are bonded in position with butyl rubber adhesive. To remove the glass, break the seal with this cutting tool.

Or cut the seal with piano wire or a bicycle brake cable core wrapped around and taped to two handles made from PVC lawn-sprinkling pipe. However, pull the wire toward you, not away as this guy did.

Glass is heavy and cumbersome. To avoid dropping it and injuring yourself or your Camaro, get a friend to help lift it out or use a pair of suction-cup handles from a tool rental store.

DECK LID

If the deck lid needs minor bump work, it can be done on or off the car. However, if it needs metal repair at its trailing edge due to a rusty hem flange, you're better off removing it. This will provide better work access.

Note that removal of the torque rods isn't necessary to remove the deck lid. The deck lid on all first-and second-generation Camaros is retained to the hinges with four bolts; two on either side. Before removing these bolts, trace the position of the deck lid hinges with a sharp china marker or pencil to aid alignment upon assembly. Do not do this with a metal scribe, as you'll scratch through the paint on the deck lid and give rust a place to start.

Camaro deck lids are not heavy, but it's not a bad idea to have an assistant support yours as you remove the last bolt. If you're working alone, place some heavy towels between the front of the deck lid and body to prevent scratching the paint. When you've done this on both sides, remove the retaining bolts and lift off the deck lid.

WEATHERSTRIPPING

Keeping water and wind out, and dry, climate-controlled air in, is the job of your Camaro's weatherstripping. It's a tough job being a weatherstrip. After decades of exposure to sun, moisture, road salt and freezing temperatures, and being squeezed and scraped with repeated window, door, hood or deck lid openings and closings, these sponge-rubber strips get flattened, torn or dried out and need to be replaced. So unless your Camaro's weatherstrips were replaced recently, plan on getting a new set. Even if the weatherstripping is in good shape, it's a good idea to replace those that cover channels that need paint or those that would suffer from paint over-spray.

Trunk Weatherstrip—This one-piece seal is glued in a channel around the trunk opening. Find the open ends near the lock cylinder and using a 1-in. wide putty knife or wide, flat-blade screwdriver, scrape the seal out of its channel. If the old adhesive bond is particularly tough and you plan to paint the car anyway, use a propane torch to loosen the adhesive and pull the weatherstrip out of its channel.

Roof-Rail Weatherstrips—These fit into channels running from the base of the A-pillar to the base of the C- pillar. They're glued to the channel with weatherstrip adhesive and retained at each formed end with wide Phillips-head screws or plastic rivets. Remove the screws or pry out the plastic rivets with a tack puller and pull the weatherstrip out of its channel. Use a narrow putty knife, if necessary, to break the weatherstrip adhesive seal. If the plastic rivets are still in good shape, save them because many weatherstripping kits do not include them.

On 1970-81 Camaros, removing the roof-rail weatherstrip from the C-pillar reveals a vertical trim molding on the leading edge of the C-pillar. It's fastened with special, flush-fit Phillips-head screws. Remove the screws and the molding and don't lose those screws!

You can remove deck lid torsion rods with locking pliers or a length of pipe. Three notches that torsion rods fit into permit adjustment for amount of assist. Lowest notch is for cars with spoilers or weak torsion rods.

If you plan to paint the car, glued-in seals such as the trunk weatherstrip can be removed with a gas torch and screwdriver or putty knife.

Roof rail and convertible A-pillar weatherstrips are retained in a channel by weatherstrip adhesive and a lip on their lower side.

The ends of some weatherstrips are secured with button-sized plastic rivets. Carefully lever out these with a tack puller and save them because they're not often supplied with replacement weatherstrips.

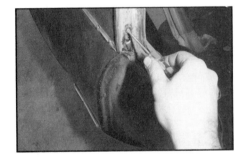

Vertical trim molding on roof pillar of 1970-81 Camaros is retained with special flush-fit, short-shank Phillips head screws. Mark these hard-to-find screws for reuse.

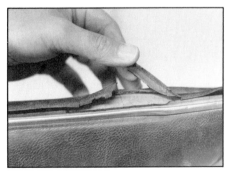

When window felts wear out, water drains down into the doors causing corrosion, and the glass scratches against hard metal parts as it is rolled up and down.

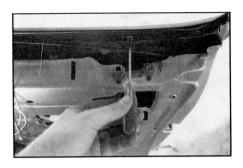

Door window outer felts on 1970-81 models are retained with plastic rivets or short-shank screws. Access to screws or rivet heads is easier with the door trim panel removed and glass stabilizers loosened.

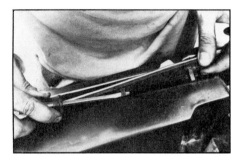

Door window outer felts on 1967-69 models are attached with screws and barb-type metal clips. To release clips, slide a screwdriver blade under each clip and gently rotate the weatherstrip bottom upward.

Door weatherstrips have steel-reinforced formed ends retained with screws or plastic rivets and plastic T-clips every 5 in. or so in between.

Quarter window inner felt on 1967 models is retained with barb-type clips and screws at either end. Also note other replaceable items: (1) U-shaped 1967-69 door jamb seal (make sure replacement has steel core) and (2) slide-on quarter window vertical weatherstrip.

T-Top Weatherstrips—On 1978-81 Camaros so equipped, it's a good idea to put fresh rubber between the top of your head and the elements. T-tops are notorious leakers! These weatherstrips are secured like the roof-rail weatherstrips—glued into the channel and fastened with screws or rivets at their formed ends.

Inner & Outer Window Felts—Otherwise known as "fuzzies," window felts are the flocked seals that press against movable glass—the door window glass and on 1967-69 Camaros, the rear-quarter window glass. When the outer window felts wear out and dry out, they allow water to drain down into the doors and on 1967-69 models, rear quarters and rocker panels. When the inner felts wear out, the glass can become scratched as it is rolled up and down.

The outer window felts incorporate a thin stainless-steel or chrome bead trim piece. On 1967-69 Camaros, these are attached to the body with barbed hooks and screws at their ends. On 1970-81 models, the outer felts are attached with plastic rivets or short-shank Phillips-head screws. To remove either type, roll down the window to its bottom stop. Remove any screws you can gain access to with a straight or offset Phillips-head screwdriver. On the early-style felts with barbed hooks, slide a thin, flat-blade screwdriver behind the molding from the topside at a 45-degree angle. Start at the end of the molding that's most accessible and gently pull the window felt up and toward the glass as you move the screwdriver along the backside, feeling and looking for the clips. When you find them, twist the screwdriver to open them. Proceed until the felt is removed. It may be necessary to remove the door or rear-quarter trim panel and remove the bottom stop for the glass so it can be lowered farther for access to the other screws. On late-style felts with plastic rivets, pry off the rivets with needle-nose pliers and get some short-shank screws to reinstall the new felts at assembly time.

Inner window felts are also incorporated into a trim molding with a stainless-steel or chrome bead. These are stapled to the door or rear-quarter trim panels with special heavy-duty, reusable staples in many applications. However, the window felts on all 1967 models are secured to the window with screws and barbed hooks. Remove them following the same procedure for outer felt removal. See page 120 in the *Door Reconditioning* chapter for details.

Door Weatherstrips—These seals run from the beltline at the rear of the door, down around the bottom and up to the beltline at the front of the door. They're fastened with a pair of wide Phillips-head screws or plastic rivets at each end and with plastic "T" clips along their perimeters. Pull the clips out with needle-nose pliers and lever out the rivets with a tack puller. See page 119 of the *Door Reconditioning* chapter for details.

Door Jamb Seals—On 1967-69 Camaros only, this U-shaped rubber piece seals the forward edge of the rear quarter windows to the rear quarter panels. Remove its two Phillips-head screws and slide the seal up and out. When buying replacement seals, ask for the type with the metal core. These cost more than solid rubber seals, but they are definitely worth it.

Rear Quarter Window Vertical Weatherstrips—These are used on 1967-69 Camaros only to seal the rear-quarter windows to the side windows. To remove them, remove the small Phillips-head screw at the lower end of each seal and pull them out of the channel.

Convertible-Top Weatherstrips—See page 160 in the *Convertible Top and Vinyl Top* chapter.

Hood-to-Cowl Seal—This one lives under the hood and as such, gets exposed to oil mist and lots of heat. It's designed to keep engine heat, oil mist and whatever water and road debris splashes up through the engine compartment out of the interior ventilation/heater air-intake at the base of the windshield. On 1967-69 Camaros, it's a one-piece weatherstrip attached to the front edge of the cowl vent grill, under the hood. The 1970-81 models have a three-piece seal, the center portion attached to the underside of the hood and two short end pieces on the cowl. The second generation seals are attached with plastic "T" clips which can be levered out with a tack puller or pulled out with needle-nose pliers.

Getting front seats out of the way first gives a lot of working space to remove interior components. Remove two rear bolts for front seats first; otherwise the seat will fall back on you as you attempt to remove the rear bolts.

INTERIOR TRIM

SEATS

Except for a few 1967-68 Camaros with the rare front bench-seat option, all Camaros have bucket front seats. All front seats are mounted to the floor with four bolts. Move the seat forward to the end of its track and remove the two rear-most seat bolts. Then, slide the seat all the way back, exposing the front bolts, and remove these. If you remove the front bolts first, the seat will fall back on you as the rears are loosened. On 1974-81 models, unscrew the Phillips-head screw retaining the front shoulder-belt guides to the seatbacks. Also check for an electrical connector for the seat-belt warning light under the seat and disconnect it. Lift out the front seat(s).

At the rear, attack the seat cushion(s) first. Kneel in one of the rear footwells and push rearward with all your might against the very bottom of the rear seat-cushion frame while lifting up. This will free the cushion frame from its hook-like floor anchor. Repeat this action for the other side. Do this in a single fluid motion and you'll feel the cushion pop up when it is free of the anchor.

On Camaros with fixed seatbacks, remove two hex screws retaining the lower outboard edges of the seat. Then lift the seatback straight up off its upper hooks, work one end clear of the rear quarter-trim panels and angle the seatback out of the car. Also remove the cardboard trunk liner if your car still has one.

On 1967-69 Camaros with a fold-down rear seat, disconnect the seatback at its lower pivots and remove it from the car.

CONSOLE

If your Camaro has one, the console is the next piece that should be removed. Items such as the radio, front speaker and dash pad cannot be removed on 1970-81 Camaros with the console in place.

The console attaches to the transmission tunnel and on second-generation Camaros, the lower part of the dash. To access its mounting hardware, you must first remove the shifter handle (1973-81 with automatic transmission), shifter plate, storage-bin lid and tray (not on 1967-69 models) and on 1967-72 models, the console ash tray. If your 1967-89 Camaro has a console-mounted oil pressure gauge, don't forget to disconnect and plug the line. Also, on first-generation Camaros, the two nuts for the console mounting bracket bolts are

accessed from underneath the car, above the driveshaft. These nuts need to be removed on '68 and '69 models to make console removal easier. Jack up the car for this.

Detailed instructions for console removal and reconditioning can be found on page 131 of the *Interior Reconditioning* chapter.

DOOR TRIM PANELS

Door trim-panel repair and replacement is covered in detail on page 135 of the *Interior Reconditioning* chapter. However, there are a few tips to keep in mind if you're just removing the panels to have a look-see or to gain access to the inner

workings of the doors and quarter-window mechanisms.

A multitude of trim-panel styles were used over the years, but almost all trim panels hang from a rail adjacent to the inner window felt. The exception to this is all 1967 models, in which the trim panels are slid under the molding just below the window sill. The widely used "hung" trim panels are affixed to the inner door panel with metal or plastic push-in clips and Phillips-head screws. But before you can remove all the screws and pop out all the clips, the window winder, inside release handle, door lock knob, arm rest and (on second-generation cars with dual sport mirrors) the driver's side remote mirror

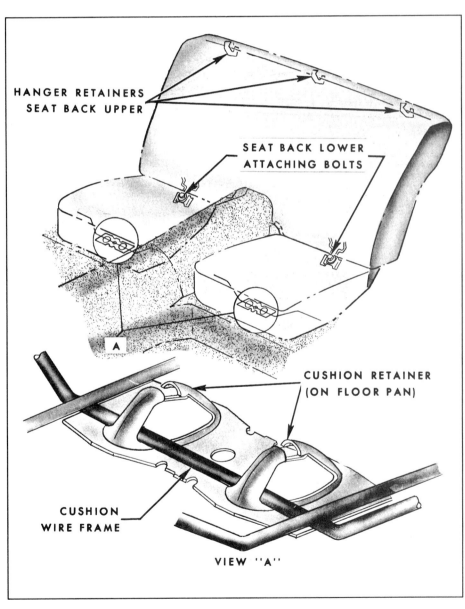

Details of rear seat cushion and seat back—1970-81 models, 1967-69 similar. Courtesy Fisher Body.

After unbolting lower corners of rear seat back, lift it straight up and off hooks.

Cardboard liner between seat back and trunk is likely waterstained and dry-rotted. If a reproduction replacement isn't available, use the old liner as a template and cut a new one from similar thickness cardboard.

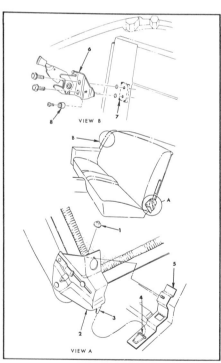

1. Link-To-Actuator Nut
2. Seat Back Link
3. Link Tab
4. Anchor Plate Slot
5. Link Anchor Plate
6. Back Lock Assembly
7. Lock Anchor Plate
8. Rubber Bumper

Details of folding rear seat back—1967-69 models. Courtesy Fisher Body.

To remove the console on some models, the shifter knob and trim plate must be first. On 1973-81 Camaros, the knob is retained by a small retaining ring. Pull off the shifter button, release the ring with point-type retaining-ring pliers and lift off the knob.

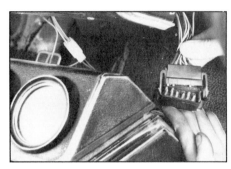

On 1967-69 models with console gauges and 1973-81 models with power windows, don't forget to disconnect the console wiring harness. Also, disconnect and plug oil line at console-mounted oil pressure gauge.

Some console mounting bolts are hidden under the shifter plate or boot and console storage bin.

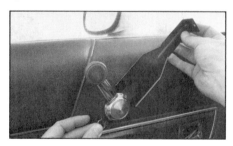

Window winder removal is virtually impossible without a special lock-clip removal tool. To disengage small horseshoe clip retaining the winder, slide the tool directly beneath the winder arm, between the winder base and door panel. On 1970-81 models with dual sport mirrors, also remove the remote-control stalk escutcheon and disconnect its bail wire retainer from the cables.

adjusting stalk and escutcheon must be removed.

Unless your Camaro is equipped with power windows, there is one tool you'll absolutely need to get the trim panels off—the lock clip tool (see nearby photo) to remove the horseshoe clip for the window cranks and 1967-69 standard trim inside door release handles. To remove the clips, merely insert the tool between the base of the handle and the trim panel directly beneath the winder arm as shown and push it against the shaft, forcing loose the horseshoe-type retaining clip.

On models with free-standing armrests, look for two or three stout Phillips-head screws running perpendicular to the trim panel. On models with integrated armrests, be on the lookout for two Phillips-head screws near the armrests running vertically into an armrest support bracket on the door.

On Camaros with pull-type inside door releases (1970-81 models), unbolt the handle from the door, pull it out a ways and disconnect the small clip behind it for the latch rod. First generation cars equipped with Custom Interiors require door trim panel removal to disconnect the door handle latch rod clip.

SAIL PANELS

These are the trim pieces covering the inside of the rear roof (C) pillars on coupe models. Removal is straightforward enough; on 1970-81 models, remove any exposed Phillips-head screws and on some models you'll have to lift the sail panels off of hooks on the C-pillar. Also, 1967-69 Camaros with Custom Trim options have rear courtesy lamps mounted on the sail panels; pop off the lenses, unscrew the lamp bases and disconnect the electrical leads at the bulb connectors. Then, pull loose the windlace molding, sneak a long, flat-blade screwdriver between the fiberboard sail panels on 1967-69 models and pop loose the three metal clips. If you aren't careful, you'll tear the metal clips out of the fiberboard.

On second-generation Camaros, the rear-most section of the sail panels lie directly beneath the slanted rear window. As a result, the hard ABS plastic end of these panels blacken, crack and sometimes crumble to dust.

SILL PLATES

At each door opening is an aluminum scuff, or *sill*, plate that protects the paint on the rocker panels from scuffs and scrapes from your shoes. It also holds down the outboard edges of the carpeting and helps keep the kick panels and rear-quarter trim panels in place. Remove four Phillips-head screws and lift each sill plate off the rocker panel. These can be cleaned up with polishing compound, rubbing compound or extra-fine (000) steel wool, but after decades of use are probably gouged, dented and in need of replacement.

REAR QUARTER TRIM PANELS

These upholstery pieces finish off the interior alongside the rear seats. On second-generation Camaros, removal is a simple matter of removing two Phillips-head screws and sliding each rear-quarter trim panel forward off of the door opening pinch-weld flange and out of the car.

On 1967-69 Camaros, you must first remove the window crank (as detailed under door trim panels previously) and the windlace trim covering the door opening pinch-weld flange. If you don't have the special horseshoe-clip removal tool, you can remove the crank retaining clip with a small hook fashioned from a coat hanger. Carefully insert a screwdriver between the back of the crank handle and the scuff washer and pry them apart. This will allow you to check the position of the retaining clip. When you've found the looped end of

the clip, slide your "hook" into the loop of the clip and pull it off the handle. Now the handle will slide off the shaft. Then remove the Phillips-head screws retaining each panel, pull the panel down (all '67 models only) or on all other first generation cars, lift the panel up and off its hanger below the rear-quarter window and out of the car. On cars with power windows, you'll also have to disconnect the leads at the rear of the switch on each panel.

First generation convertibles use three-piece trim panels; an armrest base, window regulator and folding top. Since a retaining screw is used to secure the rear edge of the folding top panel, you're better off removing the rear seat to access it. Also, the window regulator panel is screwed to the top of the armrest base from the inside, so you'll have to remove them together. Do this by removing the two screws at the front of armrest base and two more that secure the base to the folding top panel. Then pull the two-piece assembly out of the car. Finally, remove the three

Sail panels on 1970-81 models are retained with two screws. Rearmost end of panels tend to discolor and crack with long-term exposure to sun.

First-generation Camaro sail panels are hardboard-backed vinyl and retained with three metal clips. Support under the heads of the clips when popping loose the sail panels to avoid tearing the clips out of the hardboard.

On 1970-81 Camaros, remove two Phillips head screws. Then each molded plastic rear quarter trim panel can be lifted forward off door-opening pinch flange and out.

On models with free-standing armrests, remove the two or three large Phillips head screws and lift off each armrest. "Hockey-stick," armrest shown here is prone to breakage in the forward handle area. For repair tips, see sidebar on page 136.

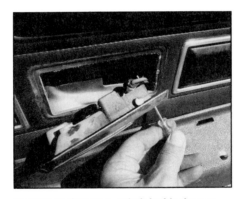

On 1970-81 Camaros, unbolt inside door release from door, pull release out a bit and disconnect clip to the latch rod. Bolts for inside door release are hidden behind a snap-in blackout trim molding on 1970-75 models.

On Camaros with integrated armrests, look for hidden screws to the armrest support.

Metal clips retaining hardboard-backed trim panels to door may tear out if not supported as trim panel is pulled away from door. Look closely and find the 8 or 10 clips holding each trim panel to the door and lever the clips out with a tack puller.

Removal of rear quarter trim panels on 1967-69 models requires that you first pull off windlace trim and door opening and disconnect clip for rear window winder as you did for the doors. On convertibles (1967 shown), folding top cover (upper left) comes out after quarter trim panel is removed.

When replacing the headliner, visors (and 1967-69 sail panels) should be re-covered with the same material. Here, retaining screws were taped to visors to help keep track of hardware. Small set screws (arrows) near base of visor can adjust tension on visor bushings.

On all 1968-73 Camaros, headliner edges are retained to the pinch flange with flexible vinyl windlace.

A-pillar posts are covered with painted metal trim. Remove two or three screws and these come out. Look for the bottom screw hidden under the roof-rail weatherstripping.

Garnish moldings over side windows on 1967 and 1970-73 models fit into roof slots with barbed hooks. Gently pry these out part way, then compress release tang on each barb with small screwdriver and finish removing molding.

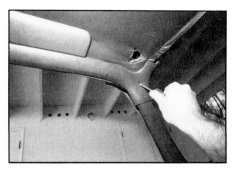

On 1973-81 models with molded headliners garnish moldings are retained with screws and plastic Christmas tree clips over the side windows which will break off when pried loose. Also note shoulder-belt holder (arrow), clipped to garnish moldings in 1972 and screwed to hardboard headliner in 1973.

screws securing the folding top trim panel and remove it.

AIR EXTRACTOR GRILLE

Ever since Camaro switched from front vent windows to *Astro Ventilation* in 1968, these black plastic grilles have been installed on the lock-pillar face of the rear quarter panels in the door openings to augment flow-through ventilation. Remove a single Phillips-head screw atop each extractor, pry the top out and slide them out. If the sponge-rubber one-way filter in each extractor is dried out or missing, or if the plastic frame is cracked around the screw heads, get a replacement.

VISORS, REAR VIEW MIRROR & COAT HOOKS

It's time to look up where the sun doesn't shine and start removing a host of items overhead. Chances are the sun visors and headliner are tattered and torn and the visor, rear-view mirror and coat hooks have to come off to get at them.

If you removed the windshield, chances are the rear-view mirror came out already. If not, remove the mirror-bracket screws from the roof (1967-69 models) or the single Allen screw at its bonded windshield mount (1970-81 models). Typically, the mirror will have sun-induced cracks around its plastic perimeter and the silver-backing on the mirror glass is going or gone. On first-generation cars with the chrome bracket, the finish is typically pitted beyond repair. Aftermarket replacements are available.

Before removing the sun visors, check to see if they will stay up flush with the headliner or tend to sag down and block your view of the road. If they do sag, try tightening the small adjustment screw at the outboard edge of the visor, next to the

visor shaft. If this doesn't work, bushing kits are available for some early-model Camaros from aftermarket Camaro specialists. Second-generation visors can be fetched from wrecking yards or new from GM. Regardless, remove the visors and their center support (if used). If you are getting a new headliner, make sure you get the visors covered with the same material as well.

Find the coat hooks and remove their single retaining screw. On 1970-71 Camaros, a total of four coat hooks were used, the extras doubling as shoulder-belt hooks to stow them up and out of the way when not in use.

HEADLINER GARNISH MOLDINGS

These moldings help retain the edges of the headliner and give it a more finished-off appearance. On 1967-69 Camaros, they consist of flexible vinyl windlace channel that presses onto a pinch flange above the windshield, backlite and, on 1968-69 models only, above the side-window openings. On 1967 models, clip-on bright metal moldings finish off the area above the side windows. After removing the A-pillar trim moldings (two or three Phillips-head screws), the windlace trim can be pulled off by hand.

The garnish moldings used with second-generation Camaros vary with the headliner type. From 1970 to mid-1973, a cut-and-sew cloth headliner was used. As the supply of these was exhausted, Chevrolet switched to a molded hardboard headliner, some time in the 1973 model run. Flexible vinyl windlace trim finishes off the areas above the windshield and backlite on the 1970-73 cut-and-sew headliner, with painted metal trim moldings over the side-window openings. After removing the A-pillar trim moldings (two Phillips-head screws, one hidden behind the door weatherstripping), you can gently pry these side moldings down enough to reveal three barbed metal clips which can be released with a flat-blade screwdriver while pulling down gently. As with the 1967-69 models, the windlace vinyl trim pulls off.

On 1973-81 Camaros with the molded hardboard headliner, the garnish moldings were beefed up to support and retain the headliner in place. For this reason, the garnish moldings adjacent to the windshield and backlite are screwed to the roof and the ones over the side windows are secured with screws and plastic Christmas tree clips. Remove the Phillips-head screws, then pop the side moldings out with a flat-blade screwdriver.

SHOULDER-BELT ANCHORS

Usually the separate shoulder belts used on 1968-73 Camaros stay in good shape because they're kept up out of the sun and few people used them. Nevertheless, to replace the headliner, they must come out. Pry open the anchor trim molding and remove the two stout hex screws and plate retaining the belts to the rear roof section. Also remove the shoulder-belt holders.

On 1974-81 models, with integrated (3-point) belts, remove the Phillips-head screws for the shoulder-belt retractor trim molding, drop down the molding and unbolt the retractor from the roof (two stout Torx-head screws).

HEADLINER

If yours is a 1973-1/2-or-later Camaro with the molded headliner, the only thing that's holding the hardboard in place now is the dome light. Snap off the dome-light lens, remove its two Phillips-head screws, disconnect its lead (make sure the battery is disconnected or sparks will fly!) and voila, the headliner comes tumbling down. It may seem inconceivable that the hardboard headliner, which is larger than the door openings, will fit through them. But with a little bending and bowing (and with the seats removed), it can be removed right out the door (or the windshield if you took that out already).

On 1967-73 Camaros with the cut-and-sew cloth headliner, you'll still have to remove the dome light. But you'll also have to cut or work the edges of the headliner loose with a screwdriver or putty knife, breaking the adhesive seal all around. Once the edges are draped down on second generation cars, unsnap the five metal listing wires from their plastic

If equipped with shoulder belts, unbolt them from their anchors in the roof. Mark these special Grade 8 bolts for reuse.

retaining clips. Save the listing wires and plastic clips because they are not supplied with new headliners.

On first generation cars, mark each listing wire from front to back as you go because the last three wires are different sizes and shapes. And be sure to mark the hole that each listing wire came out of. The extra holes at each listing-wire location allow you to adjust the tension of the headliner if necessary. On 1970-73 models, all listing wires are the same so labeling these is not necessary.

Also, if your 1978-81 Camaro has a T-top, save the old headliner to use as a template to cut the new headliner to size.

REAR PACKAGE SHELF & SPEAKER(S)

This cardboard piece (sometimes covered with vinyl trim) gets sun-baked and water-stained over the years. To remove it, slide a putty knife between its leading edge and the rear panel support, breaking the adhesive seal. It pulls right out. But first, if your Camaro has a rear speaker or blower-type defogger, climb into the trunk, disconnect the electrical leads and remove the components.

KICK PANELS

The kick panels don't have to come off to paint the car or replace the carpeting, but maybe they're scuffed up from too many kicks over the years. On 1967-69 Camaros, first remove the sill plates. Then find the Phillips-head screws that retain the panels and remove them. These screws are found encircling the vent. Second generation cars also have a screw on the outboard edge of the panel. Then remove the panels keeping the following information in mind. First generation and early second generation cars equipped with air conditioning will have a rounded cover on the right kick panel. Remove this cover and disconnect the vacuum hose and linkage from the A/C vacuum actuator. On the left side, you'll have to contend with the parking brake pedal and dimmer switch and on the right, the floor vent mechanism.

SEAT BELTS

You'll need to remove the seat belts to replace the carpeting, sound insulation or to look for rusty sections on the floorpan. Also, it's likely your Camaro's belts are faded, maybe with frayed webbing, a stuck retractor, a rusted or broken latch or whatever and should be reconditioned or

Make sure battery is disconnected before removing dome light. Unsnap connectors from bulb socket and toss wires back up through headliner opening.

Cloth headliner can be pulled loose around, the edges after all garnish moldings are removed. Don't be afraid to tug hard to break the adhesive bond.

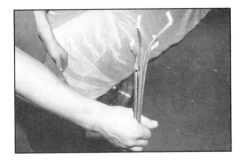
Center of cloth headliner is hung from metal wires called listings. On 1967-69 models, some of the wires are different lengths and shapes, so note which hole which listing comes out of.

After removing any rear deck speakers or rear window defogger, use a putty knife to break the adhesive seal on the leading edge of the rear package shelf and pull it out. Save it to use as a template to trim the new shelf.

replaced. On a late-model Camaro, you might get lucky at the wrecking yard and find a set of belts with good retractors. Otherwise, you might want to send your car's belts off to Ssnake-Oyl (yes, that's two esses in Snake) Products in Dallas, TX for reconditioning.

Be sure to reuse the original Grade-8 Torx-head or hex-head bolts upon installation of the belts.

CARPETING

The original carpeting used in most Camaros is a loop-type molded carpet, although a cut-pile type came into use about 1975. The loop-type carpet with its 80% nylon/20% rayon blend gives exceptional durability. And the molded shape allows the carpet to lay flat against the floor without any wrinkles. So if you're replacing the carpet, this is the best type to buy.

Cleaning—If your carpeting is just plain ol' dirty and you want to shampoo it, vacuum it first. This prevents merely moving the dirt around as opposed to cleaning it out.

Many types of carpet shampoos are available. Some are in spray form, others are liquid and require dilution with water. One type of all-around cleaner that's good for all soft trim as well as the carpeting is *Vinyl-Fab*. This cleaner works quickly and requires dilution with water for carpet shampooing. Or you can also use the same type of carpet shampoo used in the house. Although various cleaning agents are available for cleaning carpeting, there is one thing to keep in mind. Only the suds clean. Soaking the carpet doesn't clean it; it only takes longer for the rug to dry.

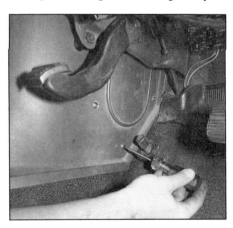

Kick panels attach to cowl with Phillips head screws. To remove the left one, you'll first have to disconnect the parking brake pedal assembly (page 62). The right kick panel may require removal of the A/C vacuum actuator on air-conditioned cars.

When using the cleaning solution of your choice, apply it with a soft-bristle brush using a circular motion. Clean a small area at a time, working from the cleaned areas into the dirty ones. This will give the carpet an equally cleaned appearance in all areas.

As you shampoo the entire carpet, wipe it vigorously with a damp turkish towel. This removes loosened dirt and any residual foam that might be left over. Finally, vacuum the carpet one more time when it's dry to remove the remaining dirt.

Removal—The carpeting will need to be removed if you're completely restoring the car, welding the car in an area directly underneath or adjacent to the carpet, or if you're replacing it. It's in two sections, front and rear, with the two overlapping under the front seats.

Typically, the front section takes the worst beating because it sees the most traffic. It's not uncommon for the carpeting to wear completely through in the driver's footwell area. Other common carpeting problems are sun-bleaching, mold and

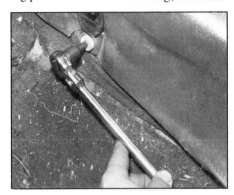

Remove the seat belts. On 1970-81 Camaros, you'll need a special Torx-head socket wrench to remove and install the anchor bolts. Save these special bolts for reuse.

Rare is the Camaro that couldn't benefit from new carpeting. With the seats, console, seat belts, sill plates and other interior trim out of the way, replacement is easy and rewarding. While the carpeting is out, check the condition of the floor insulation (replace it if cracked and flaking) and look for rust.

mildew (especially on convertibles) and cigarette burns. Regardless, it's a very rare Camaro indeed that doesn't need new carpeting after 20 years of faithful service.

To remove the front carpeting, first remove the sill plates, front seats and console or shifter boot. When all these components are removed, loosen the kick-panel retaining screws from both kick panels. Start on one side of the car by pulling on the carpet so that it slides out from under the kick panel. Then do this again for the remaining side.

On the driver's side, remove the two hex screws for the firewall carpet guard (directly beneath the steering column base) and remove the guard. Then, pull the carpet up and off the foot-operated headlamp dimmer switch. Save the soft plastic dimmer-switch grommet as some new carpet sets do not include this item. With the carpet free of the kick panels, carefully fold it up.

After the front carpeting is removed, the rear is easy to take out with the rear seat, rear-quarter trim panels and seat belts out of the way. Refer to pages 42, 43 and 131 for details on removing these items.

FLOOR INSULATION

Once the carpeting is out of the car, you should find some black mats sticking to the floor. If you're unlucky, parts of the mats came up when you removed the carpet. The factory calls this the carpeting *underlayment* and it's responsible for a number of jobs including: keeping out excessive road noise, providing a moisture barrier against snow and water that gets tracked into the interior and insulating the interior from temperature extremes.

If the mats are intact and you have no reason to suspect that rust has gained a foothold in the floor underneath them, leave them alone. Removing these petroleum-treated fibrous mats is a messy proposition.

Scrape Underlayment—To remove the mats or what's left of them, peel them off by hand. If they're sticking to the floor, you can use a paint scraper to remove them. When all the insulation is removed, take some time to carefully examine the floor for any signs of rust or rust damage.

Check for Rust—If any surface rust is evident, use a sandblaster to remove it. The sandblaster is a good way of removing the small pits caused by the rusting. If you don't have access to a sandblaster, try using a wire wheel chucked in a drill motor or a wire brush. You can also try using a liquid rust remover. *Oxi-Solv,* a reusable liquid rust remover available from the **Eastwood Company,** does a good job of removing rust and has the added bonus of leaving a zinc-phosphate coating on treated areas. Just be sure to follow the label directions.

If any areas are rusted out, now's the time to repair them using new sheet metal or one of the new floor pans available from some of the Camaro aftermarket suppliers. Methods for repairing rusted-out areas can be found under *Trunk Floor Repair,* page 87. When all signs of rust are vanquished, prime and paint the floor. Purists will want to paint the floor the same color as the body to keep that factory look.

DASH PAD & INSTRUMENTS

See pages 138 and 140 in the *Interior Reconditioning* chapter for details on removing and reconditioning these items.

RADIO & FRONT SPEAKER

If your Camaro has its original radio, it's doubtful that it is still functioning properly. Maybe you want to remove it and take it to a radio shop for repair, or perhaps you'd like to install an aftermarket stereo in its place. Or if you are restoring your car for show and want to yank out that bargain-basement Taiwanese squawk box in favor of the original article, it's time to climb under the dash.

With the console, 1967-69 heater controls and 1970-81 A/C ducts removed, gaining access to the radio is easy. Just pull off the volume and tuning knobs, bass/treble and balance adjusters, undo the shaft nuts with a deep-well socket (not pliers if you want to avoid scratching the dash!), and unbolt the rear support brace. Then unplug the antenna lead and combination power and speaker(s) connector, slide the radio back, flip it upside down and back out from under the dash.

Now you've got access to the stock front speaker, which is probably cracked and shorted-out after all these years. The 10-ohm oval factory speaker bolts to an underdash support brace via a crimped-on bracket. You'll have to transfer this bracket to the replacement speaker.

ENGINE COMPARTMENT

Probably the best thing you can do for your Camaro's engine compartment is to steam clean it before the car is immobilized and you start taking things

apart. Not only will you be dealing with far less grease and grime, but you'll be better able to ascertain the condition of many parts because you'll now be able to see them. Those items needing paint or rust removal will be readily apparent and leaks will be easier to spot without a glob of dirt and road grime stuck to them.

BATTERY & TRAY

If you haven't already done so, disconnect the battery cables, unbolt the battery hold-down bolt and lift out the battery. Place it on a block of wood to keep it from discharging into the ground. Inspect the battery cables along their full lengths—all the way to the alternator bracket on the negative cable and the starter on the positive cable. Get new replacements if there's corrosion or damaged insulation on the cables—especially the small body-ground pigtail going to the right fender. Cables for side-terminal batteries are notorious for chewed-up terminal connector screws. And before you remove it, give the battery the wire-brush and baking-soda treatment. If it's still holding a good charge and doesn't have sulfated plates, put it on a trickle charger every few weeks to keep it fresh for when you put it back in.

We've never seen a Camaro battery tray that didn't have rust on it and your's should be no exception. Light rust can be removed with Naval Jelly or sandblasting, but if the tray is perforated, replace it. Undo the two battery-tray bolts to the radiator support and two bolts from underneath the inner wheelhouse. New trays are available from numerous Camaro aftermarket parts houses. Do your tray (new or used) a favor and after applying a good coat of primer, then apply some *Eastwood Battery Tray Coating.* This low-gloss black plastic coating helps delay the onset of battery-acid induced corrosion.

RADIATOR

While your Camaro is out of commission during the restoration, give the cooling system the once-over. Unscrew the radiator cap and dab a finger around the inside diameter of the filler neck. If your finger picks up any rust, chances are the system needs to be cleaned out. Drain the coolant and unless the radiator looks like brand new inside with no build-up on the fins, send it out for cleaning and pressure-testing.

To remove the radiator, first remove the fan then unbolt the radiator shroud and set

it aside. Then disconnect the upper and lower radiator hoses, automatic-transmission cooler lines (use a flare-nut wrench) and coolant-overflow hose. On first-generation Camaros, the radiator bolts to the radiator support with four bolts. On second-generation Camaros, the radiator "floats," sandwiched between rubber mounts; remove the bolts or upper mounts and lift the radiator out of the car.

Now take a minute to inspect the shroud. Chances are it's got some cracks or road damage and should be replaced with a new or reproduction shroud at assembly time.

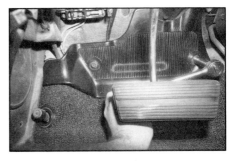

Carpet guard under steering column in driver's footwell comes out with two hex screws. Save dimmer switch rubber grommet in case new carpeting doesn't have one.

While the car is apart, now is a good time to get the radio and front speaker(s) repaired or replaced. Access is more difficult once the interior is installed.

After years of attack from battery acid, most Camaro battery trays need to be replaced. Apply a liberal dousing of penetrating oil, then unbolt the tray.

47

HOOD PAD

After two decades of exposure to oil mist, battery acid, engine heat and road grime, this pad is probably torn, crumbling and greasy to the touch. Pry off the 15-20 round-head plastic retainers with a long screwdriver or needle-nose pliers and remove the pad. As gunky as it is, you still may want to save it to use as a template for cutting a new pad, if you decide to go the original-equipment GM route. Pre-cut aftermarket replacements are available, but are not made from the original material. Original-style hood-pad retainers are available from GM.

COWL VENT GRILLE

On second-generation Camaros, this black plastic grille keeps leaves, pine bristles and other litter out of the heater/A/C plenum in the cowl. It snaps out after prying out its plastic retaining tabs and can be cleaned and detailed separately. Previous paint overspray can be removed with extra fine steel wool. You'll want it off to clean, detail and paint (if necessary) the cowl, or service the windshield-wiper linkage.

On first-generation Camaros, the cowl vent grille is a detachable sheet-metal piece. It attaches to the body with four sheet-metal screws, found at its front

Radiator in first-generation Camaros mounts to support with four bolts. Second-generation radiator floats in rubber mounts and lifts out after unbolting radiator support upper cover.

Remove hood pad retainers with needle-nose pliers or a long screwdriver. Be careful removing hood pad, its material can cause rashes. Try to keep it in one piece to use as a template for cutting a new one.

edge. Removing this panel gives you access to the windshield wiper linkage. Otherwise, this panel should be painted on the car.

RUBBER BUMPERS

These help to position the hood in its opening and keep it from rattling against the fenders and cowl. With age, exposure to the elements and engine heat, the rubber hardens and cracks. Replace them with new rubber.

On first-generation cars, there are two slide-on bumpers, one on each fender, and one adjustable bumper on a threaded steel shaft at each front corner of the radiator support. On second-generation Camaros, there are four slide-on bumpers at the sides and five adjustable ones—one at each corner and one at the center rear. Just slide out the side bumpers and pop the adjustable corner bumpers off their steel shafts.

ENGINE COMPARTMENT & LIGHTING HARNESSES

See page 100 in the *Electrical Systems & Wiring* chapter.

A/C SYSTEM

If your Camaro has air conditioning, it's wise to keep the system intact. Aside from the dangers of pressurized freon injuring you or damaging the environment, opening the system to replace a compressor or condenser invites moisture inside which will corrode the compressor or receiver/dryer in no time at all. Even if you plan to remove the engine, front-end sheet metal or subframe, keep A/C components hooked up and positioned or tied up to one side, out of the way. Then, when you've got your Camaro running and roadworthy, take it to an A/C service shop for leak-testing, evacuation and charging. It's the only way to go.

If you need to disconnect any part of the refrigerant system, have the system depressurized at a shop, and immediately cap or tape any lines or openings as you remove components.

If the system is left exposed, chances are you'll need a rebuilt compressor and a new receiver/dryer and O-rings to get it working again.

TRUNK

SPARE & JACK

Open the trunk and check to see if all of the jacking hardware—jack base, jack, handle, bumper bracket (if used), wing nut

and threaded J-hook are present and accounted for. If your Camaro still has a readable jacking decal on the underside of the deck lid , check what you have against what the decal says you should have. The "guitar-case" plastic cover used on some models for the jack is a tough item to find used. Peruse the wrecking yards for these items if you're missing any.

Inspect the spare. Does it hold air? Does it match the other tires on the car? Is there much tread left? Are the sidewalls cracked? If it's an inflatable mini-spare, is the inflater can there and in working condition? Is the spare wheel rusty? If so, dismount the tire, beadblast and paint the wheel and remount the tire with a new valve stem. Even though it's a spare, don't forget to have the wheel and tire dynamically balanced.

TRUNK MAT

Lift out the trunk mat, or what's left of it and store it away. You may need it as a template to cut a new mat later on. Look for rust under the mat. Minor surface rust is OK, but perforation, particularly in the fuel-tank anchor or rear spring perches, requires costly and time-consuming repair with a welding torch. Take your Shop-Vac and vacuum out all of the loose debris from the trunk floor.

UNDERCARRIAGE

Drivetrain, suspension, steering and brake reconditioning are covered later on in their respective chapters.

EXHAUST SYSTEM

Jack up your Camaro and support it under the subframe and rear axle with jack stands. Examine the condition of the exhaust system. If it's in fantastic, like-new shape with no dents or leaks and you don't plan any major work to the floorpan, fuel tank or trunkpan, it can stay on. You'd probably destroy it trying to take the sections apart anyway. But if the system is shot or on its last leg, it's best to remove it as it will only get in the way of any underbody repairs, is a source of head knocks and rust in the eye, and will probably suffer from overspray when you paint your Camaro.

Put on a pair of safety goggles to keep rust and foreign matter out of your eyes. Using a liberal dousing of penetrating oil and a deep-well socket, unbolt the downpipe(s) from the exhaust manifold(s). Loosen the clamps holding the sections of pipe together, and disconnect the hangers for the tailpipe(s) and muffler(s). On one-piece welded systems or old sectional systems, be prepared to get out that hacksaw, sledge hammer, chisel and torch, if necessary. Just be careful to keep the flame away from fuel lines or the fuel tank.

FUEL TANK

There are any number of reasons why you might want to remove your Camaro's fuel tank. Maybe water got inside and began rusting. Perhaps you smacked the tank on a rock or curb and the sending-unit float is giving fictional readings. Could be the tank is leaking. Possibly, curiosity has gotten the best of you and you want to see if the original build sheet from the assembly plant is to be found between the trunk-pan and fuel tank—many sheets have!

More than likely, the reason you want to remove the tank is to do some welding above or around the tank to fix sheet-metal rust holes. Unless you want to join your Camaro in that Great Restoration Shop in the Sky, NEVER DO ANY WELDING AROUND THE FUEL TANK, EVEN IF IT IS EMPTY. LEFTOVER GAS VAPORS IN AN EMPTY TANK ARE EXPLOSIVE!

For fuel tank removal procedures, see page 71 in the *Suspension & Steering* chapter.

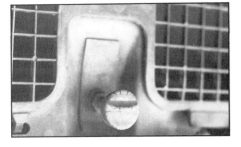

Rubber on hood bumpers should be replaced. There are two of these on 1967-69 models and five on 1970-81s. Also, black plastic cowl vent grille on 1970-81 models can be removed for cleaning and detailing.

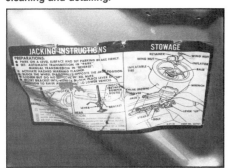

Take a good look at the jacking decal under the deck lid and see if your car's trunk is in order.

Drain plugs in first-generation cars allowed for water removal should the trunk seal or convertible top leak. Judging by the amount of sheet steel covering rust-outs in this trunk floor, several patches will have to be welded in.

When removing the exhaust system, use plenty of penetrating oil and wear goggles (not like this guy) to protect your eyes from rust and debris.

It may be unwieldy, but it's best to keep the old exhaust system for reference until you have the correct new pipes and mufflers mounted.

5 Brakes

For safety's sake, never take your Camaro's brakes for granted.

You've decided to refurbish your Camaro because it's stylish, fun to drive and compared to most cars, fast, quick and generally a whole lot of fun in the go department. But if your Camaro can't be hauled down from speed and stopped when necessary, all of that fun could come to an inglorious, expensive and possibly tragic end.

So before attempting any work on your Camaro's brake system, consider that all work must be first-rate to ensure your safety as well as that of others. If you don't have the discipline to do the job properly, let a professional do the work. It may cost a few dollars more to have someone do the work for you, but then safety is paramount. If you're proficient enough to do the work yourself, and you have or have access to the right tools, you'll find it quite enjoyable.

The number of brake system components in your Camaro is relatively small. It includes the master cylinder, brake lines and hoses, wheel cylinders and/or brake calipers, friction material and drums and/or rotors. Add the cable-operated parking brake to that list as well.

BRAKE SERVICING

Before tearing into the brakes, there's one thing you should keep in mind: **safety**. Safety can take many forms, but they're all for the benefit of you and your passengers. For instance, when you're servicing the brake shoes or pads, keep in mind that the dust surrounding them may contain asbestos.

Because asbestos is suspected as the leading cause of a number of respiratory ailments including lung cancer, keep exposure to brake dust to a minimum. Before servicing the brakes, water down the calipers, friction material and drums.

With the possible exception of the rear disc-brake option on 1969 Camaros and the four-piston, front disc brakes available on 1967-68 Camaros, replacement wheel cylinders, calipers, rotors, drums and master cylinders are widely available from GM and aftermarket auto-parts outlets and at very popular prices. For this reason, if a part is corroded, damaged or worn out, it makes good sense to replace it rather than rebuild it. Concours judges aren't likely to pull a wheel and brake drum to check for the correct casting number on a wheel cylinder. And you'll be putting your Camaro back on the road with new, unworn parts, rather than a honed-over, pitted master cylinder, wheel cylinder or caliper. Nevertheless, you should know what to look for in a brake system component when it has reached the end of its serviceable life.

We'll cover the high points of brake component rebuilding. If you have to rebuild the master cylinder, wheel cylinders or calipers, be sure to have a good reference manual on hand. A Chevrolet Service Manual or HPBooks, *Brake Handbook* are worthwhile reference materials.

If you are unfamiliar with brake system servicing, work on one side of the car at a time. That way, you can use the opposite-side drum or disc-brake assembly as a reference.

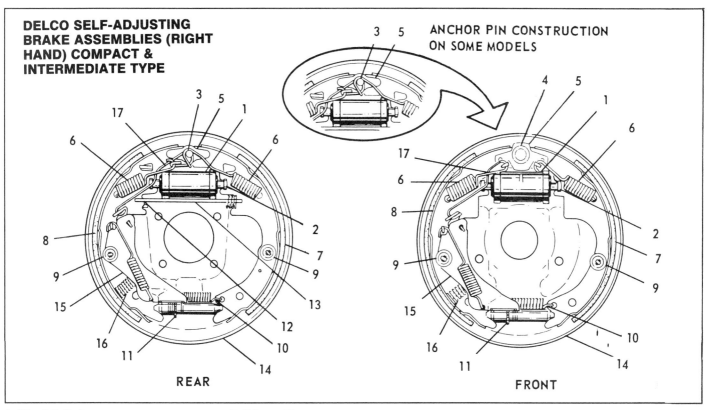

DELCO SELF-ADJUSTING BRAKE ASSEMBLIES (RIGHT HAND) COMPACT & INTERMEDIATE TYPE

ANCHOR PIN CONSTRUCTION ON SOME MODELS

REAR

FRONT

1. Wheel Cylinder
2. Cylinder Link
3. Anchor Pin
4. Anchor Bolt
5. Anchor Plate
6. Return Spring

7. Primary Shoe
8. Secondary Shoe
9. Hold-Down Cup And Spring
10. Shoe Connecting Spring
11. Adjusting Screw Assembly
12. Parking Brake Lever

13. Parking Brake Strut
14. Backing Plate
15. Adjusting Lever & Pivot Assembly
16. Lever Return Spring
17. Adjusting Link

Details of typical Camaro drum brake assembly. Courtesy General Motors.

Unlike most front drum brakes, it's not necessary to remove the spindle nut and wheel bearings to remove a front drum on 1967-69 Camaros. The drum can be pulled off independent of the hub.

Start shoe removal by disengaging return springs from center post.

Remove pins retaining each shoe to backing plate by rotating them 90 degrees.

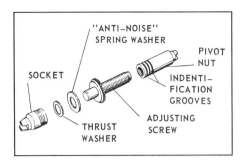

Check the star adjuster closely for corrosion, damage and wear. Lube threads upon assembly. Courtesy General Motors.

MASTER CYLINDER

The master cylinder has two jobs: providing pressure for the brakes and acting as a reservoir for the brake fluid. Its major components include the fluid reservoir, cylinder and various springs, pistons, cups and seals.

The fluid reservoir is divided into two sections, in effect making two systems out of one. The front portion of the fluid reservoir is devoted to the front brakes. The rear portion of the reservoir supplies brake fluid for the rear brakes. If the seal on one of the pistons fails, braking performance will only be lost at two wheels. This leaves braking action at the two remaining wheels.

Replacement or rebuilding of the master cylinder is required when you encounter a loss of pressure in the brake system and you've eliminated all other sources of trouble. This includes faulty wheel cylinders or calipers, brake lines, brake hoses and combination or distribution valve.

Glycol vs Silicone Fluid—Before you begin removing the brake lines from the cylinder, put some newspaper directly below them. This will help prevent brake fluid that will seep out of the lines and master cylinder from being deposited on any paint in the area. Glycol brake fluid (the standard fill in all GM cars) attacks paint, causing it to blister and peel right off the body. Only silicone-based brake fluid won't harm paint, but it has a low boiling point and other drawbacks which make it an unwise choice to use in a high-performance car that needs a firm pedal and fade-free response.

If you use glycol brake fluid, make sure it meets DOT 3 or DOT 4 specifications. On disc-brake cars, use only fluid designed to meet the higher operating temperatures of these systems.

DRUM BRAKES

A great percentage of first-generation Camaros were fitted with drum brakes at each corner. Also, every second-generation Camaro uses drum brakes at the rear. Except for minor differences between these brakes, the service operations are identical. Servicing these brakes is a simple operation using common hand tools, including a flat-blade screwdriver, a pair of locking pliers and some safety glasses.

INSPECTION

Drum Removal—After removing the wheel and tire, pull the drum off with your hands. If a rear drum won't come off, it's usually due to the parking brake being engaged or rust on the axle flange. If it's just rust, carefully and lightly tap against the outside edge of the drum with a hammer to move it outward. Work around the drum and try to avoid cocking it.

If the drum still doesn't budge, take some 240-grit sandpaper to the rust. Then try tapping the drum again with the hammer. If the drum is still stuck, you'll have to get a three-jaw puller to remove it. These are available at many tool-rental companies.

As you remove the drum, keep it openside up to prevent any brake dust accumulated therein from pouring out. Get a garbage bag, dump the dust inside and tie it closed. Brake dust is nasty stuff. If you want to clean the brake drum, don't blow it off with compressed air. Wash it with the garden hose, then wipe it dry to prevent rusting. The water will keep the dust way down.

Shoe Inspection—Nine times out of ten, you'll be replacing the shoes with new ones. But if there's plenty of lining left above the rivet heads and the shoes are free of oil or brake-fluid contamination, they can be reused. Just keep your greasy paws off the friction surface. Clean the shoes, if necessary, with a commercially available brake cleaner.

Return Springs & Adjuster—Take a minute to study the positioning of the return springs and adjuster. Make a mental note, or better yet, make a drawing or take a Polaroid picture of how things fit together. If this is old hat to you and even if it isn't, check the springs to be sure they're not broken or damaged. It's a good idea to replace these springs with new ones when replacing the linings. They're probably corroded, stretched out and/or fatigued anyway, and after all, this is a restoration,

isn't it? After decades of service, it's unlikely that the star-wheel adjuster has all of its teeth and still works like new. And self-adjusting brake mechanisms routinely need to be disassembled, cleaned and relubricated.

To remove the return springs, use your locking pliers or a stout flat-blade screwdriver. When using locking pliers, clamp them onto the straight section of each return spring near the center post on the backing plate. Pull the hooked end of each spring toward the post, disengaging it, and allow it to relax. You can also lever these springs off the post with a stout screwdriver, but be careful that the suddenly freed spring doesn't fly off and cause injury.

Use the locking pliers to remove the spring directly above the star adjuster, then spread the shoes apart at the bottom and lift off the star adjuster mechanism.

Brake Shoes—Locate the small pin and damping spring retaining each shoe to the backing plate. Using pliers, push on the cupped retainer and rotate it 90 degrees to remove the retainer and spring from the pin. You may need to hold the pin from the backside to keep it from rotating. Now the shoe and damping spring will drop to the ground. Do the same for the other shoe. Now slip the front shoe out of the parking brake strut and set it aside. Then remove the adjusting lever assembly from the rear shoe and pull the shoe out of the parking brake strut. If you're working on the rear brakes, unhook the parking brake lever from the rear shoe. Now the parking brake lever will be hanging from the parking brake cable. If you need to remove the lever, simply compress the cable spring and unhook the lever from the cable. Keep all parts from one brake assembly together to avoid confusion at assembly time.

Star Adjuster—Check the drum-brake adjuster ends, threaded portion and star-wheel teeth for corrosion or wear and replace if damaged. Otherwise, clean the adjuster with brake cleaning solvent and lubricate the adjuster threads with Lubriplate or wheel bearing grease.

Metallic Linings—Some late-'60s high-performance Camaros were equipped with metallic, instead of organic, brake linings. These are identifiable by their horizontally segmented friction linings consisting of a dark sintered-metal friction material welded to a series of bright metal pads. Metallic linings should be replaced when the bright metal backing shows through or when lining thickness is worn to less than

3/32-in. measured at the heel or toe of each pad. Note that drums must be honed to a 20 micro-in. (that's 0.00002-in.!) finish to be compatible with metallic linings. Otherwise, rapid drum wear will occur. Also, metallic linings must be seated before driving your Camaro any distance. This involves making six stops from 30 mph with moderate pedal pressure, followed by six more stops from 55 mph with one mile between stops to cool the brakes.

Rust Ridge & Scoring—Take a look at the shoe contact surface on the inner surface of the drum. It should appear smooth or only lightly grooved. Light scoring or a buildup of rust on the drum's inboard edge can be machined off by having the drum cut, or *turned,* at a machine shop.

Out-Of-Round—Another malady, evidenced by a pulsating brake pedal (but not a pulsating steering wheel), could be an *out-of-round* brake drum. This can be checked by measuring the inside diameter of the drum at four locations, 90 degrees apart, with special inside calipers. A minor out-of-round condition can be remedied by turning.

Inner Diameter—Cast into the outboard edge (wheel-mounting surface) of every drum produced after Dec. 31, 1970 is a maximum allowable *inner-diameter* (ID) specification. If the drum now exceeds this or must be cut beyond this to remove scoring, it cannot be reused. Specifications for drums produced before this time are available at local drum turning facilities. A worn drum is a thin drum, one incapable of retaining its shape or dissipating the tremendous heat energy created when you haul your Camaro down from speed. If your Camaro is 20 or more years old and still has the original drums, check them carefully. They could be paper thin.

Blueing & Heat-Checking—If your Camaro's drums ever got hot enough to blue (burn) the shoe-contact surface, they're probably not safe to reuse. Also check the ID for small stress cracks that could cause the drum to fail (basically explode) under a high load. Never weld up a cracked drum!

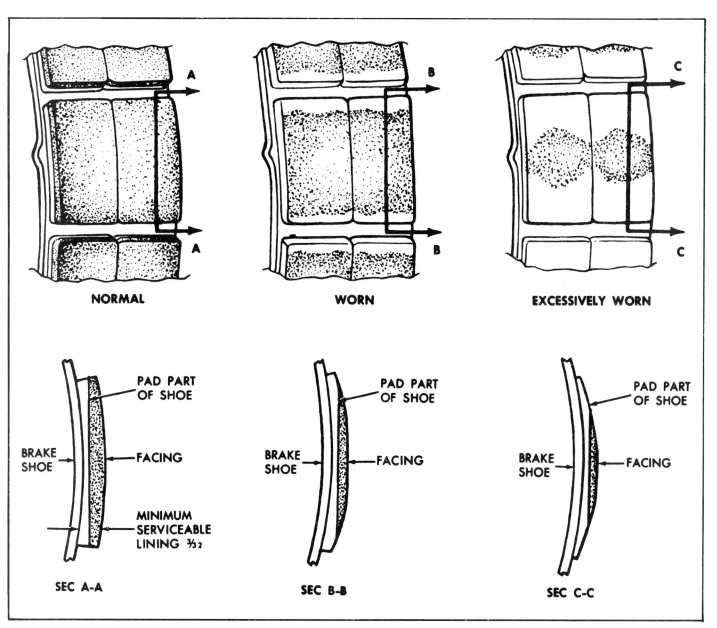

Metallic lining wear check. Courtesy Chevrolet.

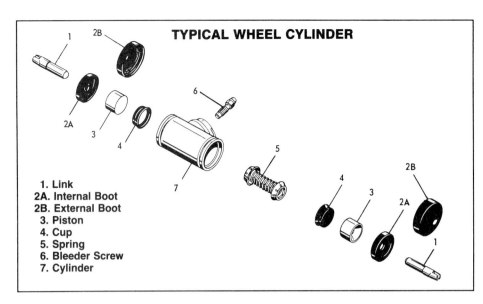

TYPICAL WHEEL CYLINDER

1. Link
2A. Internal Boot
2B. External Boot
3. Piston
4. Cup
5. Spring
6. Bleeder Screw
7. Cylinder

Typical drum-brake wheel cylinder. Courtesy General Motors.

Lugnut Studs—On the front drums, check for any stripped, cross-threaded or broken-off lugnut studs. Have damaged studs pressed out and new ones pressed in at a machine shop as necessary.

Wheel Bearing Races—Remove any buildup of grease from the hub and check the wheel-bearing races for scoring, blueing and other wear. See *Wheel Bearings*. page 60, for details.

Cleaning & Detailing—For a daily driver, you can clean the drum with a non-petroleum-based cleaner such as mineral spirits and polish the shoe-contact surface on the drum's ID with fine emery paper. Never use gasoline or kerosene to clean drums as these leave a greasy film.

If you want the drums to look good, or if

After cleaning and derusting, a good way to keep cast-iron parts looking new is to apply a coating of Eastwood Spray Gray. Courtesy Eastwood Co.

you just want to remove the rust buildup from the drum's outer surface, have them *bead-blasted or sandblasted*. That leads to another problem, though. Without some kind of coating, the cast-iron drums will rust quickly, sometimes just in humid air. You can carefully apply a light-oil coating to the outside of the drum, keeping the oil off the shoe-contact surface. But this coating will wear off quickly the first time the car sees rain. A clear lacquer coating would burn off with brake heat and a stainless-steel coating might affect brake performance. That leaves high-temperature enamel paint, a flat black being the best at dissipating heat. You might get docked points at car shows though, because Camaros left the factory with unpainted (and often rusty) drums.

WHEEL CYLINDER

If your Camaro's brake system worked just fine before you pulled a wheel and drum to inspect the linings, chances are the wheel cylinder can be reused. It's all a matter of attitude. Now is a good time to do preventive maintenance and replace or rebuild those cylinders. Or perhaps you're detailing the brakes and don't want a rusty, grimy wheel cylinder sticking out like a sore thumb. Or maybe you broke off a bleeder nipple the last time you bled the brakes. Now's your chance to make it all right.

Disassembly—If the wheel cylinder has a visible leak, rebuild or replace it. Check for the beginning of a leaky piston seal by pulling back both wheel cylinder boots. Damp is OK, but if you can see or feel any

amount of brake fluid, the cylinder must be serviced. Using a flare-nut wrench, tighten, then loosen the brake line flare nut and disconnect the brake line or hose from the rear of the cylinder. Then unbolt the cylinder from the backing plate. If the cylinder was leaking, pull off the boots and push out the pistons and spring. Run a finger through the bore. If your fingernail catches on any pits or scoring, the cylinder is a goner. If, however, it is smooth, it can likely be lightly honed out and refurbished with new rubber piston seals. Just remember to flush the cylinder with brake fluid to remove any debris after honing.

Detailing—If you want to make the exterior of a used wheel cylinder shine, remove its pistons and stuff all openings with newspaper to protect the bore. Bead-blast the cylinder until it has a uniform gray appearance, then remove the newspaper and blow the cylinder out with compressed air to remove any hidden beads. Then condition the metal with Prep-Sol or equivalent. Next, mask over the boot lips, suspend the cylinder at eye level with a coat hanger wrapped around one of its mounting bolts and spray the cylinder with Eastwood Cast Iron Spray Gray. You can give the same Eastwood treatment to disc-brake calipers and the master cylinder, if you wish.

Assembly—Install the wheel-cylinder return spring in the bore. Coat new rubber piston seals (lips facing inward) and new pistons (flat side facing inward) with brake fluid and install them in the bore. Press on new rubber boots. Thread in a new brake bleeder nipple.

BACKING PLATE

Clean the backing plate with a spray degreaser. If it's rusty, you may want to unbolt it from the front spindle or rear axle housing (after pulling the axle shaft) have it bead-blasted or sandblasted. Paint it with a semi-gloss black or have it plated with zinc dichromate (if that was the original coating).

SHOE/DRUM INSTALLATION

Installation is largely the reverse of removal, but with the following prerequisites. Keep your hands absolutely clean when handling the linings. Lubricate the six pads on the backing plate, the star-wheel adjuster threads and the parking brake fulcrum and pivot with Lubriplate grease. On a rear brake assembly, transfer the parking-brake lever and its attaching hardware to the secondary brake shoe.

Install the primary shoe (short lining) on the backing plate so it faces the front of the car and secure it with its retaining pin and damping spring. Then install the secondary shoe (long lining) in the same manner, sandwiching the adjuster and parking brake strut between the two and hooking the lower spring on both shoes. Next, position the adjusting lever and its return spring onto the secondary shoe and secure it with the shoe retaining pin and spring. Finally, hook the self-adjuster linkage, and two upper shoe return springs over the center post on the backing plate. Adjust the star wheel so the drum just barely slips on. Bleed the brakes as necessary, page 60.

DISC BRAKES

As mentioned earlier, two types of disc brakes have been used on Camaros: a four-piston, fixed-caliper design on 1967-68 models (and a few 1969s) and a single-piston, sliding caliper design from 1969 onward. A limited batch (200 or so) of 1969 models (and a handful of '68s) were equipped with four-wheel, four-piston, fixed calipers from the Corvette so that Chevrolet could legally run them in Trans-Am racing. Otherwise, disc brakes have appeared exclusively at the front, optionally on all first-generation Camaros and as standard fare on second-generation cars.

INSPECTION

You can inspect disc brake rotors and pads when the tire and wheel is removed. Take a close look at both braking surfaces of the rotor. They shouldn't have any signs of discoloration. Although minor grooving may be unsightly, it's permissible to reuse the rotor if grooving does not exceed 0.015 in. in depth. Regardless, no groove should be deep enough to catch your fingernail in. Also, if the brake pedal was pulsing up and down when the brakes were applied, check the rotor for lateral runout with a dial indicator. Lateral runout should not exceed 0.005 in. Pulsing can also be caused by improperly adjusted wheel bearings. If you discover any of these flaws, the rotors will have to be turned or replaced. Replacement of the rotor is mandatory if any cracks are apparent.

To check the pads, look between the rotor and the caliper. If the pads appear to be less than 3/16-in. thick, replace them. If they're worn unevenly, remember that uneven brake pad wear can be caused by

improperly adjusted wheel bearings. If you need to replace brake pads, remember that they must always be replaced in sets. That means if one side needs pads, you must replace the pads on the opposite side of the car as well.

1969-1981 DISC BRAKES

Caliper Removal—If you plan to rebuild or replace the caliper, disconnect the flexible brake hose now using a flare-nut wrench. To remove a sliding single-piston caliper, remove the two cap screws holding the caliper to the spindle. Use a 3/8-in. Allen wrench with a short length of pipe slid over it to remove the screws. When the caliper is off, don't let it hang by the brake

Backing plate is easier to clean up and detail after removing it from the spindle (or rear axle). Remember to install a new seal between the plate and spindle to prevent water from contaminating the brakes. Also, apply a light coating of Lubriplate to raised pads on the backing plate to keep the shoes from sticking in position.

Your assembled front drum brake assembly should look like this when you're done.

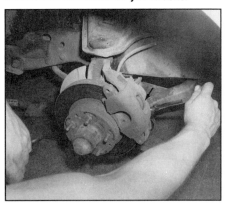

Disc brake pad wear on 1969-81 models can be quickly checked through inspection window (arrow). Caliper is easily removed after unbolting two Allen-head cap screws as shown. Don't allow the caliper to hang from its flexible brake hose.

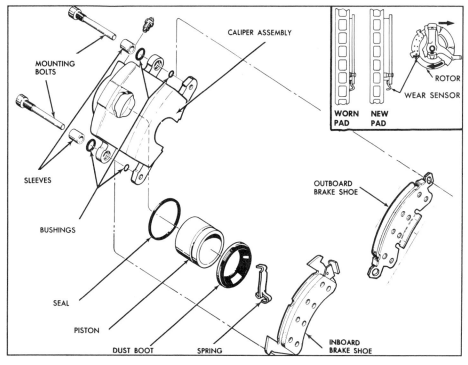

Details of single-piston, floating-caliper front disc brake used from 1969 on. Wear sensor on inboard pad first appeared in 1974, but can be retrofitted to 1969-73 models as well. Courtesy Chevrolet.

As with brake drums, disc rotors must be replaced when worn beyond a safe limit. Check rotor minimum thickness with a micrometer.

If piston is stuck, carefully blow it out with 5-7 psi of compressed air. Wood block keeps piston from flying out and injuring someone.

Worn, cracked or torn dust boot allows dirt and moisture to attack caliper piston and bore. Lever it out with a screwdriver.

Problems with corroded pistons and calipers can be minimized by using silicone brake fluid and switching to this stainless-steel piston and caliper with specially treated bore. Courtesy Stainless Steel Brake Corp.

1967-68 FOUR PISTON CALIPER

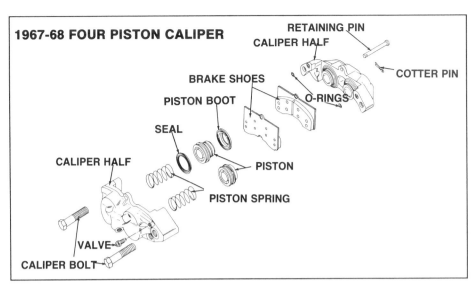

Details of 4-piston fixed-caliper disc brake for 1967-68 models (similar to modified Corvette calipers used on 4-wheel disc brake option of 1968-69). Courtesy Chevrolet.

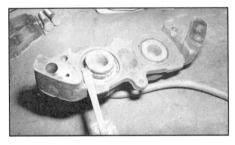

Minor caliper bore corrosion can be removed with a brake cylinder hone. If any large pits are found, caliper must be replaced.

Because 4-piston caliper halves can be unbolted, compressed air isn't required to remove stuck pistons. They can be carefully levered out as shown.

Caliper piston corroded this bad should be replaced.

hose as this can damage the hose. Use a piece of clothes hanger or a bungee cord attached to the upper control arm to support it.

Pad Removal—Remove the pads from the caliper with your fingers. Use a screwdriver if they're stubborn. Replace the pads on both front wheels if any pad is worn to within 1/32 in. of the rivet heads. As you remove the pads, you'll notice the inboard pad has a clip on its back side. This clip retains the pad to the caliper piston. Carefully pry the clip off and use it on the new inboard pad if the new pad doesn't come with one.

When purchasing new pads, make sure you get a set manufactured by a reputable company. If you buy OEM Delco pads, they'll have the sheet-metal tang on the inboard pad that acts as a wear indicator. The tang protrudes from the pad such that when lining thickness is nearing the minimum allowed, the tang contacts the rotor and makes a screeching sound that alerts the driver that pad replacement is due.

And unless you have manual brakes, it's recommended that you use semi-metallic pads for improved wear characteristics and fade-resistance. Using semi-metallic pads with manual brakes may raise the effort required to stop the car to an unacceptable level for some people. And this effect is even more pronounced when the pads are cold, such as during the first few minutes of brake operation.

Rotor Removal—To remove the rotor, pry off the dust cap with a screwdriver. When the dust cap is off, you'll see a cotter pin that's positioned through the spindle nut and spindle. Straighten the pin out with a pair of needle-nose pliers then pull it out. Now loosen the spindle nut about two turns. Then give a quick tug on the rotor to free it from the spindle. Finally, remove the spindle nut, washer, outer wheel bearing and rotor.

Flexible Brake Hose—If you're planning on removing the caliper to replace or rebuild it, replacing a flexible brake hose or are going to completely tear down the front suspension, you'll need to remove the flexible brake hose from the caliper. For this, you'll need a *flare-nut wrench*. To inspect the hose, bend the hose back on itself and check for cracks in its outer sheathing. Also check the tightness of its end fittings. If it looks suspect, replace it with a new OEM hose. Unless you're going racing or hitting the rod-show circuit, stay away from braided steel lines on a restoration car.

Opening the system like this requires that you bleed the brakes after reinstalling the caliper. This is covered later in this section.

Rotor Inspection—See drum inspection, page 52, for information on heat damage and lugnut replacement, wheel bearings and cleaning. Check the minimum thickness specification (usually 0.965 in. for a 1.0-in. disc and 1.215 in. for a 1.230-in. disc) cast into the hat, or wheel hub, portion of the rotor. Never reuse a rotor that's worn beyond this specification.

Caliper Inspection—As with drum-brake wheel cylinders, if the brakes were working fine before you took them apart, the caliper is probably OK. Nevertheless, this is a restoration. So inspect the caliper piston boot for cracks, tears or fluid leakage. A cracked boot allows moisture, road salt, dirt and grit into the piston bore, causing rapid wear and corrosion.

Caliper Overhaul—If the boot is damaged or leaking fluid, the caliper *must* be rebuilt or replaced. To do this, pry out the dust boot with a plastic or wooden stick (to avoid scratching the bore). Place a 2x4 or heavy towel inside the caliper cavity and apply 5-7 psi of compressed air to the brake hose connection. *Be careful as the piston may fly out with great force!* With the piston out, inspect the piston and bore for scoring and pitting. Minor scoring and pitting can be removed with a brake hone, but if the cost is not great, you're better off buying a new caliper. If honing is required, flush the caliper with brake fluid to remove all honing debris. If the bore and piston are OK, coat them with fresh brake

fluid, fit a new piston seal on the piston, coat it with brake fluid and carefully press the piston into the bore, making sure you don't twist or damage the seal. Then install a new dust boot. Also replace the caliper mounting bolts, bushings and sleeves if they were corroded or worn.

Replacement of the bushings is especially important to proper brake operation as they allow the caliper to slide. This sliding action centers the pads to the rotor, allowing maximum braking action and even pad wear. Remove the bushings from the grooves in the caliper mounting bosses with a small screwdriver. New bushings can be installed with your fingers. Make sure the bushings are not twisted after installation and be sure to coat them with silicone.

Rotor Installation—Rotor installation requires that you adjust the wheel bearings. This entire process is discussed in the section entitled *Wheel Bearings*, page 60.

Caliper Installation—Now turn your attention to the caliper. If the piston isn't completely seated in the caliper, you probably won't be able to slide the caliper back over the rotor. Begin seating the piston by first removing the master cylinder cover. Then, using a clean turkey baster, remove some brake fluid from the front reservoir of the master cylinder to drop the fluid level about 1/2 in. Put the cover back on the master cylinder but don't clamp it down. This will allow room for the brake fluid that will be displaced when you seat the piston. It will also prevent brake fluid from squirting all over your car and damaging the paint. But just to be careful, lay some thick towels around the base of the master cylinder to catch any overflow.

Put a C-clamp (8-in. throat or larger) squarely over the center of the piston and slowly turn the T-screw until the piston is seated. When the piston is seated, you can install the pads.

Pad Installation—Before installing new pads, be sure to use a new clip on the inboard pads. And coat the backing plates of the pads with an anti-squeal compound, or better yet, install anti-squeal shims between the piston and pad. The shims help prevent the pads from vibrating at a pitch which produces a squealing sound. They are widely available from auto parts stores.

When the pads are positioned in the caliper, slide the caliper onto the rotor. Line up the bolt holes in the caliper with those on the mounting bracket on the spindle and slide the caliper mounting bolts

through the sleeves in the inboard caliper ears. Make sure the bolts pass under the inboard shoe's retaining ears, through the holes in the outboard shoes and thread into the outboard end of the caliper. Torque these mounting bolts to 35 ft-lb. Before road-testing the brakes, top off the master cylinder then depress the brake pedal 5 times to seal the caliper pistons against the rotors. This step is important! The pedal should be hard before any driving is attempted! Install the wheel and tire, and torque the lugnuts to 80-100 ft-lb.

1967-1968 DISC BRAKES

These are the four-piston, fixed-caliper type found on 1967-68 Camaros we mentioned earlier. Much of the foregoing on single-piston disc brakes applies to the four-piston type, but with the following variations.

Brake pads can be inspected or replaced after removing a single retaining pin and cotter key. Mark the pads for reassembly if you intend to reuse them. The caliper mounts to the spindle bracket with two bolts.

To disassemble the calipers, remove the two caliper through-bolts. Take note of the location of the 0-rings for the fluid transfer passages. Compressed air is probably not required to remove the pistons. See the nearby exploded view for the orientation of components. Always use fresh brake fluid when assembling the caliper pistons and seals. Don't forget to use new transfer-passage 0-rings. Tightening torque for both the caliper mounting bolts and caliper through bolts is 130 ft-lbs.

Use a putty knife to protect the piston boots as the caliper is installed over the rotor.

BRAKE LINES AND HOSES

If you look closely, you'll find the brake lines on your car are made of double-wrapped steel tubing. That's because steel tubing is strong, durable and relatively inexpensive. Other materials such as copper and aluminum don't possess enough strength to be used as brake lines so never use these materials in the braking system. The round-wire wrapping encompassing the brake lines in certain areas is designed to prevent damage to the lines from stones or other road hazards.

When you remove the brake line that feeds the rear wheels from the combination valve under the master cylinder, you

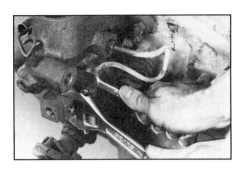

To help prevent twisting brake lines, use a wrench to keep some components, like this distribution valve, from moving.

Some retaining clips for front brake lines are on subframe.

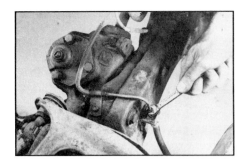

To prevent rounding off the flats, use a flare-nut wrench on all brake line connections. Tag all brakes lines for identificaticn before removing them. Spiral-wound armor helps prevent stones and debris from damaging brake lines.

must remember that it's connected to a proportioning valve. On first-generation cars, this proportioning valve is made of brass and is attached to the subframe with a metal bracket and screw.

The valve will begin to turn on its bracket if the brake line flare nut is difficult to remove because of rust or corrosion. So use a wrench on the flats of the valve to keep it from turning while loosening the flare nut for the brake line. If the nut still won't budge or begins to round off, cut the line as close to the valve as possible. This way you can remove the valve and take the nut off in a soft-jawed vise, saving the valve.

REAR BRAKE LINE

The rear brake line is made of steel tubing and stops in the vicinity of the rear-axle carrier housing. The line is connected to a flexible rubber hose. The hose, in turn, is connected to a junction block that's attached to the carrier housing of the rear axle. Two brake lines are routed from this junction block—one to each rear wheel cylinder.

Removal—To remove the rear brake line, pull it out from inside the engine compartment. However, in order to do this without bending the heck out of the line, you're going to have to remove the front parking-brake cable where it is routed through the

subframe. And this requires removing the rear parking brake cable.

The rear parking brake cable is connected to the front cable with a U-shaped metal bracket called an equalizer and a nut on an adjusting stud. This nut keeps tension against the equalizer. And they're positioned on the underside of the car beneath the driver's seat.

Begin removal by giving the threads on the adjusting stud a shot of lubricant. Then remove the nut and locknut. Don't try to wrench them off if they're stubborn because you'll only unwind the cable causing yourself extra expense. If the nut won't budge, hold the threaded end of the cable which has four flats on it with a securely fitting wrench and then remove the nut. Then pull the cable and equalizer off the stud.

Now before you start yanking on the cable housing, note that it's secured to the innermost part of the subframe with a spring clip. Pop this clip off with a screwdriver. Then pull the cable out of the subframe from the driver's side of the car.

With the cable removed, shift the brake line near the combination or distribution valve toward the center of the car and gently snake it out from between the subframe and body. It may bend a little when you're removing it but when you make a replacement line this won't be of any concern. And when you're installing the new line, you may end up bending it a little, but this won't matter. You should be able to route it where it needs to go and then bend it back into position without any trouble.

FRONT BRAKE LINES

The steel tubing for the right front brake line runs from the combination valve across the rear of the large, subframe cross-member to the flexible-hose retaining bracket. The tubing is attached to the subframe with metal clamps and bolts. Some models use a spring clip that's riveted to the subframe. Remove the clamp bolts and the compression fitting at the flexible-hose bracket to remove the brake line. When these are removed, lift the passenger side of the brake line first so it clears the subframe. Then snake the other end out and away from the steering gear.

The left front brake line can be removed by loosening and disconnecting the compression nuts. Once they're disconnected, you can remove the brake line.

Because of space limitations, probably the best and easiest way to remove the flexible brake hose from each front wheel

cylinder or caliper is to remove the brake backing plate first. To do this, remove the dust cap from the wheel hub by inserting a screwdriver between the cap and the hub. Rotate the screwdriver while moving it around the hub to free the cap.

Now straighten the cotter pin in the end of the spindle and remove it. If it's stubborn, gently place a pair of cutting pliers around the cotter pin and lever it out using the spindle nut as a fulcrum. To avoid dropping the outer wheel bearing, back off the spindle nut 1/2-in. or so then tug outward on the brake drum or disc. This will free the inner bearing from the axle shaft. Then take the nut completely off followed by the outer wheel bearing.

If you think you're going to reuse the old bearings, put them in a plastic bag, grease and all, then tag the outside of the bag. A tag on the inside of the bag will be useless when the grease attacks the ink and makes it invisible.

FABRICATING BRAKE LINES

When you've removed the old brake lines from the car, you can use them as a guide to help you bend the new brake lines. However, there's a little hitch if the brake line you're making is more than 5 or 6 feet long. If the brake line has to run farther than this, you're going to have to join two or more lengths of brake tubing to reach your destination.

To eliminate coupling brake lines together, you can order brake line by the roll from some auto supply stores. It usually comes in 25-ft.-long rolls. The only drawback to buying a roll of tubing is that it's coiled. So if you want it straight, you're going to have to straighten it yourself using a *brake line tubing bender*. It isn't difficult, it just takes a little more time and care.

The next step is to measure the length of the brake line you're replacing. You can do this by snaking a piece of string alongside the old brake line and measuring it. Be careful not to wrap the string around the brake line at any point because you'll throw the measurement off. The figure that you come up with should be used as a low side estimate. Add 1/2 in. to each foot of brake line to give you room for error. It's far easier to cut off excess brake line than it is to add it. And be sure to use a tubing cutter to avoid distorting the line.

Lay the old brake line next to the new, unbent tubing. Transfer the first bend on the old brake line to the new with a grease pencil. To make bends in the line without kinking it, you're going to need a tubing bender. K-D Tools makes a sturdy handheld bender that can be purchased at many auto part stores. Select the right size bending wheel for the diameter brake line you're using and position it on the tubing bender.

Also, be sure that the compression fittings are in place on the line before you do any bending. Otherwise, you may not be able to get them on when the line is bent.

Insert the tubing over the large bending wheel and underneath the two smaller wheels. Then gently bring the bender handles together to bend the tubing. Do a little at a time and compare it to the original to ensure the bends will be the same. If you go overboard and bend the tubing too far, it'll be difficult to bend the tubing back with the bender. And it can also fatigue the line in the area of the bend, which may lead to a leaky brake line. So work slowly and carefully.

When you're satisfied with the first bend, mark the second bend and begin working on it. Work in this manner until your new line is bent in the same places as the old.

Measure the length of the armor (the coiled wire around the brake line) using the string method discussed earlier. Armor is used to prevent stone damage to the brake lines and must always be used where it was originally installed on the brake lines. It's available from your local GM dealer's standard parts catalog in pre-cut lengths and various inside diameters.

Cut the armor to the length you need with a pair of wire cutters. Then, slide it onto the tubing to check its fit. When the armor is on the tubing, you can flare the ends. However, if you want a first class job that'll look good for a long time, spray the armor and the tubing with clear enamel or

plastic before you install the armor. When both are dry, position the armor on the tubing and give them another coat of clear.

To flare the tubing, don't get out your ordinary single-lap flaring tool and have at it. This is for two VERY good reasons. First, a single-lap flare can actually weaken the tubing. And you don't want that. Second, a single-lap flare doesn't possess the necessary strength to keep it under the compression fitting under hard braking. If the tubing slips out from under the fitting, you'll have an immediate loss of brake fluid and pressure. No pressure means no brakes. It's that simple. A double-lap flare is the only way to avoid these problems.

You can rent a double-lap flaring tool from any well stocked tool rental company or you can buy one. Buying one makes sense if you plan on using it anytime in the future because the cost of two or three rentals will pay for it.

To use a double-lap flaring tool, insert the tubing into the appropriate hole in the jaws so it extends the same distance away from the jaws as the adapter is high. The adapter you should use for this measurement must match the diameter of the tubing you're flaring. Then tighten the jaws. Slide the T-screw over the jaws so it's directly above the tubing. Now tighten the T-screw down until it's fully flared. Then back the T-screw off. Take the correct size adapter from the kit and place it over the flared end. Position the T-screw over the jaws once again and screw it down against the insert. When the tubing is folded back into itself, the job's done.

FRONT BRAKE HOSES

The front brake hoses should be replaced if they show signs of cracking, abrasion or swelling. On a Camaro that's 20 year's old or older, you're probably smart to replace these hoses regardless of how good they look.

On first-generation Camaros with drum brakes, the easiest way of removing these hoses, after disconnecting the metal brake line from them, is to remove the clip at the top of the hose where it fits into the retaining bracket. Then remove the brake backing plate and loosen the hose while the wheel cylinder remains secured to the backing plate.

Those of you with disc brakes have it easier. After disconnecting the metal brake line, remove the retaining clip from the top of the hose. Insert the tip of a screwdriver between the ear of the clip and

the metal fitting on the hose and lever the clip out. Then remove the banjo bolt that retains the hose to the caliper. Be sure not to lose the two copper washers for the banjo bolt.

MASTER CYLINDER

As with wheel cylinders, you're much better off replacing the master cylinder with a new replacement than taking a chance rebuilding one that needs scratches and pits honed out. New cylinders are widely available at popular prices and you won't have to contend with removing rust,

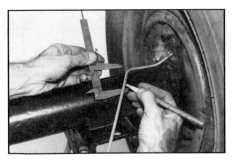

When fabricating new brake lines, transfer measurements from the old lines with a china marker.

Bend the brake line in the right direction relative to other bends. Bend just a little, then check to be sure.

If a large-radius bend is required, you may have to tweak it by hand or over your knee. Remember, easy does it.

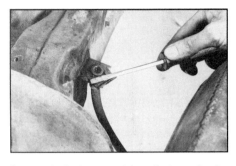

Remove brake hose retaining clip by prying it away from the hose with a screwdriver. New clip fits into step in brake hose fitting. And don't forget to check the flex hose to the rear axle.

bead-blasting, and detailing the old one. If you insist, however, follow the procedure for rebuilding, cleaning and detailing under *Wheel Cylinder,* page 54. And always use a flare-nut wrench on brake-line fittings to avoid rounding off the flats.

BRAKE BLEEDING

Whenever work is performed on brake-system hydraulics, or the system is opened to the air for any reason, it must be bled and purged of all air. Otherwise, the brake pedal will be spongy at best.

Raise the car about a foot off the ground and install jackstands under the front sub-frame and rear axle. Shake the car to make sure it is securely supported. Get a length of clear plastic hose to fit snugly over the brake caliper or wheel-cylinder bleeder nipples and a clear jar with about 2 in. of brake fluid in it. Have about 2 pints of new brake fluid on hand to replenish the master cylinder as the old fluid and air is bled out.

With an assistant seated in the car to manipulate the brake pedal, bleed all four brakes in the following order: right rear, left rear, right front, left front. At each wheel, place the hose over the nipple and submerge the end of the hose in the fluid at the bottom of the jar. On signal, have the assistant pump the brake pedal several times then hold it as you open the nipple (counterclockwise) and remove old fluid and air. Then close the nipple (clockwise) and signal the assistant to let up on the pedal. Repeat this process until a steady stream of brake fluid, with no air bubbles, is expelled. Do this for each wheel, moving from the cylinder farthest from the master cylinder to the one closest, until a firm pedal is achieved. Periodically, and after each wheel is bled, recheck the fluid level in the master cylinder to avoid running it dry and sucking more air into the system.

WHEEL BEARINGS

Wheel bearings are important components. They must be properly lubricated and adjusted to ensure their longevity and your safety. Both high mileage and vehicle age take their toll on bearings as lubrication thins out or dries up and bearing rollers and their cages wear out. So a wheel-bearing inspection should be part of any Camaro restoration.

REMOVAL

Removal of the front wheel bearings necessitates removal of the tire and wheel assembly, and brake drums, rotors and calipers. The next step is to remove the dust cap in the end of the hub or rotor. To do this, just slide the tip of a screwdriver between the lip of the cap and the drum. Then pry the cap away. As the gap widens, move the screwdriver to another area of the cap and pry again. Work in this manner until the cap is removed.

When the dust cap's off, you'll see the spindle with a castellated nut and a cotter pin running through the nut. Use a pair of needle-nose pliers to straighten the end of

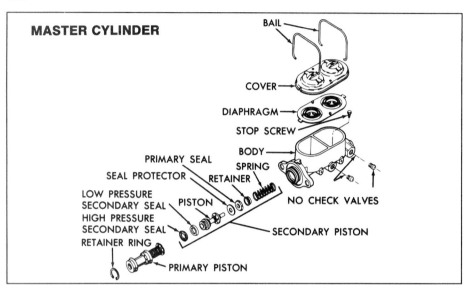

Details of typical Camaro dual master cylinder. Courtesy Chevrolet.

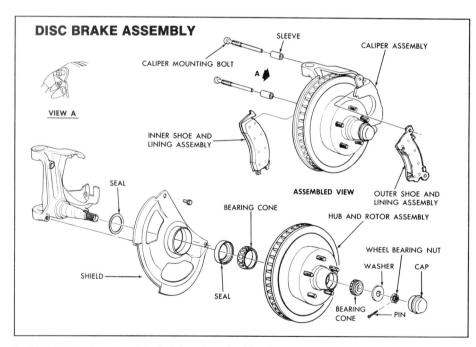

On 1969-81 disc-brake models, hub housing wheel bearings is cast integrally with rotor. Courtesy Chevrolet.

the cotter pin and remove it. Then loosen the nut about three or four turns and give an outward yank on the hub or rotor. This will free the outer bearing from the spindle but keep it from falling on the ground. Then remove the nut completely and slide the tanged washer and outer bearing off the spindle. And pull the hub or rotor (after removing the caliper) off the spindle.

If you plan on reusing the bearings, keep them with their respective hub or rotor. Bearings and races wear into a matched set, much like an engine's valve lifter and camshaft lobe. If the bearing is used with a different race, it'll start wearing in a different manner, creating friction, heat and possibly bearing failure.

Inner Bearing & Dust Seal—To remove the inner wheel bearing, carefully flip the hub or rotor over. If you look through the spindle cavity, you'll see the inner bearing. This bearing is covered by a metal ring with a rubber lip known as a dust seal. You have to remove the dust seal to remove the bearing.

The dust seal can be easily removed by inserting the tip of a stout screwdriver under the inside of the seal and prying it out using the edge of the cavity as a fulcrum point. Then remove the bearing.

Bearing Outer Races—If you're replacing the bearings, you have one more step to complete: removing the outer bearing races. This is accomplished by placing a long drift against the portion of the race that faces the inside of the hub or rotor and tapping on it with a hammer. Just remember to tap lightly and work your way around the race to prevent cocking it in the bore.

CLEANING

To do the job right, both old and new bearings should be cleaned in clean solvent such as *Gunk Parts Cleaner*. Pour the solvent into two suitable containers. The first container will be used for initial cleaning and the second for final cleaning.

Place the bearings in the first container and let them soak for about five minutes or so. When this time has passed, use a toothbrush to clean out any residual grease. Then place the bearings in the second container of solvent and clean them again.

When the bearings are scrupulously clean, place them on some newspaper and let them dry completely. If you care to, you can dry them with compressed air. Just direct the air nozzle so it's parallel to the long side of the bearings. And whatever you do, don't spin the bearings with

compressed air. If you do, the bearings can spin so fast they'll literally fly out of the cage like a hand grenade. People have been killed doing this. So don't try it. When the bearings are thoroughly dry, you can pack them with fresh wheel-bearing grease, but inspect them first.

INSPECTION

When the bearings are clean, take a close look at them for signs of damage or wear. Carefully check the cage for any dents and for signs of scratching around its outside diameter. Then look at the bearings themselves. They should appear bright and smooth with no signs of discoloration, pitting, flaking or smears. Now look at the inner and outer races. They should have no signs of smearing, pitting, indentations or cracking. Any bearings that can't pass all these tests must be replaced. Your safety depends on it.

PACKING

Before packing the wheel bearings, make sure you have the proper type of wheel-bearing grease on hand. First of all, you must use wheel-bearing grease, not regular chassis grease. Wheel-bearing grease has a more fibrous composition than other types of grease. Second, if your Camaro has front disc brakes with their resulting higher operating temperatures, they require a grease with a higher melting point. Grease that's formulated for wheel bearings used in disc-brake equipped cars fits this requirement just fine. And it can also be used for drum-brake applications as well.

Bearings can be easily packed by hand if you know the proper technique. Starting with clean hands, dip the edge of the bearing into the grease and place it in the palm of your hand. Now push the edge of the bearing into the grease until you see it extruding out the other side. This indicates the grease has filled the bearing correctly. Work your way around the bearing using this technique until it's completely packed with grease. And be sure to smear a thick coating of grease on the outside of the rollers.

INSTALLATION

Bearing Outer Races—If you're using new bearings, you'll have to install new outer races. You can do this with a press, brass drift or old bearing race. First take a look at the entire bore where the old race used to be. If you see any nicks or gouges, remove them with fine emery cloth. When

On Camaros with front drum brakes and 1967-68 disc brake models, separate bolt-on rotor is used.

the bore is free of any nicks, position the new bearing race at the beginning of the bore with the tapered side facing out.

The trick is to drive in the new race without cocking it in the bore or damaging it. You can do this with a properly prepared old bearing race. To prepare the old race, grind about 0.010 in. off the outside of it. Then position the side of the old race with the thick shoulder against the new race and carefully drive it in using the side of a hammer. Work slowly to prevent cocking the race in the bore.

If you prefer to use a drift to seat the race, be sure it's made of softer metal (brass, for instance) than the race or you'll damage it. And be sure to work slowly.

Inner Bearing & Dust Seal—When the outer bearings races are installed and clean, apply a small coating of grease to the inner bearing race. Then install the inner bearing in the race. Next, position the inner-bearing dust seal into the bore and drive it in using an old seal and the same method you used for installing the bearing races. However, you don't need to grind the old seal. Drive in the new seal until it's flush with the hub.

Hub or Rotor & Outer Bearing—Before positioning the hub or rotor onto the car, make sure the spindle is clean and free of any old grease. Apply a light coating of new grease onto the spindle, on the lip of the dust seal and on the outer bearing races. Now slide the inner bearing and hub or rotor over the spindle and into position.

At this point, install the outer bearing over the spindle and into place so it supports the hub or rotor. Then slide the tanged washer over the spindle and screw the spindle nut on by hand. For drum brake applications, slide the drum over the hub. Now you're ready to adjust the bearings.

ADJUSTMENT

Adjust the bearings by tightening the spindle nut to 12 ft.-lbs. while spinning the drum or rotor. This step is important in that it seats the bearings and helps prevent excessive wheel-bearing play later on.

Then back off the nut until it's just loose. Now hand-tighten the nut until one of the holes in the spindle line up with the slots in the nut. Be sure not to rotate the nut more than half the distance of one of the flats on the nut.

Next, slide a new cotter pin through the nut and spindle. And bend its ends against the spindle nut making sure they won't touch the dust cap. Finally, install the dust cap and you're finished.

PARKING BRAKE

If a brake drum won't come off, check to make sure that the parking brake is disengaged. If the parking-brake mechanism and cable are free, rust may have formed around the axle-shaft flange, preventing drum removal. The cure for this is to use some 240-grit sandpaper to sand the rust off the flange. When the rust is removed, you should be able to slide the drum off. If it still won't come off, you'll have to use a three-jaw puller to remove it. These are available at many local rental shops.

Cable Removal—Disconnect the center parking brake cable (it's the bare one underneath the car) from the parking brake cables that run through each brake backing plate. First, loosen the adjuster at the "Y" underneath the floorpan, practically under the driver's seat. Use two wrenches to break the lock nut free. Then remove the lock nut and adjusting nut.

Follow both sides of the cable back toward the rear axle. You'll find the cable sheath is attached to small brackets in the floorpan with a clip that has two round ears. Remove these clips and put them aside. Then remove the cable from the "U" shaped metal connector and remove the rear cables from the axle.

Cable Installation—Feed the parking brake cable from each brake backing plate through the cable brackets attached to the side rails. Begin installing the center parking brake cable by feeding the cable through the hole in the "U" shaped metal connectors near both rear brake backing plates. Then loop them into the connectors.

Now hook the cable guide (the long steel rod with hooks on both ends) into the rear hole on the side of the right-hand subframe rail. And slide the cable over the free end. Next, slide the cable into the equalizer and slide the equalizer over the threaded end of the front parking brake cable. Then screw the adjusting and locknuts onto the stud. Make sure that the cable isn't kinked or binding at any point.

Adjustment—After installing the parking brake cables, you'll need to adjust the parking brake. Do this by releasing the parking brake. Then slowly push down on the parking brake until you hear it click twice. Now tighten the adjusting nut until the rear brakes drag slightly. You can check this by rotating the rear tires by hand.

When you've adjusted the brake, tighten the lock nut against the adjusting nut. Then try to apply and release the parking brake. If the parking brake doesn't release completely or it's hard to apply the parking brake, make sure the cables and equalizer are lubricated. Use *WD-40* for the cables running through sheaths and *LubriPlate* or motorcycle chain oil for the exposed bare cable. And be sure to lubricate the floorpan where the cable rubs against it with *LubriPlate*.

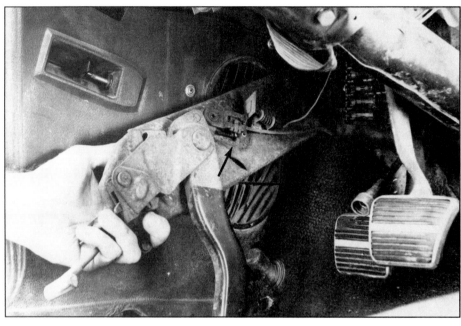

If necessary, disconnect parking brake cable from pedal assembly by removing clip (arrow). The complete pedal assembly comes out after removing bolts to the firewall and lower instrument panel.

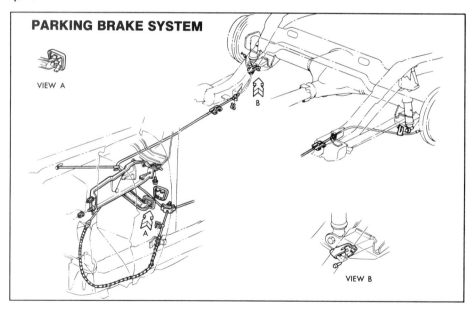

Typical Camaro parking brake linkage—1967-69 models with rear drum brakes shown. Courtesy Chevrolet.

After decades of use and abuse, your Camaro's independent front suspension and steering linkage will likely have a lot of worn parts, causing clunks, shakes and imprecise steering. Kanter's low-cost front-end rebuild kit (Kanter Auto Products, Boonton, NJ, 201/334-9575) shown here provides the most commonly replaced parts except the idler arm.

FRONT END

Camaro's independent front suspension uses upper and lower control arms of the wishbone design. A coil spring is positioned between the lower control arm and fits into a seat in the subframe rail. And the shock absorbers pass through the inside of the coil springs. This is true of both first- and second-generation cars. However, there is one item that's different: the steering linkage.

The recirculating-ball steering gear and parallelogram-type linkage (consisting of a Pitman arm, idler arm, center link, tie rods and integral steering knuckles and wheel spindles) on first-generation cars is positioned behind the front suspension and crossmember. With high-g cornering loads, deflections in the control-arm bushings act on the steering linkage, causing an oversteer condition and "squirrelly" driver response. When engineers began designing the second-generation Camaro, it was found that the car could be made to handle more predictably if the steering linkage was moved forward of the front suspension. With this set-up, any control-arm deflections would result in an understeering condition, allowing the driver to simply crank in more lock to compensate rather than constantly over-correcting as with the 1967-69 design. And so it was changed with the introduction of the 1970 Camaro.

What Wears Out?—It doesn't take a rocket scientist to figure out that after decades of use and abuse and tens, perhaps hundreds of thousands of fun-filled miles, your Camaro's front suspension and steering linkage may be due for some attention.

Imprecise handling, shimmies and shakes at speed, clunks and groans as you drive over "speed bumps" or up driveway ramps, and wrenching noises as you turn the steering wheel from lock-to-lock, are all cues that the time has come to refurbish the front end. Tires worn asymmetrically is another indicator. And aside from the obvious safety implications of a "loose" front end, returning the front suspension and steering to like-new condition should be the cornerstone of any Camaro restoration. After all, Camaros are good-handling cars provided their underpinnings are kept up to snuff.

Having recommended a complete front-suspension teardown, keep in mind that the two most frequent wear items (shock absorbers and anti-roll-bar bushings) don't require suspension disassembly to replace them. If your Camaro has had recent front-end work or is a low-mileage or late-model car, tracks straight and has even tire wear, it may not need to be gone through. It should, however, be inspected.

Front-end items to be scrutinized closely, in descending order of their failure rate, are: the idler arm, upper control-arm pivot bushings, lower ball joints and the tie-rod ends.

INSPECTION

Shock Absorbers—If your Camaro tended to float over bumps and cycle through its suspension travel more than once afterward, the shock absorbers may need to be replaced. With the car parked, press down sharply on the front bumper and release abruptly. If the front end bobs up and down several

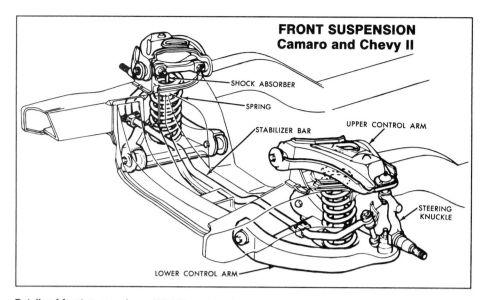

FRONT SUSPENSION
Camaro and Chevy II

SHOCK ABSORBER

SPRING

STABILIZER BAR

UPPER CONTROL ARM

STEERING KNUCKLE

LOWER CONTROL ARM

Details of front suspension—1967-69 models. Courtesy Chevrolet.

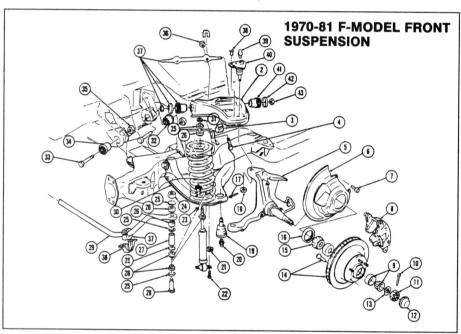

1970-81 F-MODEL FRONT SUSPENSION

1. Shim
2. Arm Asm, Upr-LH/RH
3. Bumper, Strg Knuckle
4. Bolt
5. Knuckle, Steering
6. Shield, Front Disc
7. Bolt
8. Caliper Asm, L/R Fronts
9. Bearing Asm, Frt Whl Otr
10. Pin, Cotter
11. Nut, Steering Knuckle
12. Grease Cap
13. Washer
14. Disc, Front w/Hub
15. Bearing Asm
16. Seal Asm
17. Pin, Cotter
18. Nut, Upr Ball Stud
19. Ball Stud Lwr Cont Arm
20. Stud, Straight
21. Nut "U"
22. Screw
23. Nut, Hex
24. Nut, Stabilizer Link, Hex
25. Retainer, Stab Link
26. Grommet, Stab Link
27. Spacer
28. Link, Front
29. Shaft, Stabilizer
30. Arm, Lower
31. Nut, Shock End
32. Bushing Asm
33. Bolt, Strg Knuckle
34. Bushing Asm
35. Nut
36. Nut
37. Shaft Unit
38. Rivet
39. Fitting
40. Stud Unit
41. Bushing
42. Retainer, Bushing
43. Nut (5/8"-18)

Exploded view of front suspension—1970-81 models. Courtesy Chevrolet.

times through jounce and rebound, the shocks' internal fluid seals are shot. Sometimes, a visual check will spot a shock absorber with an external fluid leak. Replace any leaking shock absorber. Also, visually check for worn, dry-rotted or missing shock-absorber mounting bushings. Remember that shocks should always be replaced in axle sets (pairs).

Anti-Roll-Bar Bushings—Camaros tend to be pretty flat-cornering enthusiast's cars, so if yours goes around sharp corners seemingly on its door handles, check the bushings in the front anti-roll bar, particularly those on the upright links at the lower control arms. Smog, sunlight and age gang up on the original-equipment rubber bushings, dry-rotting them in no time at all.

Idler Arm—If you notice any response lag when turning the steering wheel as you enter a corner, suspect a worn steering component, most likely the idler arm. Jack up the car with the front wheels off the ground and support it on jack stands. The idler arm is bolted to the right-front frame rail. With the front wheels pointed straight ahead, grab the bottom of the arm at its center-link connection and rock it up and down with about 25 lbs of force. If you notice any appreciable play—more then 1/8 in. at its ball socket—replace the arm.

Control-Arm Pivot Bushings—These metal-sleeved rubber components live a harsh life. On the lower arm, they're bombarded by road debris, the elements and road salt. On the upper arm, they're somewhat protected from the elements, but get baked in engine heat, particularly that of the exhaust manifolds. And the bushings in both arms experience tremendous deflection loads when you go into a corner hard. If you notice some unsettling wander at high speed, investigate these bushings. The uppers can be viewed from inside the engine compartment; the lowers from underneath. If they are cracked, dry-rotted or oil-soaked, remove the control arms and press in new bushings.

Ball Joints—Also a tough job is the one the upper and lower ball joints perform. As the main links between your Camaro's steering and front suspension, these hefty pieces locate and support the spindles through full steering lock and full suspension travel. It is little wonder then, that they wear rapidly, especially when people forget to lubricate their ball joints once a year or so. Also lethal to ball joints is a torn rubber boot, which lets grease out and water and dirt in. Don't fool around with worn ball joints. Lose a ball joint and you could lose a front wheel-spindle, brakes and all—while you're moving down the road.

Because the lower control arms act as seats for the front springs, they take the weight of the car—and so does their ball joints. So they tend to wear faster than upper ball joints. Realizing this, General Motors began installing lower ball-joint wear indicators in Camaros starting with 1974 models. These are actually spring-loaded ball studs that retract into their sockets as they wear. After scraping off any rust or dirt build-up from around the grease nipple on the lower part of the ball joint, check if the flats on the nipple are above or below the surface of the joint using a ruler or tape measure. If the nipple's flats have receded below the surface, the joint has worn past its 0.050-in. limit and needs to be replaced.

Whether it's an upper or lower ball joint, replace any that won't accept chassis grease. Aside from the fact that the joints need grease to function, non-acceptance means the joint's ball socket is worn internally to the point the spiral grease passages are closed off. Worn ball joints

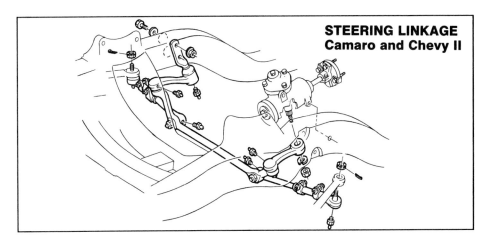

STEERING LINKAGE
Camaro and Chevy II

Steering linkage details—1967-69 models. Courtesy Chevrolet.

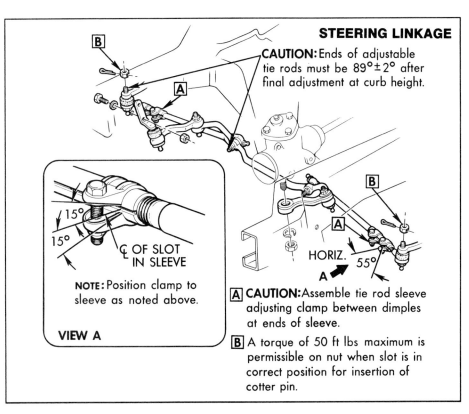

STEERING LINKAGE

CAUTION: Ends of adjustable tie rods must be 89°±2° after final adjustment at curb height.

NOTE: Position clamp to sleeve as noted above.

VIEW A

A CAUTION: Assemble tie rod sleeve adjusting clamp between dimples at ends of sleeve.

B A torque of 50 ft lbs maximum is permissible on nut when slot is in correct position for insertion of cotter pin.

Steering linkage details—1970-81 models. Courtesy Chevrolet.

Cracked bushings and broken strut have rendered this 1967 Camaro's front anti-roll bar useless.

When upper control arm bushings are shot, washers on end of shaft will be offset compared to control-arm flange. Bushings in this shape will cause car to wander at highway speeds.

Steering column attaches to steering box at rag joint. Remove 2 bolts and brass ground strap to disconnect.

Metal bracket and retaining nuts secure column to bottom of dash. Wedge plate above top of bracket can be shimmed to adjust column height.

65

sometimes sound off with a groan when the wheels are turned from lock-to-lock.

If you suspect a worn ball joint, temporarily remove the cotter pin and tighten the castellated nut on the end of the affected spindle to about 20 ft-lb. Doing so will remove any wheel-bearing play which could be mistaken for ball-joint play (Just don't forget to readjust the wheel bearings after the check or you'll destroy the roller bearings). Jack up the front end under the lower control arm and support the car on

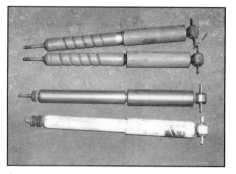

Spiral-groove Delco shocks (top) were original equipment on many Camaros, but are no longer available.

Monroe Formula GP gas-charged shocks have high damping efficiency without being overly stiff.

Larger-diameter anti-roll bars and urethane bushings will help your Camaro corner flatter and more predictably. But urethane will squeak if not lubed periodically and concours judges may dock points for using non-stock bushings.

jack stands. While observing the fit of the lower ball joint in the spindle, grasp the tire at the top and bottom and attempt to rock it inward and outward. Some play (about 1/16 in.) is normal, but if wear exceeds this, the joint is worn.

Repeat for the upper ball joint, only this time jack under the subframe and allow the wheel to hang free. Usually, the upper joint will be OK.

Then, if you are going to drive the car anywhere, readjust the wheel bearings, page 61.

If you notice some play between the ball-joint stud nuts and the spindle boss, the nuts may need to be tightened. Remove the cotter pins and check that the lower nut is tightened to 65 ft-lbs. for '67-69 models and 85 ft-lbs. for 1970-81 vehicles. The upper nut must be torqued to 50 ft-lbs. for '67-69 models and 65 ft-lbs for second generation cars. Then reinstall the cotter pins.

Tie-Rod Ends—A side-to-side shimmy felt in the steering wheel is good cause to inspect these steering components. There are actually four tie-rod ends, two inner and two outer on your Camaro. For some reason, the outers tend to wear out first. Since these components are not loaded to a great extent, the only reason for a premature demise is lack of periodic lubrication. Look for a torn boot that lets grease out and water and dirt in. If the boot's torn, the tie rod's a goner.

Springs—As a rule, Camaros have relatively high-rate springs to begin with and don't get overloaded towing trailers or moving your neighbor's refrigerator. So spring sag is not a big problem. But if your Camaro is drooping particularly low or is tilting to one side, keep these parameters in mind. The distance from the center of the lower control-arm pivot bolt to the lowest point on the inboard corner of the lower ball joint should be about 2-5/8 in. to 2-7/8 in. on first-generation Camaros and 2-1/8 to 2-5/8 in. on second-generation cars. These distances are computed by drawing a horizontal line through the center of the pivot bolt and another horizontal line from the ball joint and measuring the distance between the lines. Also, on all Camaros, the distance from the forward end of the rocker panel to the ground should be about 8-1/4 to 9-1/4 in. If the car is sagging more than 3/4 in. or side-to-side variance exceeds 1/2 in., you'll need new front springs. Be sure to rule out decreased ride height due to ultra low-profile 50-series tires if you're looking at a modified car.

RESTORED VS. RESTIFIED

Before tearing apart your Camaro's front suspension and steering, have a good fix on what you want to do with the car. If the future itinerary will be filled with car shows and perhaps the prying eyes of sharp-penciled investors, go the bone-stock route with rubber bushings and original-style tires. If yours is a pre-1974 Camaro, that means no radial tires. And be prepared to detail the heck out of the control arms, spindles, steering box and so on.

On the other hand, if you're building this Camaro as a driver, there's no good reason to avoid the aftermarket suspension goodies that have been developed for Camaros over the past two decades. At least, you have a choice when it comes to tires, shock absorbers, anti-roll-bar bushings and control-arm bushings. If you're not fussy about originality, you can have a *restified* Camaro that looks stock on the outside, but has aftermarket pieces underneath to improve handling. Some of these improvements can be dramatic.

Gas-Charged Shocks—The spiral-groove AC/Delco twin-tube shock absorbers that were original equipment on most Camaros are no longer available from Chevrolet—as of this writing. It seems that GM realized it couldn't compete in the aftermarket with original-style shocks. So unless you're so picky about originality that you'd rather drive around on shot shocks that are correct in appearance than non-correct replacement shocks, you've got a dilemma.

This problem is actually an opportunity to install shocks that greatly out-perform their predecessors. Single-tube, gas-charged shocks are pressurized with nitrogen to keep the shock's hydraulic fluid from overheating and aerating (and thereby losing its effectiveness, as the shock is cycled rapidly up and down. Even at low speeds and on relatively smooth roads, the damping effectiveness of gas-charged shocks are superior to their conventional double-tube counterparts. Because gas-charged shocks look much the same as conventional shocks on the outside, there is no reason to draw attention to your, uh, modification. If the gas shocks are a non-conventional color, paint them semi-gloss black prior to installation.

Urethane & Solid Bushings—The stock rubber bushings used on the anti-roll bar and control-arm inner pivots go a long way toward reducing road-surface-induced vibration and harshness from your Camaro's front end. But comfort has its price, and Camaro drivers pay with imprecise handling (at the limit) caused by *compliance* and *deflection* of these rubber-bushed components. On the other hand, if you don't probe the outer limits of handling with your Camaro, you might never notice the imprecision of a response lag during a transient maneuver.

One way to improve response is to eliminate or at least minimize the degree of compliance or deflection by installing solid or hard-plastic bushings in place of the rubber. Understand that Concours judges will sneer at this practice and you'll likely feel that pea in the road in the steering wheel and the seat of your pants. But geometry changes will be minimal and your Camaro will zig when you want it to zig and zag when you want that, too.

Solid bushings available for control-arm inner pivots may be steel-sleeved with hardened ball bearings (strap on your kidney belt first) or composite Delrin and aluminum Del-A-Lum bushings. The latter is our pick and are available from **Global West Chassis Components in Montclair, CA (714/946-7828).** You'll need to send them your control arms to have the special bushings installed. A benefit is the Global West bushings are lubed for life.

A compromise solution to the ride vs. handling bushing dilemma is to use urethane bushings. These are available for both the control arms, and the anti-roll bar. A major drawback is urethane's propensity to squeak—don't try to sneak up on anyone; they'll hear you coming a block away. Frequent spraying with silicone can minimize the squeaking or you can pop for graphite-impregnated urethane bushings. Not only are these black (and therefore less likely to arouse attention), but the dry graphite lubricant virtually eliminates squeaks—and the squawks of your peers.

Larger Anti-Roll Bar—Was it Einstein or Newton who theorized that for every action, there was an equal and opposite reaction? Well the same is true for your Camaro's suspension dynamics. Start fooling around with suspension geometry or roll stiffness and you'll have to make other changes to achieve a new balance. The "bigger is better" philosophy has you thinking that a larger (thicker cross-section) front anti-roll bar will reduce your Camaro's body roll during hard cornering. It will, but it will also increase the tendency of your Camaro to understeer (plow or swing wide in a turn). An attempt to reduce understeer by installing a larger rear anti-roll bar could result in excessive oversteer (tendency for the rear end to swing wide or spin out in a turn).

All of which isn't to say that you can't experiment with different diameter anti-roll bars. But be conservative and seek out the advice of aftermarket companies with sound technical departments. You see, the engineers at the GM Technical Center know a little bit about Newton (or was that Einstein?) too.

Low-Profile Radial Tires—Unless you plan to turn your Camaro into a museum piece or trailer it to and from car shows and not much else, there's no good reason to overlook 20 years of dramatic improvements in tire technology. Radial-ply, 60-series (meaning the tire's sidewalls are 60% as large as its tread) tires weren't even on the Camaro option sheet until 1974, but what a difference they've made since. A radial tire is designed with more flexible sidewalls that allow a flatter contact patch with the road. No Camaro should be without them, but here's a compromise: Equip your Camaro with radials for day-in/day-out driving pleasure, but have a set of original-style tires mounted on a set of spare wheels for car shows and the like. Reproduction raised white-letter Firestone Wide Ovals and red-stripe Uniroyals are available from the **Classic Camaro Parts and Accessories in Huntington Beach, CA, 714/894-06511** and other aftermarket suppliers.

Fast-Ratio Steering—As amazing as it may sound, all first- and second-generation Camaro steering boxes interchange, 1967 to 1981. This creates some interesting possibilities, because whereas horsepower saw a steady decline in the 1970s, Chevy really got its act together on steering gears. Late-model Z/28 power-steering boxes with fast ratio, high effort and variable assist can be installed in earlier models with tremendous results. Just remember to install the pitman arm backwards on 1967-69 models. And if you want to fool the car show judges, transfer the steering-box cover from your old box to the new one.

DISASSEMBLY

Disassembling the front suspension can be safe and enjoyable if you have the right tools on hand. This includes a pickle fork or gear-puller-type ball-stud separator tool to separate the steering linkage, another pickle fork to separate the steering knuckle from the ball joints, a coil-spring compressor, a floor jack and a sledge hammer. If your tool box isn't stocked with these items, rent them as they're needed or even

Modern steel-belted, low-profile radial tires make a tremendous improvement in handling over bias-ply tires originally available when 1967-73 Camaros were new.

If you're a stickler for originality, reproductions of Firestone Wide Oval, Goodyear Polyglas, Uniroyal Tiger Paws and other "vintage" rubber are available from the aftermarket.

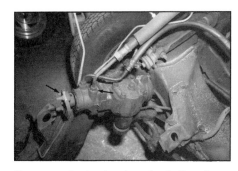

If a power-steering box has slop in it, replace it with a new or rebuilt unit—don't attempt to adjust it yourself. Common wear items are rag joint (left arrow) and high-pressure hose to pump (right arrow). You'll need a crowfoot wrench to loosen fittings at box.

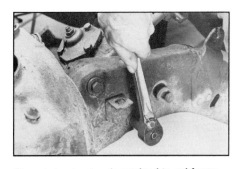

Manual steering box is retained to subframe with three Grade 8 bolts. Sector-shaft screw (arrow) can be adjusted to set steering wheel free-play.

If you plan to replace the tie-rod ends anyway, small-diameter pickle fork can be used to separate steering linkage components. To save good tie-rod ends for reuse, separate them with a screw-type gear puller instead.

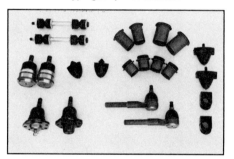

New parts to have on hand for a front-end rebuild include upper and lower ball joints, control-arm bushings, outer tie-rod ends, anti-roll bar bushings and bump stops. Also necessary, but not shown, is a new idler arm.

Large-diameter pickle fork and hammer are used to separate spindle from ball joints.

The amount and placement of these shims determines front-end caster and camber. When removing upper control arm shaft, note the location and number of these shims so they can be replaced in the same position upon assembly. Otherwise, front-end alignment will be way off.

buy them if you wish. But don't try to use one pickle fork where the other is needed. The slot in the ball-joint separator is larger than the slot in the steering-linkage fork. Using the larger size won't provide the wedge action needed to separate the linkage pieces, and the excessive force you'd probably apply to try to force the pieces apart will likely cause damage to steering components. And if you want to reuse the existing tie-rod ends without damaging their rubber boots, disconnecting them from the spindles with a gear-puller-type ball-stud separator tool is best.

You'll also need the services of a machine-shop with a hydraulic press to remove and install the control-arm pivot bushings and lower ball joint. Don't try to hammer these pieces in yourself as you'll end up trashing the ball joint and bushings and maybe damaging the control arms as well.

Clean First—Remember that years of dirt and corrosion build-up on these parts will make it a nasty place to work. If you have the opportunity beforehand, take your Camaro and have the undercarriage steam-cleaned. You'll be better able to see what you're working on and less dirt and crud will be raining down in your face and eyes.

Hardware Counts—The steering linkage is most easily separated while it's on the vehicle. This is because the linkage is solidly supported which makes it much easier to separate with a pickle fork. As you remove the idler arm from the right-hand subframe rail, note that its retaining bolts use prevailing-torque nuts. The outside end of these nuts are compressed slightly to prevent them from loosening unintentionally. Wherever you remove these nuts (all prevailing-torque nuts for that matter), they must be replaced with the same type of nut which provides a preload. This is a safety item that must not be overlooked. Also of utmost importance to safety is the quality of any replacement nuts or bolts you use; they must be Grade-8 or better—your life may depend on it!

There are two ways of separating the steering-relay rod from the Pitman arm on the steering gear. They differ in that one method uses a pickle fork to separate the two and the other uses a gear puller. Why a gear puller? To prevent damaging the steering box internals. Ditto for the Pitman arm. Never use a hammer to knock the Pitman arm off the steering box, unless you want to buy a new Pitman arm and steering box.

It may be tempting to loosen stuck parts, but never use the old "blue wrench" or any

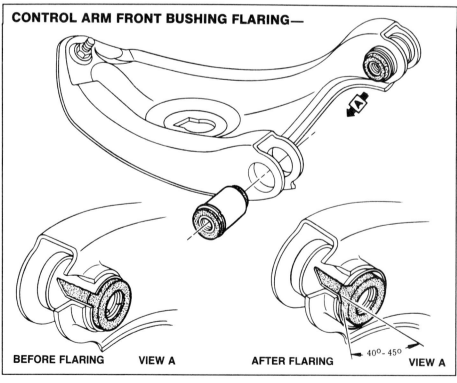

CONTROL ARM FRONT BUSHING FLARING—

BEFORE FLARING   VIEW A   AFTER FLARING   40°- 45°   VIEW A

New control arm bushings must be installed using a press, not pounded in with a hammer. On 1970-81 models, front bushing must be flared as shown. Courtesy Chevrolet.

other type of heat to loosen steering-linkage components—unless you plan on throwing them out anyway. The heat takes the temper out of the steel forgings and weakens them, making them prone to bending forces.

Alignment Tips—If, during the course of rebuilding the front end, you replace any components (and it's likely you will), the alignment will have to be reset and the steering wheel realigned. But you can keep things in the ball park so the car is at least driveable to the nearest alignment shop. As you remove the tie-rod ends, count how many turns it takes to screw them off the shaft and install the new ends an equal amount of turns. This will keep toe-in within reason. And as you remove the upper-control arm from the subframe, count the number and location of shims under the cross-shaft bolt heads. Tape them together and mark them as to location. This will allow you to duplicate the previous caster and camber settings.

Safety Always—We've said it elsewhere, but it bears repeating. Never venture under your Camaro unless it's firmly supported. Shake it a bit to make sure it's sitting as solid as the Rock of Gibraltar on jack stands. And by all means, remember that the front coil springs can be lethal missiles if they get loose when you're unbolting the lower ball joint from the control arm. The spring must be compressed with a spring compressor or safety-chained to the subframe as the lower control arm is lowered, or both.

Likewise, when you've got the lower control arms off the car, inspect them closely for cracks, particularly near the ball-joint socket and inner-pivot eyes. These steel stampings can be checked for cracks by the Magnaflux process at any automotive machine shop. Never weld up a cracked control arm; new replacements are available from Chevrolet.

Cleaning & Detailing—You'll likely be reusing the control arms, coil springs, cross-shafts and spindles, so a good cleaning is in order. Each can be degreased in a *Safety-Clean* tank with fresh solvent and a stiff bristle brush. The asphalt-like undercoating on some Camaro lower control arms can be difficult to remove; these may need the wire brush treatment. After most of the dirt and grease is removed, have these pieces bead-blasted. Cover the machined taper of the spindles with four or five wraps of duct tape to prevent abrading this surface. The control arms and springs can get a few light coats of semi-flat black paint or *GM Reconditioning Paint*, part

no. 1050104. Don't paint the spindles, but if you want to delay the onset of rust, give them a light coat of oil. The cross-shaft gets a light coat of grease or silicone.

With these tips in mind, you can begin refurbishing your Camaro's front end. Or, you can go to an expert as we did. We went to John Costa, and the Camaro experts at **Guldstrand Engineering, Culver City, CA (213/391-7108)**.

REAR SUSPENSION

Early first-generation Camaros (1967 only) used mono-plate leaf springs that measure 56 in. between the centers of the spring eyes. The use of single-leaf springs reduces the amount of unsprung weight the suspension has to handle thereby offering a more comfortable ride. It also does away with friction between leafs which occurs with multi-leaf springs. This friction retards the action of the suspension to surface irregularities. These springs are splayed (spread out) at the rear to accommodate the muffler and fuel tank.

Shortly after the first Camaro was introduced and high-horsepower V8s were nestled into the engine compartment, the problem of *axle hop* reared its head. It seems the torque reaction of the rear-axle assembly arising from the high-output engines overburdened the springs, causing the rear wheels to skip on the road surface during acceleration. The problem was traced to the leaf springs winding up then quickly unwinding. It was cured in 1968 by repositioning or staggering the rear shock absorbers so that one was in front and the other behind the rear axle. This damped axle windup and solved the tramping problem; well almost. Also beginning in 1968, multi-leaf springs, which tended to have a higher load rating, were made standard on all V8s with a horsepower rating above the base 327's 210 hp and optional with smaller-output engines. From 1970-on, multi-leaf springs and stagger-mount rear shocks were standard equipment on all Camaros.

COMPONENT REMOVAL

While certainly not as sophisticated as the independent rear suspension designs of the Corvette and expensive European sports cars, the Hotchkiss Drive set-up (live rear axle and longitudinal-mounted leaf springs) at the rear of all 1967-81 Camaros does have several advantages. Aside from the low manufacturing cost (remember, the Camaro has traditionally been an affordable performance car), the

Coil spring can be a lethal weapon if not controlled when lower control arm is lowered. Use a spring compressor to compress spring slightly before disconnecting lower ball joint from spindle. Support lower arm on a floor jack as arm is lowered.

Remove original-equipment upper ball joints by drilling holes in rivet heads, then chiseling off as shown. Remaining portion of rivets can be knocked out with a punch. New ball joints must be installed with Grade 8 bolts. Don't forget the lock washers!

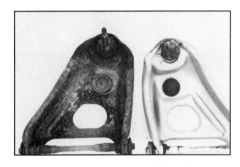

Control arm at right was sand-blasted to remove rust and old paint. Detail arms by spraying with GM Reconditioning Paint, part no. 1050104.

Finished front end, ready to install. When painting front end, don't paint spindles as this enthusiast did. Leave them unpainted as factory did. Nevertheless, it's a good idea to tape over the wheel-bearing surfaces of the spindle for protection.

REAR SUSPENSION COMPONENTS

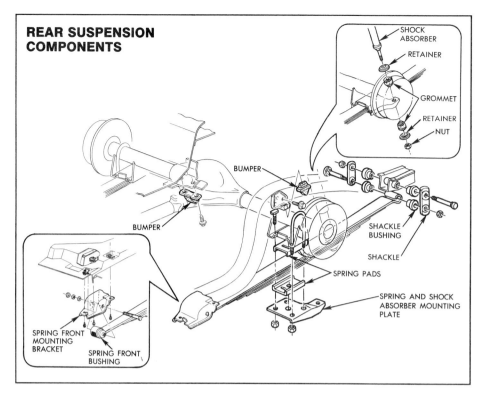

Details of 1970-81 rear suspension; 1967-69 similar. Courtesy Chevrolet.

If you are working alone, rear axle, shocks and leaf springs can be removed as a unit and disassembled later. Leaving the wheels and tires on makes it easy to wheel around the garage.

FUEL TANK INSTALLATION

Details of 1970-81 fuel tank assembly. Courtesy Chevrolet.

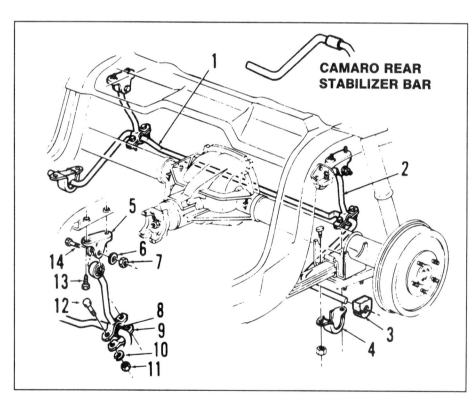

CAMARO REAR STABILIZER BAR

1. Shaft
2. Support ASM
3. Bushing
4. Bracket, Shaft
5. Bracket, Support
6. Washer (1" OD x 1/32"Thk)
7. Nut
8. Bushing
9. Clamp
10. Washer (5/16)
11. Nut, Hex (5/16"-18)
12. Bolt, Hex (5/16"-18 x 1")
13. Screw (3/8"-16 x 1 1/8")
14. Bolt (3/8"-16 x 2 1/4")

Details of 1970-81 rear anti-roll bar. Courtesy Chevrolet.

On 1967-68 models, fuel filler neck and shield are screwed to both tail panel and trunk floor.

Camaro's rear suspension is rock-solid durable and decidedly easy to work on. About the only routine wear items are the rear shock absorbers, and if you're preparing an older Camaro to be a driver, these may be the only rear-suspension components you'll likely have to replace.

In the course of a ground-up restoration, however, you'll want to remove, inspect, clean, detail and install the rear-axle assembly and leaf springs as well. Along with the rear axle, you'll have to attend to details such as leaky axle shaft seals, noisy rear wheel bearings, cracked flexible rear brake hose and seized or broken parking-brake cables. Also, second-generation Camaros born as Z/28s, SSs or with the F-41 suspension option will have a rear anti-roll bar that will likely need new rubber or urethane bushings.

So if it's just shocks you need for that driver Camaro, and the axle seals are tight, spring-eye bushings are not cracked and the springs are not sagging, skip ahead to the headings for shock-absorber removal and installation. Otherwise, let's get on with dropping the rear axle and springs.

Leaf Springs & Axle Assembly—The rear leaf springs and rear axle assembly can be removed together if you wish. The advantage here is that by leaving the tires on, the assembly can easily be wheeled around like a pushcart until you're ready to disassemble it. Leaf spring removal can be accomplished with the proper 1/2-in.-drive hex sockets, breaker bar (sometimes with a length of pipe inserted over their handle), lots of penetrating fluid and box wrenches. And maybe a little extra persuasion in the form of a torch or so-called "blue-tip wrench." Sometimes the nuts rust to the shackle bolts so tightly that heat is the only thing that will free them. If you need to use a torch, first remove the fuel tank. If you don't, you probably won't be around to talk about it.

Fuel Tank—To remove the fuel tank, disconnect the negative lead of the battery. Then siphon the fuel from the fuel tank into a suitable container, such as a gas can. Remove the clamps from the fuel line(s) and evaporative control system (1970 and later models). Disconnect the battery feed wire from the sending unit. Also, be sure to remove the sending-unit ground wire (which is usually black) where it's secured to the body. And remove the neoprene filler-neck coupler if the tank has one.

The fuel tank is held in the car with two metal retaining straps that fit on the underside of the tank. The front of each strap is folded over a hanger in the body and has a small sheet-metal screw holding it together. Don't remove these screws until the tank is out of the car. The rear of each strap is fitted over a long bolt located between the rear of the tank and the rear valance panel. This is where you begin the tank removal.

Squirt some penetrant on each bolt before you begin. The long rusty threads on the bolts will make you glad you did. Sup-

port the fuel tank with a floor jack or milk crate to keep it from falling after you remove the retaining straps. Fit a deep-well socket of the correct size over the bolt and onto the nut. Now remove the two nuts, then the fuel tank. And be sure to place the fuel tank in a cool area, out of the sun. But do not store the now-vented tank with gasoline still in it in a confined garage or basement near a furnace or hot-water heater. The vapors can be explosive.

Brake Line & Anti-Roll Bar—Now you can turn your attention to the rear axle. However, if the axle is already out of the car, skip this next step and go directly to spring removal.

But before you start removing anything, remember that the brake line is attached to the axle. And so is the anti-roll bar (if your car has the fortune to have one) via the U-bolts that retain the axle to the leaf springs. Whatever set-up you have, give the threaded ends of the axle retaining bolts a good shot of rust penetrant and allow it to work into the threads. In the meantime, turn your attention to the brake line T-fitting near the differential inspection cover.

Use a flare-nut wrench to break the nut loose on the brake line that runs from the front of the car to the T-fitting. Remove the anti-roll bar by loosening the nuts at the top of each anti-roll bar strut. These struts are found near the ends of the bar and link it to the body. Then remove the bolts.

The ends of the anti-roll bar are secured to the axle with sheet-metal brackets. And these brackets are held in place with the nuts for the inner axle retaining bolts. Remove these nuts with a breaker bar and a hex socket.

Shock Absorbers—Also be sure to remove the shock absorbers. The easiest way to do this for the rear shocks of first-generation cars is to remove the two bolts that secure the upper shock mount. Then remove the lower attaching nut and shock absorber. Also, if you plan on removing the rear axle, unbolt and remove the tires and wheels, then slide off the rear brake drums. This makes the axle a whole lot lighter when the time comes to remove it.

Driveshaft—Next, remove the driveshaft U-bolt retaining nuts. When these are removed, insert a screwdriver between the universal joint and the pinion flange and pry them apart. This will help pull the U-bolts out of the pinion flange. With your free hand, be sure to hold onto the driveshaft so that it doesn't drop on the ground or your head. Tape the U-joint

Before removing rear axle, disconnect rear brake hose at T-fitting.

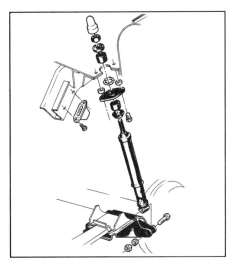

Rear shock absorber mounting—1967-69 models. Courtesy Chevrolet.

After disconnecting U-bolts at rear of driveshaft, pull driveshaft out of transmission. Have a bucket on hand to collect transmission fluid that may leak out.

Rear axle can also be removed independently of the springs. When all attaching hardware has been removed from axle, hoist it over the springs with a couple of floor jacks and a jackstand, or get a helper to lend a hand.

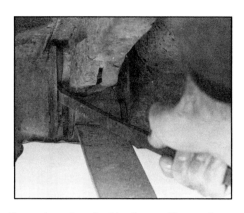

Pry rusty spring shackles loose with a pry bar or large screwdriver.

Check condition of axle bearings and install a new seal to keep gear oil from leaking onto the rear brakes.

bearing caps so they won't fall off.

Axle Removal—Support both ends of the axle with jackstands to ensure your safety. This will prevent the axle from falling on you or onto the ground if the leaf springs are broken. Don t be surprised if most of them are. It's common.

The rear axle is secured to the rear leaf springs with bolts. First-generation cars use "T" shaped bolts that pass through the spring perches while second-generation cars use "U" shaped bolts that fit over the axle tubes. Be sure to use a six-point hex socket and a breaker bar to break them loose.

When all the nuts are loose, you can use a ratchet to remove them the rest of the way. At this point the axle will only be resting on the springs and jackstands. If you care only to remove the springs, you can jack the axle up and then support it with jackstands. To get the axle out of the car without removing the springs, you'll need a floor jack and a helper to keep it balanced on the jack.

Position the cup of the jack underneath the inspection cover so it rests against the carrier housing. Lift the axle with the jack until the brake backing plates clear the springs. Then move the axle to one side of the car so that the carrier housing is near the leaf spring.

To get the differential carrier over the springs, first support the axle with a pair of jackstands on the axle tubes and lower the jack. Then position the cup of the jacks (yes, you'll need two floor jacks at this point) underneath the backing plates. Now pump the jacks up high enough so that the differential carrier will clear the top of the springs. And pull the axle out from under the car as far as you can.

At this point, support the end of the axle that's not under the car with a jackstand and remove the floor jack from the same

end. Finally, position the jack underneath the differential carrier and jack it up until the remaining backing plate will clear the spring. Then carefully pull the axle out from underneath the car. That wasn't too bad was it?

Rear Springs—Take a wire brush and clean as much rust and debris off the threaded portion of the rear-spring shackle bolts as you can. Then spray them with penetrant. Ditto for the front spring-perch retaining bolts. Use a six-point hex socket and breaker bar on the shackle nuts. If that doesn't work, try using the 'ol blue-tip wrench. With the torch, heat the nut until it's cherry red in order to free it. And if you use a torch, chances are very good that you're going to burn the rubber spring bushings. So work in a well ventilated area and take appropriate safety precautions.

When the nuts are off both top and bottom shackle bolts, you're ready to begin removing the shackles. But be warned, the shackle bolts are almost always rusted to the bushings. So you're going to need a lever to get between the shackle and the spring eye to pry the shackle off.

To remove the remaining part of the shackle from the bushing in the frame rail, rock the shackle plate back and forth while using your lever to push against the end of the bolt. Be sure to use lots of penetrant to ease removal.

To remove the bushings from either the frame rail or the spring eye, pry on the edges of the bushing until it's free of the rust. The rust will make a crunching noise as it breaks. Then twist the bushing out of the hole. If the bushing is especially hard to remove, insert a flat-blade screwdriver an inch or so into the bushing and pry in all directions until it's free.

Removal of the spring eye bolt at the front of the spring takes some patience because you must first remove the spring perch to access it. This front spring perch uses retaining bolts that are screwed into clip nuts. And unfortunately, it's very common for the clips on the nuts to shear as you're trying to remove the retaining bolts. If this happens, you'll have to cut the head off the affected retaining bolt and fish out what remains of the bolt and nut from inside the body pocket. You may even have to cut the bolt and nut to get it out and this is probably best done with a torch.

When the spring perch is free of the car, remove the spring-eye bolt and retaining nut with a six-point hex socket. If the bolt is frozen to the spring eye bushing, you may have to use a sledgehammer and heat to remove it.

RECONDITIONING

REAR AXLE

Reconditioning of the rear axle can range from sandblasting and painting the axle housing to installing a new differential assembly, depending on the axle's condition and your enthusiasm.

If all you want and need to do is clean the axle and paint it, consider having it sandblasted first. You can do it yourself if you have the right equipment such as a sandblaster and an air compressor. Both siphon and pressure- feed sandblasters can be used, but be aware of their advantages and disadvantages. These are covered in *Chapter 7*, page 88. When the surface is free of rust and conditioned with a metal conditioner, you can begin priming it.

If for some reason you don't get a chance to paint a part and it begins rusting, first clean off the surface rust with a metal conditioner. Then paint the part.

Select a primer which offers good corrosion protection such as zinc-chromate or an epoxy-chromate primer. You must use primer first to ensure good adhesion of the paint. If the surface is deeply pitted, consider using a primer-surfacer with fast build properties such as *Kondar* over the primer. Kondar will fill the pits quickly, saving you re-coating and sanding time. Ditzler also sells a zinc-chromate primer (DPE-1538) that doesn't require a finish coat that's ideal as a black undercarriage finish. Just keep in mind that it won't fill deep pits. Whichever primer you choose, be sure that it's compatible with the topcoat you're going to use and follow the directions on the can faithfully.

Adjustment and replacement of the internal parts of the rear axle is straightforward if you have a good mechanical background. If you want to recondition the axle yourself, be sure to get a factory shop manual to help guide you. However, in the manual you'll find that some specialized tools are required which aren't readily available from tool rental outlets. You may need some connections to get the tools, if you don't want to buy them. For those of you who'd rather have a pro do it, there are shops that specialize in reconditioning rear axles. And they can usually be found in the Yellow Pages.

For information on reconditioning the brakes and brake lines, refer to *Chapter 5, page 50*. With the exception of the brake lines, it's easiest to recondition the brakes with the rear axle in the car.

LEAF SPRINGS

If you're on a tight budget and the springs have sagged, have them re-arched at a spring shop. The only drawback to this is that it's a temporary fix that won't last for many miles. Another alternative is to buy a good set of used springs. Or you can go first class and buy a new set of springs. The choice is up to you and your checking account.

If the spring appears to be in good condition but needs new spring-eye bushings, have new bushings pressed-in at a spring shop since bushing replacement requires special tools.

DRIVESHAFT

The driveshaft in your Camaro is adversely affected by extreme working angles. If the difference in the working angles between the universal joints is more than about 4 degrees, the chances of your universal joints failing increases proportionally with an increase in working angle. A special tool called an *inclinometer* is required to check the working angles, so it's beyond the scope of this book.

However, a good rule of thumb if you want your U-joints to last a long time, is to avoid raising the ride height of the rear end of your Camaro. When your car is at its designed ride height, the U-joint angle is quite small. This allows the U-joints to live a long and happy life. Replacement of the universal joints is covered in the shop manual.

Cleanliness of your driveshaft is also important. Any undercoating or foreign material that gets on it can cause a vibration problem.

FUEL TANK

As you progress with the restoration, don't forget the fuel tank. If the tank had been leaking, have it pressure-tested. Many radiator shops can do this type of work. They can also boil out the tank to remove any residue which accumulates over time. This reduces the chance of corrosion, clogging or other fuel system problem later on. These shops may also offer a sealing process to stop minor leaks. Carefully consider any guarantee and your safety before agreeing to this.

Any dirt that exists on the exterior of the tank can be cleaned off with plenty of soap and hot water. Use a scrub brush or fine steel wool if necessary to remove stubborn spots but take care not to damage the galvanized exterior of the tank. When the tank is free of dirt, wash it down with *Prep-Sol*. When the Prep-Sol has dried, coat the tank with clear urethane paint.

Sending Unit—Another component of the fuel tank is the sending unit. It's responsible for providing your fuel gauge with feedback as to the amount of gasoline in the tank. And a wire-mesh screen incorporated into the sender also filters contaminants out of the fuel.

The sending unit is secured to the tank with a large steel locking ring. This ring has a cam-like action that tightens down on the sending-unit flange as it is tapped into place. If the sending unit needs to be replaced, it can be removed from the tank with a brass drift. You can also use a block of wood if you're careful not to let any wood chips fall into the tank when you remove the sending unit. Don't use any metal objects to remove or install the lock ring as a spark may be produced. And if any residual gas fumes are inside or around the tank, the tank may explode.

To remove the lock ring, tap against any of its protruding tabs so the ring rotates counterclockwise. If the ring is stubborn, sometimes it helps to keep downward pressure on the sending unit. This compresses the 0-ring underneath the sending unit and in turn relieves tension on the lock ring. When the slots in the lock ring line up with the tangs in the fuel tank, the lock ring can be removed. Now you can snake the sending unit out of the tank, taking care not to damage the float and filter sock.

Before installing the sending unit, make sure you're using a new 0-ring to seal it. Otherwise, the tank may leak. With the new 0-ring in place on the tank, snake the sending unit through it and into the tank. Then make sure the sending-unit tubing is properly positioned in relation to the tank so you can install the fuel hose later. Now slide the lock ring over the tubing and onto the top of the sending unit. Have an assistant push down on the sending unit while you tap the lock ring into position. This entails rotating the lock ring approximately 120 degrees after it fits over the tangs on the fuel tank.

COMPONENT INSTALLATION

LEAF SPRINGS, AXLE & ANTI-ROLL BAR

If both the springs and axle are out of the car, the easiest way to put them back in is to position the axle underneath the car first. Raise the axle with a floor jack until it's about six inches from the underside of the car. Then support it with jackstands.

To prepare the springs for installation, place the spring perches over the ends of each spring and slide the retaining bolt through it. Install the lock washer and nut onto the bolt and torque the nut to 75 ft-lbs. Also, lubricate the two-piece rear shackle bushings with silicone and insert them into the rear eyes of both springs.

Next, lubricate the remaining shackle bushings with silicone and position them into their holes in the frame rail. Now for installing the shackle plates. As you install these plates, remember that the flanged portion of the plate must face away from the leaf springs. Then position one of the shackle plates so that it's next to the outer bushing in the frame rail and the square hole is at the top. Next, slide the shackle

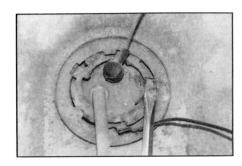

To service fuel tank float, sending unit and in-tank filter, remove lock ring. Be careful not to create any sparks if fuel vapors are present in tank.

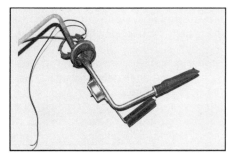

After installing new sock-type filter and 0-ring on lock ring, install sending unit in tank.

Biggest problem installing rear axle is lifting it over springs. Get a helper.

On 1970-81 models so equipped, install new bushings on rear anti-roll bar assembly.

bolt through the plate and bushing. Now fit the round hole in another shackle plate over the end of the bolt and loosely install the nut. And perform these same steps for the remaining side. Remember, the threads of the upper bolt have to face the trunk pan.

Finally, install new clipnuts into the pockets provided in the spring-perch pocket where the front of the leaf spring will be located. Next, install the springs.

Position the spring perch on the end of the spring to the pocket in the body. Do this by first locating the tab on the spring perch into the slot in the body. You may need to use a floor jack to lift and compress the spring. Then loosely install the retaining bolts. Now swing the rear of the spring up into position and slide the lower retaining bolt through the shackles and bushings. The threads for the lower shackle bolts must point toward the outside of the car. Finally, loosely install the shackle nuts. Follow this operation for installing both springs.

If your Camaro was equipped with a rear anti-roll bar, install it now. Be sure to use new bushings.

After all bolts and nuts are loosely installed, you can torque them to specifications. Begin with the spring perch bolts that are screwed into the side rails. This ensures that the perch will be correctly positioned in the body.

SHOCK ABSORBER INSTALLATION

Position a jack under the rear axle and raise the car slightly. This loads the leaf springs, reducing the distance between the axle and body, making it easier to install the shock absorbers and rear brake hose. After you've raised the car, support it with jackstands and proceed to install the shock absorbers.

First-Generation Cars —Begin by placing a washer and grommet onto the threaded portion of the shock. With these parts in place, slide the threaded end of the shock through the mounting plate in the body. Then install the upper grommet, washer and nut but don't tighten the nut just yet. This "hanging" of the shock will make it easier to install the lower retaining bolt. And when you're installing the lower retaining bolts, make sure their heads point toward the front of the car.

Second-Generation Cars—On 1970-81 cars, compress the shock before you position it to the body. Then slide it into the car so you can loosely install the upper attaching bolts. Now install a washer and grommet onto the threaded end of the shock.

Hold these pieces in place so you can pull on the lower part of the shock while threading it through the hole in the lower shock mount. Then install the remaining grommet, washer and nut.

With all the bolts and nuts installed, torque the bolts to specifications. When tightening the nut on the threaded end of the shock, only tighten it far enough to compress the grommet so that the grommet is the same diameter as the washer. This is important for a smooth ride.

Finally, attach the rear brake hose to the T-fitting on the axle. Be sure to use a flare-nut wrench to tighten the hose nut; otherwise you may round it off.

FUEL TANK

First, make sure that you've got the fuel-tank sending unit securely mounted in the tank. And use new anti-squeak material on the ribs of the tank where it will contact the tank bracing in the floorpan. This material is available in roll form from your local Chevy dealer. Just measure the length and rib on the tank and cut the identical size piece from the roll with a pair of scissors. With the anti-squeak liner in position on the tank, raise the tank against the body with a piece of plywood on a floor jack. Now wrap the retaining straps around the tank and slide them over the new tank retaining bolts. Install the nuts and torque them to specifications.

Connect the ground strap from the sending unit to the frame. And be sure to connect the fuel gauge wire(s) to the top of the sending unit. Install new, reinforced fuel hoses and evaporative control hoses and clamp them in position. Fuel leaks near the exhaust system can be dangerous.

DRIVESHAFT

Next, connect the driveshaft to the pinion flange of the axle. Do this by lubricating the slip yoke (splined end) of the driveshaft with transmission fluid. This helps prevent damage to the seal on the end of the extension housing.

Slide the slip yoke over the output shaft of the transmission until it's in as far as it will go. Then line up the caps on the universal joint with the recesses on the pinion flange and slide the driveshaft almost completely into them. However, just before you seat the U-joint caps completely, slide the U-bolts over the caps and into the holes in the flange. Now you can seat the driveshaft into the flange. Then install the lock washers and retaining nuts.

Front-end sheet metal primed and ready for paint.

In this section of the book we will discuss removal, reconditioning and assembly of the front end sheet metal. This consists of all the body sheet metal and various supports from the front of the doors forward.

COMPONENT REMOVAL
HOOD

The hood is bolted to two hinges which in turn are bolted to each fender. When removing the hood, be sure the hood is opened up as far as it will go as this reduces tension on the hood hinge springs. And be sure to have an assistant or two on hand. You wouldn't want the hood falling on you or your car after removing the bolts.

If you plan on just removing the hood or hinges, mark their positions beforehand. That way they'll be in the ballpark as far as alignment to the fenders and header panel when you reinstall them. You can use a fine-point felt-tip marker or a china marker to

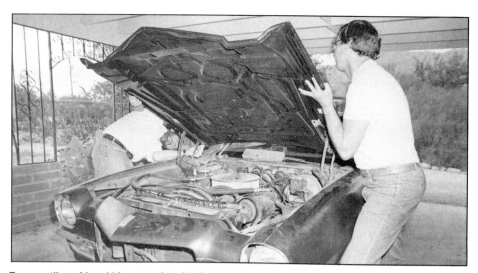

Trace outline of hood hinges on hood before removing hood bolts. Get a friend to help with removal; the hood is heavy and unwieldy. Don't drop it.

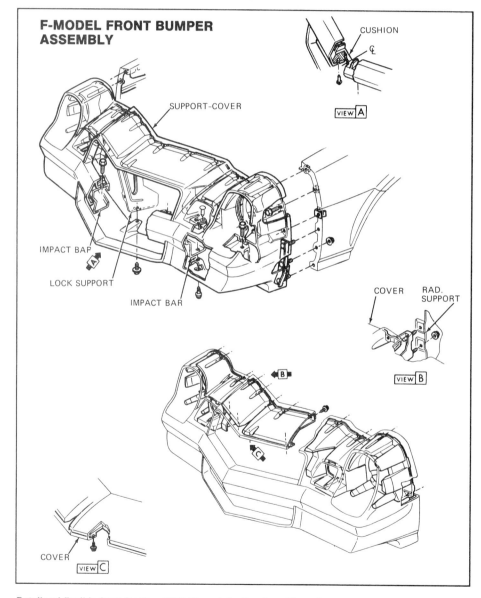

F-MODEL FRONT BUMPER ASSEMBLY

CUSHION

$\mathbb{C}$

VIEW A

SUPPORT-COVER

IMPACT BAR

LOCK SUPPORT

IMPACT BAR

COVER RAD. SUPPORT

VIEW B

B

C

COVER

VIEW C

Details of flexible front fascia—1978-81 models. Courtesy Chevrolet.

transfer their outline to its mating part. Don't use anything sharp such as a scribe because you'll cut into the paint, creating a place for future rust.

Removal—Begin hood removal by laying an old blanket across the cowl and front fenders. This will help prevent damage to your car if the hood slips onto it. On 1970-81 models, disconnect the windshield-washer hoses from the rear underside of the hood. If your car is equipped with a cowl-induction hood, disconnect the wire leading to the flapper-valve solenoid. Then loosen the hinge-to-hood retaining bolts. Now have both of your friends hold onto the front and rear sides of the hood while you remove the bolts. Once the hood's off, you can remove the hinge-to-fender retaining bolts.

Latch & Support Bracket—The hood latch is situated between the radiator-core support and the header panel. A support bracket is bolted to the bottom of the latch and also to the core support and front valance panel. Remove the latch bolts and latch first. Then you can access and remove the support bolts and support much more easily.

FLEXIBLE FRONT FASCIA (1978-81)

Beginning in 1978, Camaros use a flexible front fascia that combines the bumper, valance panel and header panel into a single unit. It consists of a soft urethane outer skin over an impact absorbing fiberglass skeleton (see nearby exploded view). If you're going to remove this heavy and cumbersome assembly, have a helper assist with supporting the fascia as you disconnect its attaching hardware.

With the upper and lower grilles and the headlamp and parking lamp assemblies removed, begin front fascia removal by removing two nuts for the fascia's fiberglass support at the hood latch. Through the headlamp bezel openings, remove three bolts that hold each fender to the fiberglass support.

Through the upper grille opening, remove the two bolts on either side that retain the fiberglass support to the radiator support. Also push in three plastic tabs retaining the soft urethane cover to the bumper support and one tab retaining the cover to each headlight support. There are two more plastic tabs in the lower grille opening holding the cover to the support—push these in.

With the front fascia firmly supported, reach underneath and remove two bolts retaining the cover to the support and one bolt securing the fascia to the hood latch center support. The fascia should now come free.

If the soft fascia was in a collision greater than 10 mph, it's likely that the inner supports and fiberglass skeleton is collapsed or cracked, requiring replacement of the assembly. Minor damage to the urethane skin can be repaired by following the procedures in the *Body Prep & Paint* chapter.

FRONT FENDERS, HEADER & VALANCE PANEL

The front fenders are bolted to the cowl, core support, header panel and valance panel. The fenders are also bolted to the fender skirts (inner fender liners). Steel shims are used between the fender, cowl and core support mounting point to align the fenders. On first-generation Camaros, fender extension panels connect the lower front portion of the fenders to the valance panel. Fender extension panels are also used on 1978-81 models to extend the fender line down along the rear edge of the flexible urethane front fascia. On 1967-68 models, brackets that retain the ends of the front bumper are sandwiched between the fender and fender extension.

On first-generation cars only, the bolts that attach the fender extension to the fender are the bolts that are most likely to give you trouble when you have to remove them. Because of their location, they tend to rust quite readily. And if the rust is severe enough, the bolts have a nasty habit of breaking the clip on the clipnuts they fit into.

If this happens, use an open-end wrench to hold the clipnut when you're removing each bolt. This is a tough job because the bolts are difficult to access. So, give 'em a good shot of penetrant before removing them and be sure to use a six-point socket to avoid rounding the head. Ditto for the bolts that secure the fender extension to the valance panel.

Fender Skirts—The fender skirt (inner fender liner) is attached to the outer fender by those nasty clipnuts that are used on the fender extensions. So, the same precautions are in order here. However, if the clipnuts start spinning, there's no way to remove the bolt except by cutting the bolt head off with a cutting wheel or hacksaw. Or you can remove the inner and outer fender together.

In addition to being attached to the wheel opening of the outer fender, the

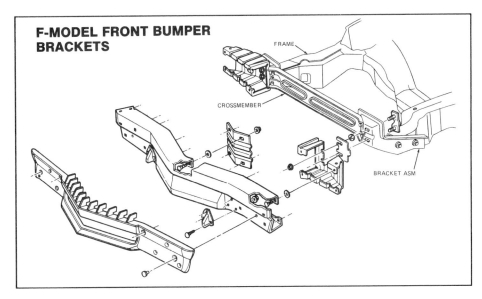

F-MODEL FRONT BUMPER BRACKETS

FRAME

CROSSMEMBER

BRACKET ASM

Front bumper support and brackets—1978-81 models. Courtesy Chevrolet.

fender skirt is attached to an extension of the main body at its rear. It's also attached to the bottom of the radiator-core support and the firewall. The fender-skirt rear and core-support mounting bolts are found at their respective areas. You're better off removing these bolts first so the fender skirt doesn't fall on you.

The fender skirt is attached to tabs on the firewall with a metal bracket and two

Bolts for fender skirts fit into sheet-metal clip nuts. If these start spinning, only way to remove fender is to cut off bolt heads.

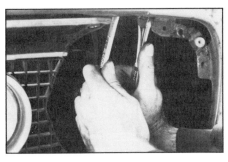

You'll need two wrenches to separate the header panel from the front fenders because the nuts are free-floating.

As you unbolt fenders, keep record of the number of shims at the four rear fender mount locations.

On 1967-77 models, spread fenders apart slightly at the front and lift out the header and valance panels.

On 1967-69 convertibles, unbolt these liquid-filled vibration dampers, or "cocktail shakers," from the radiator support.

After stripping core support of all hardware and wiring, squirt penetrant on the mounting bolts and wait 10 minutes. Then, using a large breaker bar and socket, remove the bolts and rubber cushions for the radiator support.

With time and exposure to engine heat, oil and the elements, subframe cushions crack and decompose. New cushion (right) helps restore integrity to front end.

bolts. These are found at the bottom rear of the fender skirt near the firewall and are accessible from inside the wheel well.

RADIATOR-CORE SUPPORT

The radiator-core support provides a mounting surface for the radiator and the front fenders. It also supports the leading edge of the front sheet metal and ties it all together. A variety of components are attached to the core support depending upon the model and options. This includes the radiator, battery tray, voltage regulator, cocktail shakers (1967-69 convertibles only), wiring, hideaway headlight wiring or hoses (1967-69 Rally Sport only), horns and other items. Keep in mind that the core supports for 1967-69 models are interchangeable as are supports for 1970-73, 1974-77 and 1978-81 models.

Before removing anything, it's a good idea to take photographs of these components in position on the core support. What also helps a great deal is tagging all wires and hoses. This will help you remember what went where when it comes time to install these pieces. There really

isn't any specific order to follow when removing these components. However, sometimes it's helpful to remove the radiator first to ease access to other items. Have suitable boxes and bags on hand to store the components as you remove them.

When all components are removed, you can remove the core support from the subframe. It is secured to the subframe with two large-diameter bolts, lockwashers and nuts. Two rubber cushions are positioned between the support and the subframe to help isolate noise and vibration from the passenger compartment.

First, squirt some penetrant over the exposed bolt threads and nut. Allow the penetrant plenty of time to soak in because the breakaway torque needed to loosen these bolts can be super high. Then slide a 3/4-in. box-end wrench through the front of the subframe and place it over the nut. Now position a breaker bar with a hex-socket over the bolt head and break the bolt loose. Do this for both retaining bolts. Then remove the core support from the subframe.

COMPONENT RECONDITIONING

When reconditioning the front-end sheet metal, it's a good idea to use new clipnuts. Even though the old clipnuts might look good, the clip may be weakened. And when you torque the new bolts into them, the clip may crack allowing the nut to spin. Which in turn creates a removal problem. So play it safe and use new clipnuts.

FENDER SKIRTS

Rust usually starts at the joint at the forward section of the fender skirt. Light to moderate rust can be removed chemically or by sandblasting.

Extensive rust is another matter. If a flat section is rusted out, you can probably replace it with new sheet metal. However, if a rust hole larger than what can be filled with weld appears on the rounded section, you might be better off purchasing a new fender skirt or finding a good used one in a Sunbelt wrecking yard. Otherwise, you'll have to form the metal to fit the curvature of the fender.

If you decide to buy new fender skirts, take a careful look at them beforehand. Because the metal used to form the skirt must be stretched quite a bit, it can split in the creases above the splash apron.

Be aware that new fender skirts for first-

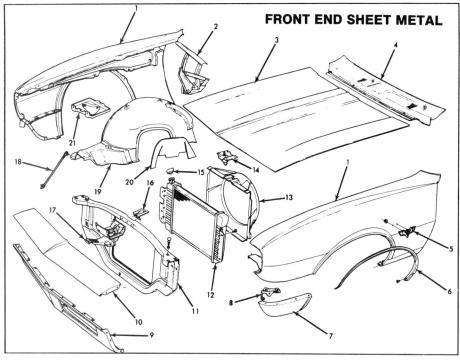

FRONT END SHEET METAL

1. Fender	8. Bumper Bracket	15. Radiator Cap
2. Hood Hinge	9. Valance Panel	16. Bracket (V8s only)
3. Hood	10. Header Panel	17. Hood Catch
4. Cowl Vent Grille	11. Radiator Support	18. Brace
5. Trim	12. Radiator	19. Skirt
6. Molding	13. Shroud	20. Seal
7. Fender Extension	14. Hood Lock	21. Battery Tray

Details of front-end sheet metal—1967-68 models, 1969 similar. Courtesy Chevrolet.

generation Rally Sports are almost impossible to find. The only alternative is to buy a standard fender skirt and modify it to accommodate the headlight door linkage.

Making A Template—You can use a piece of paper to make a template to transfer the linkage holes from the old fender skirt to the new one. Do this by taping a large piece of paper (butcher paper or newspaper) over the holes in the skirt from the underside. Then carefully cut the perimeter of the paper with an X-acto knife so that it fits in the depressed area of the skirt exactly. Working from the top side of the skirt, trace the linkage holes onto the paper. Now you have a pattern from which to work.

Next, remove the pattern from the old skirt and position it into the depression on the new skirt. Then securely tape it into place. If it slips, the hole locations will be off. Take the X-acto knife and cut around the holes you drew on the pattern. This will scribe the outline of the holes onto the new skirt.

After transferring the holes onto the skirt, find the middle of them in order to position the drill bits correctly. You can do this with a circle template. Just place the template on the hole and scribe small lines where the centerlines of the hole are. Then take a small straightedge and draw two lines across the hole, 90 deg. apart. The point where the lines intersect is the center of the circle.

Mark the center of the circle with a prick punch. Then drill the hole using the proper size drill. For holes larger than 1/2-in., you'll have to use a hole saw. After drilling the holes, take a small file and remove any burrs from them. Finally, condition, prime and paint the holes. You wouldn't want rust to start in on your new fender skirts would you? This can be done with a small artist's brush.

SPLASH APRONS

Rubber splash aprons seal out water and dirt from the engine compartment. They're located at the bottom edge of each fender skirt and secured to it with wire staples. If they're still in good shape without any apparent rips, clean them with soap and water. Then coat them with some protectant.

If the splash aprons are beyond hope, or you just want new ones, reproduction splash aprons are available from aftermarket sources such as **Classic Camaro Parts & Accessories in Huntington Beach, CA** and **Arizona GM Specialists in Tucson,**

AZ, just to name a few.

Removal & Installation—Before removing the old splash aprons, trace their outline onto the skirt with a china marker. This will help you position the new aprons. Next, straighten out the bends in the staples with a screwdriver or needle-nose pliers. Then pull the splash aprons off the staples and remove the staples.

To install new splash aprons on old skirts, place them in position against the skirt. Then gently clamp them in place with C-clamps or locking pliers. Next, place a 2x4 into a vise. This will give the apron support during the drilling operation. Position the skirt and apron over the

2x4 so that the engine compartment side of the skirt faces you. And slide the skirt so that the first staple holes are directly above the end of the 2x4.

Slide a 1/16-in. drill bit into one of the staple holes in the skirt and drill through the apron. As you complete each set of staple holes, move the skirt so that the next set of staple holes are directly above the 2x4. Work in this manner until you've drilled all the staple holes.

With all the staple holes drilled, you can install the staples. Slide the ends of each staple through the holes in the skirt and apron. Now bend the ends of the staples over with needle-nose pliers. Then posi-

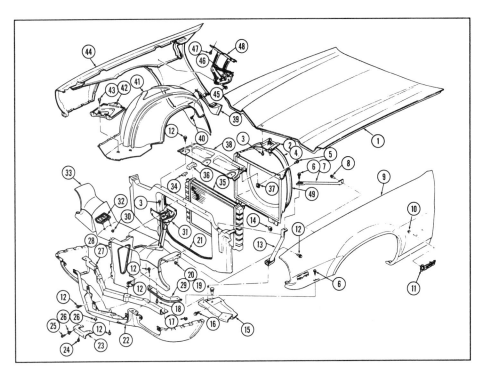

1973 CAMARO FRONT END SHEET METAL

1. Hood Panel	17. Nut	33. Header Asm
2. Plate Asm	18. Screw (5/16"-18)	34. Cap Asm
3. Screw Asm (5/16"-18 x 7/8")	19. Spacer	35. Body Asm, Radiator
4. Shroud Upper	20. Nut (5/16"-18)	36. Pad
5. Screw	21. Support	37. Nut
6. Screw (5/16"-18x 5/8")	22. Bumper Asm	38. Panel
7. Brace	23. Bracket	39. Brace Asm
8. Screw (5/16"-18 x 5/8")	24. Screw (1/4"-14 x 3/4")	40. Screw & Washer Asm
9. Fender Asm	25. Screw (1/4"-14 x 5/8")	41. Skirt Asm
10. Nut	26. Nut	42. Tray Asm
11. Nameplate	27. Reinforcement	43. Screw (5/16"-18 x 5/8")
12. Screw (5/16'-18 x 13/16")	28. Support Asm	44. Fender Asm
13. Brace Asm	29. Panel	45. Screw (3/8"-16 x 13/16")
14. Clip	30. Nut (3/16")	46. Spring
15. Bracket	31. Catch Asm	47. Screw (3/8"-16 x 13/16")
16. Screw (3/8"-16)	32. Emblem	48. Hinge Asm
		49. Shroud

Details of front-end sheet metal—1973 models, 1970-72 similar. Courtesy Chevrolet.

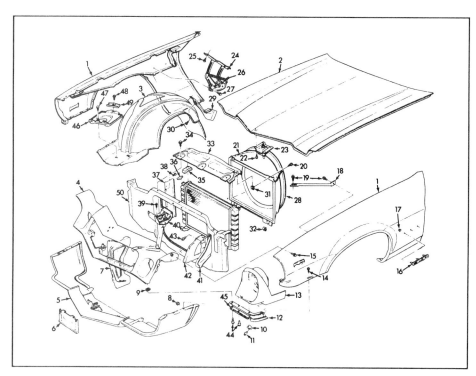

1. Fender Asm
2. Hood Panel
3. Skirt, Frt Fdr
4. Panel Asm
5. Panel, Valance
6. Bracket, License
7. Support, Hood Latch
8. Nut, "U"
9. Nut, "U" (3/8"-16)
10. Retainer, Filler Pnl
11. Screw
12. Panel, Filler
13. Housing, Headlamp
14. Screw
15. Screw
16. Plate
17. Nut

18. Brace, Frt Fdr To Rad
19. Screw
20. Screw
21. Shroud, Fan/Lwr
22. Screw Asm
23. Plate, Hood Lock
24. Hinge Asm
25. Screw
26. Spring, Hood Hinge
27. Screw
28. Shroud, Fan/Upr
29. Brace, Fdr Skirt
30. Screw
31. Nut, "U"
32. Clip, Fan Shroud
33. Panel, Rad Upr
34. Screw, Hex

35. Pad, Rad Retainer
36. Cap, radiator
37. Baffle, Rad
38. Screw
39. Screw, Hex
40. Catch Asm, Hood Latch
41. Support, Jacking
42. Support, grille
43. Screw, Hex
44. Screw (#8-18 x 1/2")
45. Retainer, Frt Bpr Filler
    Pan Inner
46. Tray Asm, Battery
47. Screw
48. Screw, Hex
49. Clamp, Battery
50. Support, Rad

Details of front-end sheet metal—1974-77 models. Courtesy Chevrolet.

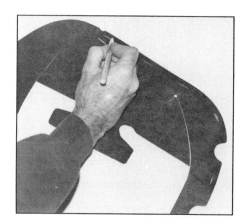

Transfer splash-apron staple locations with a china marker. Then drill new holes with a 1/16-in. drill bit and reuse the staples.

Jim Altieri of Arizona GM Classics shows how to transfer mounting hole locations to new fender. Attach large sheet of copier paper to definable character line of old fender and punch holes with ice pick. Then use template to mark hole locations on new fender.

tion the skirt over the 2x4 in the vise the same way you did earlier. Finish the job by peening the ends of the staples so they hold the apron securely.

If you're installing new aprons onto new skirts, you'll also have to drill through the skirt. So use the old splash apron as a pattern to drill the new holes.

FENDERS, HEADER & VALANCE PANELS

Fenders can be repaired either on the car or off. If they just have a few small dings in them, they can be repaired by a little hammer and dolly work, filling them in with body filler, then repainting. However, if rust is extensive, you'll have to weld in new metal before filling and painting. This type of rust usually occurs on first- generation cars at the joint between the fender and fender extension and on all Camaros at the lower rear of the fender, just behind the wheel well.

The header panel of 1967-73 models and valance panel of 1967-77 models are another home for rust. That's because they're in the direct line of fire from stones and rocks.

Minor rust can be removed chemically or by sandblasting. Major rust may require repair or replacement, depending on its severity. Header panels of 1974-77 Camaros are fiberglass and the entire nose section of 1978-81 models is flexible

Altieri applies asphalt-based undercoating to rust-prone areas such as inside lower rear of front fenders.

Another trick is to cut in hard-to-reach areas with color coat before installing fenders and other front-end sheet metal.

urethane, so rust is no problem there. But different maladies, such as cracked fiberglass (or sheet molded compound) or gouged urethane can occur.

For more information on rust repairs, refer to the section in this book on Trunk Floor Repair, page 87, or HPBooks' *Paint and Body Handbook.*

The really nice thing about front-end sheet metal is that it can be unbolted and replaced with new or good used pieces. If you're in the market for new fenders, keep in mind that first-generation Rally Sports use different fenders than models without this option. This is due to the deeper bracketry at the front of the fender for the hideaway headlights. If you can't find Rally Sport fenders, you can use standard fenders equipped with Rally Sport conversion brackets. These are available from Classic Camaro and other Camaro parts supply houses. Also, 1970-73 Rally Sports have a specific header panel to accommodate the soft Endura nose and European-style round park/turn signal lamps.

Once the fenders, header panel and valance panel are in good shape, prime them, lightly sand, then cut in the edges and hard-to-reach areas with four to six color coats. This will help ensure more uniform paint color.

CORE SUPPORT

The time you'll spend reconditioning the core support is directly related to its condition. If the core support only shows signs of minor rusting, sandblast those areas then condition them with metal conditioner. After conditioning, you can proceed to prime, sand and paint the affected areas using the same technique that you'd use for any other sheet metal repair. The same goes for rust holes, sometimes a problem adjacent to the battery tray or at the bottom of the support. If the rust holes are small and few, you can cut out the affected sections and patch them with new metal. Then finish them by conditioning, priming and painting the metal. If the core support has severe rust, such as around the mounting holes where it attaches to the subframe, or any sort of collision damage, you'll be time ahead by purchasing a replacement core support.

HOOD

The hood is a durable component, double-paneled for rigidity, that seldom requires major work. The heat of the engine and a light oil mist that inevitably seems to form in the engine compartment

keeps moisture from attacking the paint and metal. The biggest enemy is battery acid or acid fumes, followed by the occasional splash of paint-removing glycol brake fluid or a backfire through the carburetor and resultant fire. Most rust can be removed by sandblasting. Minor fender benders often spare the hood from any sheet-metal damage.

But if the hood does require sheet-metal repair, it's often difficult if not impossible to do a good job. Unless the damage is in an area not covered by the hood's inner panel, hammer and dolly or welding torch access is the limiting factor. Don't even bother trying to restore a hood that's been cut for an aftermarket scoop or hood pins. The required welding or brazing and lead or plastic filler work won't hold up over time. The paint will eventually crack due to the heat cycling (heating and cooling) of the engine compartment and the slamming of the hood by ham-fisted brutes.

So if the hood is damaged, try to find a new or good used replacement before investing a lot of time and money in a restoration that won't stand the test of time.

Insulation Pad & Retaining Clips—The hood insulation pad is constructed of loose-woven fiberglass sheet similar to a home heater filter pad. It is retained to the underside of the hood with about 15-20 retaining clips with barbed heads that fit into corresponding holes in the hood frame. Some clips have holes surrounding the barbed heads. These clips can be removed by inserting a pair of needle-nose pliers through the holes and pulling them free of the hood. The clips with solid round heads can be removed by sliding your fingers or an opened needle-nose pliers underneath the heads and pulling them out.

Cleaning—Because of its fibrous surface, the only way to recondition the hood pad is to clean it with a solution of soap and hot water. To do this on the car, cover any exposed sheet metal with plastic first. The idea here is to protect the paint from the cleaning agent you'll be using on the insulation. Then spray the insulation with a strong cleaning agent such as Fantastik. Let the cleaner soak into the insulation for a few minutes to give it time to work. Remember that you're dealing with a porous item that's been exposed to 15-20 years of road grime, gunk and oil mist.

Now rinse the insulation with buckets of clean hot water. If this doesn't clean all the grime off the first time, let the insulation dry then try cleaning it again. If after

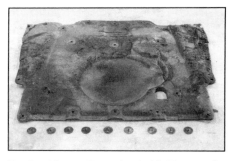

Hood pad is usually retained with 15 or so of these snap-in, round-head plastic clips. Use the old pad as a template to cut a new pad if necessary.

cleaning it, it still doesn't meet with your satisfaction, or it starts to fall apart, you'll have to replace it.

New insulation blankets are available from GM dealers and many aftermarket suppliers. The GM pad is a universal type that fits all hoods, large and small. So you'll need your old pad or a good used one to use as a template to cut it to shape and cut out the numerous holes for the retainers—a messy and time-consuming task. The aftermarket pads are made from the fiberglass furnace-filter material—not exactly the original stuff—but are precut to size with all the holes cut out for the retainers and include new retainers (although not always the round-head factory part) to replace all of the old retainers you'll break off removing the old pad.

COWL-INDUCTION TESTING

1969 Models—Camaro hoods have gone through various levels of trim through their first fourteen years. And the cowl-induction hood available on 1969 models was one of them. The cowl-induction hood was one of those great options that you could see working. Put your foot to the floor and watch the flapper valve at the rear of the hood open up as you hear the moan of the distant Holley or Quadrajet carb. It's fantastic! If you miss this sensation because your system is on the blink, read the following to find out how it works.

All that actually happens when you put the pedal to the metal is that a switch on the throttle linkage closes. When the switch closes, 12 volts is provided to the flapper-valve solenoid. This solenoid then opens the valve as long as it's receiving 12 volts. When you release the accelerator, the contacts in the switch open, breaking the circuit to the solenoid.

If the flapper valve in the hood isn't working, first check the condition of the fuse in the fuse block. If it's blown, you've

Subframe is sandwiched between two rubber cushions. Lower cushion and washer will come out when you remove subframe retaining bolts. You'll have to lower the subframe to access the upper cushions.

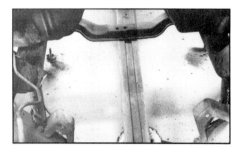

To remove the subframe from the car, support the car on jack stands. Then support the subframe with a 2x4 running underneath the rear crossmember and underneath the center of the main cross rail. With a floor jack positioned underneath the 2x4 that runs from front to back, carefully pull the subframe out from under the car.

found the cause of the problem. However, if it isn't blown, you'll have to check the system.

System Check—To begin, first make sure the engine isn't running. Then rotate the ignition key to the **ON** position, but don't start the engine. Have an assistant put the accelerator pedal to the floor and keep it there. Start at the wire attached to the solenoid. Insert the positive probe of a volt-ohmmeter into the back of the connector. And attach the negative lead of the ohmmeter to a good ground. Make sure the ground is good, otherwise the test results will be wrong. With the ohmmeter properly connected, you should get a 12-volt reading.

If you do, clean the terminal on the solenoid and the connector with a piece of fine sandpaper or a small wire brush. Now reconnect the solenoid feed wire to the solenoid. If the solenoid still doesn't work, you'll have to spring for a new one. Replacement parts are not available to rebuild the solenoid.

The above test can also be performed with a test light. If the bulb lights, 12 volts

is being supplied to the solenoid. If the bulb doesn't light, work your way back in the system until you find the point at which the light lights up. When you find this point, you'll have isolated the problem to the area between the glowing light and the solenoid

The switch that controls the 12-volt feed to the solenoid is located on a bracket directly to the rear of the accelerator linkage. If this switch isn't working, the solenoid won't be working either. You can check the switch operation with a test light. With the ignition still in the **ON** position, connect one lead of the test light to the wire leading to the rear of the switch. Connect the other lead to a good ground. If the test light doesn't light, and the connections are good, backtrack the wire to find out why it isn't providing 12 volts to the switch. However, if the test light lights, you can proceed to test the switch.

Now remove the test-light lead from the ground and connect it to the remaining terminal on the rear of the switch. Then push the switch plunger in. If the switch and its connections are in good condition, the test light will light. If it doesn't light, clean the switch terminals and try testing again. If the test light still doesn't light, the switch is defective and must be replaced.

REMOVAL

The subframe is usually removed for one of three reasons: you're doing a ground-up restoration, it requires rust repair or because it was damaged in an accident and is being replaced.

Although the subframe can be removed with the engine, transmission, front suspension and sheet metal on it, it's very awkward to separate from the car this way and should be avoided if at all possible. We'll discuss the removal in terms of the ground-up restoration and rust repair.

After the front end sheet metal, core support, engine, transmission, front suspension, parking brake cable, fuel lines and brake lines have been removed, removal of the subframe is straightforward. Information on removal of these components can be found elsewhere in this book.

Support Body & Subframe—Begin subframe removal by supporting the body with jackstands along the front and rear of both rocker panels. Then spray some penetrant on the core-support-to-subframe bolts and the subframe bolts that secure the subframe to the body.

Support the subframe by placing a 2x4 underneath the back of the subframe so that it runs from side to side. Have a couple of friends hold it in place. Then position a 5-ft.-long 2x4 on the cup of a floor jack.

Now roll the jack under the subframe so the 2x4 runs from the front of the car to the back. And make sure it intersects the first 2x4 at a right angle like a cross and also fits under the portion of the subframe that runs underneath the engine. With this setup, you can roll the subframe out from under the car when you've removed the subframe-to-body retaining bolts.

If you don't have a floor jack, just support the subframe at all four corners with jackstands. When you remove the subframe bolts, you'll have to drop the subframe and drag it out from under the car.

Now use a socket on the end of a breaker bar to loosen and remove the core-support bolts. Don't be surprised if these bolts break when you try to loosen them. Use the same tools with a short extension to remove the subframe-to-body bolts.

When the subframe is separated from the body and out from underneath it, you can begin the reconditioning process.

RECONDITIONING

First, carefully inspect all six subframe mounting points. The metal must not be rotted or the mounting holes enlarged. The rear holes on first-generation cars should measure 1-11/16 in., center holes 1-1/2 in. and front holes 1 in. in diameter. On second-generation cars, these measurements are 1-9/16 in. at the middle and rear, and 7/8 in. front.

If any of these areas are rotted, you must repair them or obtain another subframe in good condition. To repair these areas, cut the rotted section out with a grinder equipped with a cutting wheel. A Dremel Moto-Tool with #409 cutting wheels works pretty well but it's rather slow. When you've reached good metal, place a piece of paper under the area you just cut out and trace its shape onto the paper.

Now cut out the pattern and transfer it to an appropriate size piece of 1/8-in. sheet steel or large steel washer of the same thickness. Cut the steel with your favorite hacksaw and dress with a grinder. Don't cut with a torch because the steel may warp and be more difficult to fit and finish later.

The next step is welding the piece into position. This is best done with a MIG welder or arc welder. If you don't have access to or experience with either of these welders, trust your work only to a professional welder. And don't try brazing because it doesn't possess the strength needed in this high-stress application. If you elect to arc-weld, use SAE 6013 or 7011 contact rod to join the pieces.

After everything's welded up, grind the weld flat with an 18- or 24-grit sanding disc. Next, cut the subframe-mount hole to size with a hole saw. Then sand the area one more time with a 120-grit sanding disc and prep, prime and paint the area if that's all that needs to be done to the subframe. If you need to remove rust from the remainder of the subframe, do the painting later.

To derust the subframe, and the front suspension for that matter, you can have it done chemically or mechanically with a sandblaster. When the subframe and suspension is free of rust, prep the metal with metal conditioner following the manufacturers directions. When you've conditioned all the bare metal surfaces, shoot some primer on them. Primer must be used before primer-surfacer for proper paint adhesion.

Although many of the two-part catalyzed primers have superior durability over "regular" primers, don't use them unless you have a respirator to wear that filters out *isocyanates*. Use these respirators if breathing properly for the rest of your life is important to you. After applying the primer, scuff-sand it with 240- or 320-grit sandpaper. This will give a tooth to which the primer-surfacer you'll be applying next can adhere to.

The primer-surfacer you'll use is dependent on the surface of the metal. If the pits are shallower than 1/32 in., just use a normal primer-surfacer. However, if the pits are deeper than this, use a primer-surfacer with a high build rate such as *Kondar*. This material fills in most pits with only two coatings and saves time compared to the shoot, sand, shoot, sand, process you'll follow using regular primer-surfacer.

And as you sand, be sure to use 150-grit paper if you're using an air-powered orbital air sander. If you're sanding by hand, use 240- or 320-grit paper and a lot of water for the preliminary sanding and 400- or 600-grit paper with water for the finish sand.

When you've finished sanding the primer, tack the surface so it's free of any dust or dirt. Then apply four to six coats of color—in this case semi-gloss black. If you want to use the real stuff, get some GM Reconditioning Paint from your local GM dealer, part #1050104. Or if you're

not so picky about things out of sight and out of mind, you can use acrylic lacquer or enamel as the topcoat.

But if you plan to drive your Camaro a lot, and especially if that involves Rustbelt winters or gravel roads, consider using *acrylic urethane paint*. This paint is made up of two components. When these components are mixed together, they react with each other generating heat, much like an epoxy adhesive. And like epoxy, this paint is very durable, resisting all types of stone chips and the like. But be careful if you plan to use this type of paint, since it contains isocyanates. Isocyanates can cause many respiratory problems if inhaled. So be sure to follow the manufacturer's recommendations on the use of proper respirator equipment when spraying this material.

ASSEMBLY

When both body and subframe are painted and ready to go, you can join them using new subframe mounting cushions (also known as biscuits), bolts and washers. These hard-rubber mounts are available from GM dealers and many Camaro aftermarket parts suppliers.

In the typical aftermarket subframe mounting kit, you'll find three pairs of cushions. One pair is to be used for the front mounts, one for the center mounts and one for the rear mounts. And each cushion comes in two halves: a cushion with an integral washer and bushing and another cushion with a step molded into it. When these two halves are assembled, a 1/8-in. void is created between the two. It's this void that will be filled with the subframe metal.

If you have to separate these two halves for any reason, keep them together with a string or piece of wire because they're made of different hardnesses of rubber. If you don't, you may have problems with noise, vibration and harshness being transmitted from the subframe to the body.

Before installing the cushions, clean the bushings and washers with metal conditioner and spray them with a clear acrylic. This will help prevent them from rusting. Also, spray the cushions with a coat of silicone or rubber preservative to keep them pliable. Now to determine which cushions go where.

Subframe Cushion ID—There are two methods to identify the subframe cushions on first-generation cars. One way is to ask the company or parts person which cushions go where. The second alternative is

for those of you who don't have the foggiest idea of which cushions to use at what location. This entails carefully examining and measuring the cushions.

When you know which cushions go where, you can install them in the subframe and position the subframe to the body.

The rear cushion set is the smallest; the outside diameter of both cushions measure about 2-1/8 ins. The center cushion set is a little different. The upper cushion (with the washer and bushing) has an outside diameter of 2-1/2 in. And the outside portion of this cushion can be identified by its height of 7/8 in. This height is measured without the washer. For all practical purposes, the cushions that don't have a washer for the rear and center cushion sets are interchangeable with one another. If you're not sure which ones they are, trial fit them onto the step on washered cushions. They should fit easily but snugly.

Rusted subframe cushion mount is typical of cars exposed to salt and snow. Damage can be repaired by cutting out rust and welding in a steel plate. However, if subframe shows much rust elsewhere, you're better off getting another used subframe in better shape.

SUBFRAME
With the front-end sheet metal off the car, now is a good time to inspect the six subframe mounts. It's likely that they're dried out and cracked and in need of replacement. Two are exposed by removing the radiator support. You can replace the other four mounts easily after supporting the body on jack stands cushioned with stout 2x4 boards.

If you want to totally recondition the subframe, you'll have to remove the engine, transmission, front suspension, steering and brakes. If that's your intention, then skip the assembly and go on to the next section of *Subframe* before assembling the front end sheet metal.

Subframe ready for installation. Note 5/8-in. alignment holes outboard of center mounts that should align with similar holes in body ahead of footwells. With wheels and suspension in place, subframe is easy to wheel around.

After radiator support is installed with new mounts, install fenders using same amount of shims as you took out.

Leave fender bolts loose as you do preliminary alignment of sheet metal. Here, Jim Altieri uses a 2x4 to coax a stubborn fender into alignment with its lower bolt hole.

Install the header panel bolts finger-tight. . .

...then offer up the valance panel. This job is easier if you have an assistant. Only when all sheet metal is in place and in reasonable alignment should you tighten bolts.

The cushion that's used for the radiator-core support also has an outside diameter of 2-1/2 in., just like the center cushion. However, it can be differentiated from the center cushion by its height; it's only 9/16-in. tall excluding the washer. And because of its unique (smaller bushing diameter, only one of the washerless cushions will fit it.

INSTALLATION

Begin subframe installation by installing the cushions onto the subframe. Working from the rear of the subframe forward is the easiest. Insert the rear and center cushions which contain the bushing into the top of the subframe. Make sure that you have the right cushions in the right holes.

The next step requires raising the subframe to meet the body. This can be done by running a 2x4 between the rear of the subframe rails. Then position another 2x4 so that it rests against the underside of the first 2x4 and underneath the engine cradle portion of the subframe. Now position a floor jack underneath the second 2x4 so that the subframe is balanced when you raise it. Having a friend around to help you at this time is an asset.

Before raising the subframe, slide a 4-in. long, 1/2-in. diameter bolt through the top of the subframe mounting-bolt cage nuts on both sides of the firewall. This will help line up the subframe to the body as you raise it. Now raise the body so that the 1/2-in. bolts line up with the bushings in the center mounting cushions.

When the subframe is in position, install the remaining half of the rear cushions. Then screw the bolts with accompanying flat and lock washers into the rear cage nuts in the body. Next, remove the 1/2-in. bolts from the center cushions, and install the lower half of the center cushions.

Complete the body attachment by installing the center-cushion retaining hardware.

Align Subframe—The next step is to align the subframe to the body. This requires using 5/8-in. diameter bolts about 2-in. long for alignment pins. These bolts need to fit through the alignment holes in the subframe mounting ears and their corresponding holes in the body. The shank portion of the bolts should fit through these holes with minimal clearance. If they won't quite fit through the holes, use some 240-grit sandpaper on them until they just slide through the holes. The tighter the fit, the better the subframe will be aligned to the body.

Slide the "alignment pins" through the hole in the firewall just outboard of both center cushions. The pins should fall through the holes in the center cushion mounting ears in the subframe. If they don't, shift the subframe until they do. With the pins in position through the body and subframe, torque the center and rear mounting bolts to 75-95 lb-ft. Then remove the alignment pins.

Final Assembly—Now turn your attention to the front cushions. The front cushion with the washer is inserted through the subframe rail into the hole in the bottom of the core support. The washerless cushion fits over it on the top side of the subframe. Slide the bolt and washer through the bushing from the top and secure it with a flat washer, lockwasher and nut.

With the subframe in place, you can install the engine, steering linkage and gear and front suspension. You're better off installing the engine now instead of later when the sheet metal is in place. This could save some possible damage to the sheet metal as the engine is hoisted over it into position.

ASSEMBLY

Now comes the time when you see your dream car coming together. If you plan on painting the car after you've assembled the front end, skip the steps for installing exterior trim such as the grille, moldings and side marker lamps. Install these components after you've painted the car. The reason for this is it'll give you the opportunity to do a more thorough paint job and save you masking time.

GENERAL TIPS

As you prepare to install some of the sheet metal, you may have a little trouble lining up the bolt holes. To get around this, carefully slide a Phillips head screwdriver through the holes to align them with each other. Then hold the parts in position as you withdraw the screwdriver and replace it with the retaining bolt.

Basically speaking, assembly is the reverse of disassembly. Start by installing the radiator-core support if it has been removed. This is covered in the *Radiator Core Support* section of this chapter. Then install the fender skirts, fenders, fender extensions, header panel and valance panel. This will position the front end sheet metal for painting.

Whatever you do, don't tighten any fasteners for front-end sheet metal parts until all parts have been installed and aligned. Once the fenders, hood, valance and header panels have been aligned, you can tighten their bolts and nuts to specs. Now, let's take a closer look at how each component is installed.

Also, get a friend to assist you. It isn't that some of the sheet metal pieces are heavy, they're just awkward to handle. You don't want to drop or scratch any of the panels after you've reconditioned them. Wear gloves to protect your hands.

Fender Skirts—Position the fender-skirt braces to the tabs on the firewall and loosely install the retaining bolts. Next, have a friend hold the skirt against the brace while you install the skirt attaching bolts. Now loosely install the remaining skirt hardware except for the fender retaining bolts.

Fenders—As with the fender skirts, installation of the fenders is much easier if you have a friend to help you. They can hold the fender for you while you install the bolts and shims. Also, if you intend to apply rustproofing to the inside of the fenders, now is the time to do it. It's more difficult to thoroughly coat the fender after it's installed.

Begin installing the fenders by having your friend hold the fender in its approximate position just a few inches away from the car. Then push up on the outside edge of the fender skirt while your friend slides the fender into place. This will help prevent the skirt from scratching the fender.

And as the fender is being positioned, make sure that it fits over the core support and cowl so that the bolt holes are lined up. Then install the fender bolts beginning with the upper bolts at the top of the firewall, cowl and core support. This helps support the fender while you install the remainder of the retaining bolts. Then, on 1967-69 models only, install the fender extension.

Header & Valance Panel—When the fenders are installed, carefully spread the front section of the fenders away from each other slightly. Make sure the sheet metal is secure so that it can be prepared for painting.

If you have a 1978-or-later Camaro, you can install the urethane front fascia now as detailed in the photo sequence below.

3. Skin is retained to fiberglass with numerous hex screws and these plastic rivets.

1. Gary Nori of Z&Z Auto, Orange CA (714/997-2200) begins assembly of 1978-81 front fascia by mounting bumper brackets to subframe.

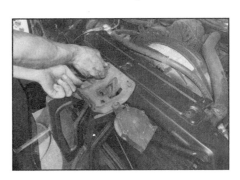

4. Bolt hood latch and support to radiator support.

2. Fiberglass support fits inside soft urethane nose.

5. Bolt bumper support to brackets.

6. Slip nose over bumper support and raise into position with a floor jack. Nori then finishes job by bolting the nose to the fenders, fender extensions and hood latch support.

Now it's time to delve deep into the innards of your Camaro. Find any rust-outs in the floor, tail panel, rear wheelhouses, rocker panels or quarter panels and you'll be needing a welder. To help with the search, take your Camaro (if it is still mobile) to a steam-cleaning facility and have the undercarriage degreased.

Chances are that your Camaro has suffered its share of parking-lot dings and maybe a fender-bender over the years. And if it has spent much time in Northern climes where they use calcium chloride (road salt) to de-ice the roads, or lived near the coast (Atlantic, Pacific or Gulf) where high humidity and salt air deliver a potent 1-2 punch to mild steel, rust and corrosion have gained a foothold. Then, there's rust damage caused by water getting past a trunk, T-top or door weatherstrip or leaky convertible top and standing in a puddle on the floor.

That's the easy stuff. How about designed-in rust problems: poor drainage of the cowl, front fenders, rocker panels, rear quarter panels, chafed-through windshield and backlite trim-molding clips and the granddaddy of them all, creeping roof rust beneath a vinyl top? Camaro's got 'em all!

Or maybe a previous owner decided he liked those J.C. Whitney hood pins or fiberglass pro-street-look hood scoop. Perhaps your Camaro went under the knife to install an aftermarket sunroof or fender flares.

These and other maladies will have to be remedied in the course of a restoration. If worse comes to worst, everything that unbolts—namely the hood, header and valance panels, fenders, doors and deck lid—can be easily replaced with new or good used replacements. The hood in particular is tough to repair due to its double-panel construction, proximity to the engine and frequent opening and closing.

But unless you're building a car around a rare VIN number or are very handy with a welder and have lots of time on your hands, do not underestimate the amount of work required to repair the basic structure of the car—namely, floor, roof, cowl and rear quarter panels.

For knocking out everyday door dings or filling that small rust hole, we recommend you pick up HPBooks' *Paint & Body Handbook*. For some of the more labor-intensive and Camaro-specific maladies facing the restorer, we recommend you read on.

DEGREASING

But before you get started banging, cutting, grinding and welding, it's best to start with a clean piece of metal. If you're repairing the firewall or cowl, fender skirts, radiator support or floorpan, chances are it will have a layer of grease and grime covering it from years of leaking oil and grease mixing with dirt.

The easiest way to loosen and remove most of this buildup is by *steam cleaning*. And many coin-operated car washes have high-pressure detergent or kerosene-spray booths that remove most of the big chunks. The best steam-cleaning operations have a lift, so the operator can raise your Camaro to working height and get his wand underneath for direct access to the transmission tunnel area and other buildup areas.

For engine compartments, cleaning with an

aerosol petroleum-based solvent such as *Gunk* and followed by rinsing with high-pressure water works quite well. Heavy deposits can first be scraped off with a putty knife.

Small parts can be degreased at your local engine- rebuilding shop's Jet Spray tank. The tank cleans the parts with alkaline solution followed by a high-pressure water spray.

DERUSTING

Aside from fixing rust holes or welding in a new sheet-metal panel, you may just want to remove surface rust from an otherwise straight, structurally sound and usable part. Generally speaking, there are three ways to remove rust from metal: sandblasting, acid treatment and alkaline treatment.

Sandblasting—This is a good derusting method for large pieces of sheet metal, such as the body, and heavy, cumbersome objects such as the subframe. It offers the benefit of also removing old paint and providing a tooth (surface irregularlity) for new paint to adhere to. Its major drawback is that if not done correctly, the operator can distort and stretch the metal. Sand, of course, gets everywhere, so never use sandblasting to clean engines, transmissions or anything with moving parts. For more details, see the nearby sidebar, page 88.

Acid—Derusting with acid solutions is perhaps the most common method around the garage. Over-the-counter products such as *Naval Jelly* and other jelled phosphoric acid products do a good job of removing minor surface rust buildups and are safe to use provided gloves and eye protection are worn.

More potent acid pickling solutions can be used to derust small parts such as bumper brackets. But never use acid pickling solution to remove rust from mechanical parts that are subjected to stress, such as wheel spindles. The acid chemical reaction causes excess hydrogen to be deposited in the metal. The resulting hydrogen embrittlement can cause a metal part to fail (break) under stress.

Alkaline—Yet another way to remove rust from iron and steel parts is with an alkaline solution, most often lye or trisodium phosphate. Never use this solution on non-ferrous metals such as brass or aluminum as it will corrode rather than clean them. Provided you take careful precautions to cover all exposed skin, your eyes and wear a protective hood, a lye solution (similar to

that found in household Drano) can be used to clean small parts.

If you need to derust a complete body, the best way is to haul it off to a so-called metal laundry or dip and strip facility such as Redi-Strip. Here, trained operators will dip your Camaro, stripped of all chrome trim, weatherstripping, non-ferrous metals (including trim and VIN tags), soft trim, wiring, electrical and drivetrain pieces beforehand, into an electrolyte-charged caustic bath. When you get your Camaro body back, it will be rust-free, but also missing its body caulking and any plastic filler or fiberglass repair sections it had previously.

PAINT STRIPPING

If you just want to change the color of your car and not damage the factory-applied primer underneath, use Ditzler DX-525 paint remover. Removing the color coat will help avoid excessive paint film thickness, which can lead to checking and cracking later.

If rust is not a problem, but you just want to remove the old paint to gain access to the metal or want to switch from an enamel to a lacquer-based paint, you can use paint stripper. Commonly available paint strippers are hydrocloric-acid based and in gel form. This is very nasty stuff to work with. Wear rubber gloves, a heavy-material long-sleeve shirt, full-length denim jeans, high-top shoes, eye goggles and a respirator mask. Make sure there's plenty of ventilation.

To aid cleanup, place newspapers around the car body to catch the acid jelly and old paint that falls to the ground. Cover the car's tires and any weatherstripping you want to reuse. Tape over the

Quickie enamel paint job had to be removed to repaint with acrylic lacquer. Hydrochloric-acid based paint remover works best if applied in thick coats. Ventilate the work area well to prevent inhaling toxic fumes. Tape over seams to keep remover from seeping into doors and trunk openings. When paint bubbles up, scrape it off with plastic squeegee.

Afterwards, hose off remaining residue with water. Then treat bare panels with metal conditioner.

body seams (doors, hood, trunk, etc.) to keep the jelly from oozing through the cracks and removing paint where you want to keep it.

Carefully pour some stripper onto a 3- or 4-in. bristle brush you'll not mind throwing away later. Spread the stripper over the painted surface, the thicker the better. You'll be able to see the paint bubble, buckle and lift after about 5 min. Scrape off the stripper while it's still wet with a plastic squeegee used to spread plastic body filler. Then, reapply to any areas that still have paint. Don't let the stripper dry on the painted surface. Afterward, roll the car outside and hose off all stripper. Use a *3M ScotchBrite* pad if necessary to remove any traces of the stripper.

TRUNK FLOOR REPAIR

Trunks. They separate the bad Camaros from the good. If your trunk is an untouched original in good condition, chances are the rest of your car is in sound shape also. On the other hand, if your Camaro's trunk looks like a giant rusty sieve, don't fret. You can make it look as good as new. All it takes is a little work.

A common place for rust in Camaros is the *trunk pan*. That's the depressed area of the trunk. Another spot where you may find rust is in the *trunk extensions*: the metal that runs from the edges of the trunk pan to the quarter panels. Both of these areas can be repaired. The quantity of work required is dependent on the extent of the rust. Let's take a look at each area.

PATCH REPAIR

Minor rust holes about 1/8-in. diameter and less can be filled with a variety of materials including lead, plastic body filler and epoxy. And if you have access to an

SANDBLASTING

Sandblasting is a good way to remove rust and paint in a hurry. Pressure-feed sandblaster shown here is efficient and effective. Mask glass and chrome to prevent pitting and tape over seams to prevent sand buildup inside panels.

Sandblasting is one of the most common methods of removing rust from automotive components. Since the sand is fine, it can remove rust pits in metal surfaces without damaging the underlying good metal. And it has the added bonus of roughing up the surface, providing a "tooth" for excellent bonding of the primer-surfacer. Basically speaking, there are two types of sandblasters; siphon-feed and pressure-feed, each having their advantages and disadvantages.

Siphon-Feed—A siphon-feed sandblaster uses air to suck sand out of a container and blow it out the nozzle. This type of sandblaster is relatively inexpensive to purchase but results take longer and it requires more sand as compared to a pressure-feed unit. If you're thinking of tackling other restoration projects in the future, automotive or otherwise, you're better off purchasing a pressure-feed sandblaster.

Pressure-Feed—As its name implies, a pressure-feed sandblaster uses sand in a pressurized tank. The air pressure in the tank forces sand out of the nozzle at high speed, enabling it to cut through years of rust and scale in seconds. It also uses considerably less sand in the process than a siphon-feed unit.

Rent Or Buy?—If you'd rather not purchase a sandblaster (which requires a suitable size air compressor), you can rent one. Many of the larger rental shops will rent commercial type sandblasters with compressor for as long as you need it. However, because of the minimum rental time on these units, it's best to have quite a few parts lined up that require sandblasting to offset the cost of doing it *yourself*. Another alternative is to have a shop that specializes in sandblasting do the work for you. However, as with the case of renting a sandblaster, you can usually get a better price if you have a few parts that require sandblasting rather than only one or two.

Cautions—If you've decided to do the sandblasting yourself, keep in mind that sheet metal can't withstand the pressure of a sandblaster without reducing the spraying pressure and holding the spray nozzle at a 45-deg. angle to the surface. Disobey these rules and you'll end up with distorted sheet metal that has the surface of a rolling golf course.

Safety—This is another aspect of sandblasting that's often neglected. No matter where you're sandblasting, you must wear a respirator designed to filter out sand from the air

you're breathing. An incorrect respirator or the lack of one can allow minute sand particles to enter your lungs, causing a condition called *Silicosis,* leading to coughing, shortness of breath and other more serious respiratory problems.

A sandblasting hood and gloves are also necessary to prevent injury to your eyes and skin. In addition, it's a good idea to wear a long-sleeved jacket and long pants. Due to the safety equipment and clothing necessary for your protection, things can get pretty hot so try to schedule the sandblasting for a cooler part of the day such as early morning. Sandblasting early in the day leaves time for priming and painting parts later that day to prevent rust from starting again.

Also be sure to follow the manfacturer's operating recommendations closely. They know their equipment best and can tell you ways of doing the highest quality job in the least amount of time with a minimum of hassles.

Sand—If you've lined up a sandblaster, all you need is sand. Although you can use sand from the beach or even play sand, these types of sands tend to be very damp, requiring straining and drying before they can be used. Silica sand is the most common abrasive used in sandblast-

For safety's sake, wear a respirator, hood and gloves when sandblasting. It's also a good idea to wear long pants and a long-sleeve shirt. Direct stream of sand to metal at a 45-degree angle.

sandblasting, apply metal conditioner to ... rust from attacking bare metal before it ...inted.

equipment and is just the ticket for ...sting the parts from your car. ... type of sand can be found in ...y builder's supply stores listed in *Yellow Pages* and is inexpensive. ...oring sand is another matter of ...rtance. Sand used in a ...dblaster has to be almost entirely ... of moisture to eliminate clogged ... and nozzles. Wet sand is far and ... the leading cause of ...daches when you're sandblast-... Few things are as frustrating as ...ng aside a day to sandblast parts ... all that happens is you spend the ...disassembling the sandblaster to ...n out wet sand that prevented ... from sandblasting. So keep the ... in a dry area away from any ...sible sources of moisture.

... the way, sand can be reused. ...ever, as the sand coming out of the ...le hits the surface being stripped, ...ugh edges, which provide the cut-...action, are smoothed out. So, the ... time you use the sand it doesn't cut ...ell as the first time and so on. But if ...absolutely have to, used sand can ...eused after straining it through a ...le layer of window screen.

...al Prep—After sandblasting off ...e rust and scale, you're ready to ...int the part. But before you do, ...ent any future rusting by apply-...a metal conditioner with water in ...atio recommended by the manu-...urer and wipe the metal while it is ...wet with conditioner. If you let the ...ditioner dry without wiping it, this ...romote rusting. An alternate to ...al conditioning is to use a primer ...contains metal conditioner, such ...uPont Vari-Prime.

...for some reason you don't get a ...ice to paint a part before it begins ...ng again, clean off the surface ...with a metal conditioner and then ...t it. And remember that a freshly ...blasted piece of metal can rust ...quickly if not primed and painted ...st immediately.

oxy-acetylene torch, you can braze the holes.

If the rust holes are larger than 1/8 in. but don't decimate an entire panel, you can repair them with sheet steel, provided the panel is flat. However, if an area has lots of holes in it, or is an odd-shaped stamping, you're better off cutting away the whole section and welding in a new preformed panel. The process is the same for both types of repairs.

Required Tools—To do this type of job, you'll need a cutting wheel, nibbler or aviation snips. An advantage to using a cutting wheel or nibbler is that you can cut out the rusted section without bending the surrounding metal. However, similiar re-sults are obtainable using two pair of snips: a left-hand pair and a right-hand pair.

Cut Out Rust—If you only need to patch an area, begin cutting from the inside of the rusted area. Cut out all the rusted and thin areas until you get to solid metal. Then trim the area so that you can easily make a patch to fit it. The fewer number of angles the better.

Make Template—Take a piece of light cardboard and press it against the sheet metal directly under the hole. Five or six sheets of paper stapled together will do in a pinch. Take a pencil and follow it around the perimeter of the hole so its shape transfers onto the cardboard. Be sure to follow the pattern of the hole exactly.

Proceed by removing the cardboard and cutting out the shape you drew with a pair of scissors. This is the pattern you're going to use to cut out the patch panel. Lay this pattern down onto the new sheet metal. Sheet metal is widely available from auto-motive paint stores.

Now transfer the shape of pattern onto the metal with a pencil. When you've traced the pattern completely, remove it. Then color over the pencil line with a broad-tip magic marker. Be sure to use a dark color. Lay the pattern down in its original position and hold it securely. Then trace its outline again. However, this time use a scribe or sharp nail to transfer the shape onto the new metal. The outline of the pattern must be clearly visible in the magic marker. This is the pattern you'll follow to shape the patch panel.

Cut Patch Panel—You can cut the pattern out with a nibbler, cutting wheel or a pair of left-hand and right-hand aviation snips. To use the nibbler you'll have to cut a small hole next to the pattern. This allows the lower jaw of the nibbler to get un-derneath the metal. Then simply follow your pattern. Nibblers don't distort the metal because of their forked jaws. The

Dremel Moto-tool is just the ticket to cut out rusted sections of severely rusted trunk pan. Use Dremel 409 cutting wheels and wear gog-gles.

When using metal snips, take care not to dis-tort the metal. Snips are a good choice where one sheet metal panel lies directly beneath the one you're cutting.

Raised spot weld areas can be leveled off with a grinder, sanding disc or file.

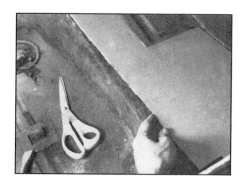

Cut a piece of cardboard to act as a template for the repair panel. Take your time and make the template accurate.

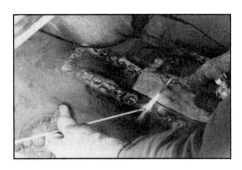

When you're satisfied with the fit of the replacement panel, tack it into position. Then weld the seams, taking care to cool the surrounding sheet metal with wet rags to prevent distortion.

After welding, grinding and priming, replacement panel blends into trunk well.

adjacent metal is securely held by the upper jaws while the lower jaw pushes the metal up between the upper jaws, cutting it cleanly.

Cutting wheels are available for drill motors as well as Dremel Moto-tools. Just follow the pattern with the wheel until it's cut out.

If you'd rather use aviation snips, you'll encounter some trouble as you try to go farther than the depth of the cutting jaws. However, you can get around this with a pair of left-hand and right-hand snips. As you reach the end of the cutting action with one pair of snips, begin another cut with the other pair of snips about a 1/2-in. away from the first cut. This gives a relief cut. The relief cut will let you cut again with the first pair of snips. Repeat this process of alternating between snips until you have the pattern cut out. Any small raised areas created by the cutting action of the snips can be straightened out on the anvil of a vise with a flat bumping hammer.

Trial Fit—With the patch panel in hand, trial-fit it to the hole. If it needs to be bent to fit the hole, work it with your hands or with a bumping hammer or dolly. When it's properly contoured, check that it fits the hole with little or no gap. The closer it fits the hole, the better. If it's slightly

larger than the hole, trim or file it. Then check the fit again. Do this as many times as necessary until you get a good fit.

Tack-Weld—The next step is to hold the metal in place while you tack-weld it. If some other metal part of the car is supporting it, you're all set. However, if you're welding in a piece, say in the middle of the trunk pan, it'll have to be supported before you can tack-weld it. And there is a trick you can use.

Cut a piece of brazing rod about 4-in. longer than the patch panel. Carefully braze the center of the rod to the center of the patch panel. The idea is to have the brazing rod hanging over the edge of the panel by 2 in. on both sides. As you braze the rod, try to use as little heat as possible to prevent the patch panel from warping.

Now position the patch panel on the car so the brazing rod is on top of the sheet metal surrounding the hole. Then carefully check the fit of the patch panel to the hole. If it's misaligned with the hole, bend the brazing rod until the misalignment is eliminated. Then lay a brick over the ends of the brazing rod to stabilize the patch. Check the patch alignment one more time.

When the patch is in position, tack the patch every 2 in. or so with brazing rod. However, if the patch is in a structural area, use welding rod. It possesses much more strength than brazing rod. When the patch is tacked into position, tack it again between the first tack welds.

Final Welding—At this point you can re-heat the brazing rod "hanger" to remove it from the patch. Then finish welding the patch in its entirety. When the patch is completely welded, grind down the weld to the surface of the adjacent metal. An 18- or 24-grit disc on a grinder works great. Just take care not to cut through or burn the metal as you grind.

Dressing the Patch—When the weld is flush, fill any minor surface irregularities with lead or plastic body filler. When the surface is smooth, prime it and paint it. If you can't paint it right away, at least prime it to prevent rust from forming. *Du Pont 215S* primer is a good choice.

PAN REPLACEMENT

Now if the majority of the trunk pan is rusted out, you're better off cutting it out and replacing it with a new one. These are available from many of the Camaro aftermarket suppliers.

Preparation—Before attempting to cut out the old trunk pan, support it from underneath with a floor jack or some wood.

Just make sure that it's supported so that it doesn't bend the surrounding metal as you cut it out of the car.

From underneath the car, take a look at the fuel tank rails (also known as trunk floor reinforcements). These rails provide a means of securing the fuel tank to the trunk pan. They're spot-welded to the underside of the trunk pan and to the inner tail panel. Although these rails don't have to be removed from the car, they have to be removed from the trunk pan. And then they have to be welded to the new trunk pan. So unless you're a good enough welder to do overhead welding, which is quite difficult, you're better off removing the rails from the car. However, if you prefer, new fuel tank rails are available from the aftermarket.

Before removing the fuel tank rails, measure their location on the trunk pan. Otherwise, you'll have problems later on determining where to locate them on the new trunk pan. So take measurements from the edge of the rails to the reinforcing ribs. Do this at both the front and back. And be sure to write these dimensions down.

Remove Rails—To remove the rails, drill out the spot welds that attach them to the inner tail panel. An alternative method is to use a spot-weld cutter to cut through the welds. Another alternative is to cut the rails just behind the slotted holes for the fuel tank bolts, moving toward the front of the car. The only drawback to this method is that you'll have to do some finish work on the cut when the rails are welded back together.

Trial Fit—Place the replacement pan on top of your old pan in exactly the same location. This is important. If you're off now, you'll have gaping holes to fill later where the old trunk and new trunk pan don't meet. Mark the area around the edge of the trunk pan with layout dye or a black magic marker. Then follow the shape of the trunk pan with a metal scribe so it transfers to the car.

Cut Out Pan—To cut the trunk pan out, use the aforementioned Dremel grinder with cutting wheels. Using a nibbler to cut the pan out will create a large gap unless you adjust for the width of the cutter. Whichever method you prefer, remember to cut to the outside of the scribed line. If you don't cut enough of the original metal out the first time, cut or file it out later. But once too much metal is cut out, you'll have to weld in another piece of metal, adding to task at hand. So be careful.

Remember that the rear frame rails are just to the outside of the trunk pan. If you cut much farther outboard than the very edge of the trunk pan, you'll probably end up cutting the frame rail. So again, be careful.

Transfer Fuel Tank Rails—With the pan out of the car, you can remove the fuel tank rails from it. Do this by drilling out the spot welds. With the rails off the trunk pan, grind or file off any remaining spot welds. This will allow the rails to sit flat against the trunk pan when you're welding them, easing the job.

Next, locate the fuel rails on the new trunk pan. Use the dimensions that you recorded earlier to do this. With the rails in position, plug-weld them using the holes that you drilled previously to remove them from the old trunk pan. And to obtain the necessary strength, use welding rod, not brazing rod.

Weld New Pan—Now position the new trunk pan assembly to the car from underneath. With the pan temporarily in position, support it from underneath as before. To get the pan perfectly in position, insert shims between the supports and the pan until the edges of the trunk pan are perfectly aligned with the hole in the car.

At this point, you can tack-weld the pan. And again, to ensure the structural integrity of the pan, use welding rod, not brazing rod. Tack-weld the perimeter of the panel every 2 in. on the first pass. The second time around, tack weld halfway between each previous weld. When you've finished tack-welding the pan, run a continuous bead around it.

Finish the job by grinding the weld flush on both sides of the pan. Then fill any surface irregularities with plastic body filler or lead.

DASH & COWL REPAIR

An area often prone to rusting on Camaros (and many other cars for that matter), is the leading edge of the dash panel where it meets the cowl. And the rust can often be seen spreading at the base of the windshield. Repairing this type of damage requires removing the windshield, cutting out the rusted area and welding in new metal if rusting is severe.

After removing the windshield, as outlined on page 38, scrape away all traces of windshield sealer. Carefully inspect the area where the dash panel overlaps the cowl panel. If only minor rust exists, grind or sandblast it away and prime and paint the metal.

If rust is a major problem here, you'll have to cut out the affected areas and weld in new metal. Or you can install a new dash panel. These are available for some Camaros.

UPPER COWL PANEL

Take a careful look at the upper cowl panel. If it's severely rusted, you'll have to repair it before repairing the dash panel. That's because the dash panel overlaps the cowl at this point.

To repair rust in this area, carefully peel back the dash panel with an air chisel. You can also use a cold chisel and a hammer, it just takes longer. The key word here is care; don't try repairing this area only to destroy something else. When the dash panel is removed, you can carefully examine the cowl panel to determine how much rusted metal you have to remove to get to sound metal.

Also remember that whatever rust shows on the surface is probably just as bad or worse on the side you can't see. So if the rust is quite bad, you're probably better off removing the cowl panel and repairing it thoroughly.

Cut Out Cowl Panel—This entails using a drill or spot-weld removal tool (such as that available from the **Eastwood Company, Malvern, PA**) to remove the spot welds that secure the cowl panel to the car. A spot-weld removal tool is best because it cuts only the uppermost weld, leaving the underlying metal virtually intact. It also makes it easier when plug-welding the panel back into position. When installing the repaired panel, you'll have to plug-weld it into position using the holes as a guide.

With that in mind, cut out the rusted metal with a Dremel Tool and cutting wheels. Save as much of the rusted metal as possible because it will serve as a pattern for the new sheet metal.

Make Patch Panel—Now you can make your patch panels. Begin by making a pattern out of thin cardboard stock. Use a good section of the cowl that most closely resembles the area that was rusted out. Then use this area to draw a pattern on the cardboard taking into consideration the bends of the metal.

Transfer this pattern onto an adequately sized piece of 18-gage mild-steel sheet metal. This type of metal is best because it can be easily formed and welded. When

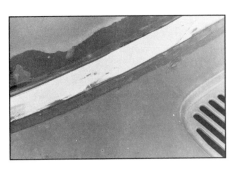

Typical problem with 1967-69 cars is rusting at the forward edge of the dash. This is caused by a leaky windshield and requires windshield removal for proper repair.

Cutting wheel comes in handy when you have to remove a section of the dash without damaging the sheet metal underneath.

you've laid the pattern, carefully bend the metal using a vise or locking pliers and a hammer. At each stage of the forming process, compare your work to the pattern and old cowl panel. That's the best method for ensuring a good fit.

When you're satisfied with your patch panel, weld it to the cowl panel. Don't braze the metal at this point because the cowl acts as a structural component and is subjected to high loads. After welding, grind all the welds flush. Then condition, prime and paint the cowl.

Weld Cowl Panel—Position the repaired cowl panel in the car. To align it properly, use the holes you drilled to remove the spot welds. When everything's lined up, clamp the cowl panel into position. When you're satisfied with the alignment of the panel, weld it into place. Now turn your attention to the dash panel.

DASH PANEL REPAIR

If the dash panel exhibits a considerable amount of rust, cut it away as you did the rust on the cowl panel. Be sure to use the cutting wheels or a nibbler to avoid distorting the metal. Otherwise, you'll have to do a lot more finishing work to get the panel looking good.

Form Dash Panel Patch—Next, make a pattern using cardboard as discussed earlier and transfer it onto 18-gage sheet metal. After cutting out the dash patch panel, bend the leading edge of it to fit the curve of the windshield. Do this by measuring the distance from the front edge of the dash to the bend. Then transfer this measurement onto the patch panel.

To put this bend into the metal, put a couple of 6-in. long 1x3-in. pieces of wood (such as floor boards) into the jaws of a vise. Then place the patch panel between the wood. Position the panel so that the bend mark is about 1/8 in. above the boards. Now start to bend the panel by

Repair of a compound curve, like this outboard dash section, requires two panels: one for downward angle and one for the flange.

Lower dash section was repaired with lead filler. Special file is needed to smooth the lead without clogging.

Water trapped in air plenum at bottom of lower cowl rusted through. A hole here admits water, road grime and engine compartment fumes into a car's interior via the ventilation system.

pulling down on it with your hands. When the angle of the bend is close to that of the original dash, use a rawhide mallet to form the remaining bend angle. Proceed in this manner until the bend in the patch panel is formed entirely. Just keep in mind that if you run into trouble with a stubborn bend, you may need to put relief cuts in the metal, shaped like a "V." This will allow you to bend the metal more easily.

Weld Dash Panel—Clamp the patch panel to the dash and check your work. When pushing down on the leading edge of the dash, the patch panel should closely fit the contour of the cowl. A variation of less than 1/8 in. is tolerable because drip-check sealer can compensate for the difference. If the fit is good, tack-weld the patch panel in place. Then recheck the fit.

When the cowl panel fits well, tack weld its trailing edge, then run a continuous bead. Now tack the leading edge of the dash every inch or so. When you've tacked the entire panel, lay a continuous weld bead across it. And remember to plug-weld the holes in the cowl.

Finish the dash by grinding the welds flush with the surface. Then fill the surface imperfections with body filler and finish it off as you would any body panel.

LOWER COWL REPAIR

The lower sides of the cowl are another favorite spot for the rust bug. Holes in this area can allow water, dirt and air into the interior. And this can lead to even more rusting.

Rust in this area can be repaired by cutting out the affected sections with a cutting wheel or nibbler. If the rust has moved into the seams, drill out the spot welds and remove the bad metal.

Make a patch panel using cardboard as explained earlier. Then weld it into position and finish it with body filler.

QUARTER PANEL REPAIR

Repair of the quarter-panel sheet metal can take many forms depending on the condition of the quarter panel and your checking account. If the panel exhibits very small holes either in front of or behind the wheel opening, they can be repaired by welding or with plastic body filler or fiberglass. Larger holes in these areas and in the arch of the wheel opening might require the use of a small patch panel.

Replacement panels vary widely in quality once you stray from original-equipment GM sheet metal. Beware of bargain-priced imported panels from across the Pacific.

Patch Panels—Patch panels are offered in sizes to fit the areas forward and rearward of the wheel opening, the wheel opening itself and most of the quarter panel. Some of these panels, generally those imported from Pacific Rim countries, are made very cheaply and don't fit well. So make sure that you buy your panels from an established source who will refund your money if the panel doesn't fit.

You can check the fit by holding the new panel against the same area on your car. All the body lines and folds should line up. They should also have the same sharpness. Folds that are more rounded than the original indicate the parts were made on substandard dies. If the patch panels don't pass all of these tests, forget buying them.

Original Vs Reproduction Panels—If you're planning on replacing the entire quarter panel, you're better off buying GM original-equipment sheet metal. The factory panels fit the best although they're not perfect. For instance, they may be missing part of a flange. So you'll have to form it yourself. And the dimples inside the wheel opening that are supposed to coincide with the screws for the wheel-opening molding may not be in the right location. Also, many of the GM panels one of the authors purchased came pre-dinged and pre-dented, still inside their original boxes. But even in light of these deficiencies, the factory pieces are far better than some cheap reproductions from across the Pacific. Have you ever tried hammerforming an entire quarter panel?

QUARTER PANEL REMOVAL

Before you begin cutting away the quarter panel, there are some important things to keep in mind. For instance, the car must be sitting on its rear suspension. If you've removed the rear axle, supporting the rear leaf springs where the rear axle would normally sit will work just fine.

Supporting the car helps simulate normal loads and stresses in the chassis. If the car isn't supported in this manner, the chassis could shift when the quarter panel is removed. And if a new quarter panel is installed onto a shifted chassis, it could buckle when the rear wheels are on the ground once again.

Also, the front subframe should be installed and supported directly under the coil-spring cavities. An alternative is to have the car supported by the front suspension. As with the rear suspension, this helps load the chassis correctly for the best possible alignment of the quarter panels. And all of this talk about loading the chassis is especially critical on convertible models because they're more prone to flexing than coupes.

Panel Alignment—Another item to keep in mind before removing the quarter panels is their *alignment*. To help you install the quarter panels in proper registration, two items on your car can help you: the deck lid and the doors. Take time to align these components to the quarter panel as they'll help when it comes time to install the new panel.

The deck lid should be aligned so that the gap between it and the quarter panels is even from front to rear. Also, the leading edge of the deck lid should be parallel to the tulip panel. The tulip panel fits in front of the deck lid and behind the rear window. If the deck lid isn't bent but the gap between it and only one of the quarter panels is slightly uneven, the quarter panel may not be properly aligned in the first place. A small adjustment to correct this situation can be made when the new quarter panel is installed. However, you can't correct for a gap that's 1/4-in. out of parallel.

The door should be aligned so that its trailing edge is parallel at all points to the front of the quarter panel. When you've done this, take a look at the bottom edge of the door. It should be evenly spaced from the rocker panel along its entire length. If it isn't, the chassis may be bent. This would necessitate a trip to the frame shop to get things right before you do any work to the quarter panels.

One way to check the alignment of the body is to use the specifications listed in the *Fisher Body Service Manual*. The specifications given there allow you to check the distance between various points on the underbody of the vehicle, crossing side to side and front to back. If your car doesn't meet these specifications (allowing for a tolerance factor of plus or minus

Before removing quarter panel from car, deck lid and door should be properly aligned to it. This provides reference points to align new quarter panel.

1/8 in.) chances are the frame is bent or twisted and needs professional help.

Gaging Holes—The only real tool you need to check these specifications is a tape measure. The measuring points are usually from gaging hole to gaging hole or between a bolt hole and a gaging hole. A gaging hole is a hole in the chassis which has no bolt or stud running through it. These holes are nominally 3/4-in. diameter and can be found in a number of places.

For instance, a gaging hole is located in the front of each rear frame rail in first-generation cars, just inboard of the spring perches. Another gaging hole is located on each subframe mounting ear, where the center subframe mounting bolts are found. By measuring the distance between the centers of gaging holes such as these, you can determine if the chassis is bent or not.

The tape should be horizontal when checking these dimensions. If the gaging holes you're checking aren't on the same horizontal plane, you can tie a couple of large nuts on one end of a piece of string and a small nail on the other. The string will provide the location of the hole and the nuts put tension on the string so it's taut. Then thread the nail and string through the gaging hole. Make sure the string is centered in the hole and then check the dimension. If you have any doubts, check with a local frame alignment shop.

Required Tools—If you live in snowy climes where they use salt on the roads, chances are at least one entire quarter panel needs to be replaced. Although this isn't a job to be taken lightly, you shouldn't consider it beyond your reach if you've had some experience with bodywork (with accompanying good results), welding, and have the necessary tools available. This includes oxy-acetylene torches, an air hammer with a forked panel-cutting bit,

flanging pliers (for '67 and '68 convertibles, sharp metal snips, a grinding bit and body filler material.

You'll need the air hammer with the panel-cutting bit to cut away the old quarter panel with minimal distortion to the surrounding metal. And this is especially important when adapting a coupe quarter panel for a convertible application (necessary on 1967-68 convertibles) since you'll need to leave a portion of the old quarter panel attached to the body. This serves as a blend area for the new quarter panel.

Also, if you look carefully at the angle of the window sill on the coupe quarter panel, you'll find that it's canted at a steeper angle than the window sill on the ragtop. So you'll have to cut off this portion of the coupe quarter panel and blend it into the ragtop window sill.

And if you look carefully at the top portion of the convertible quarter panel, you'll see a ridge that follows the down sloping curve of the panel alongside its inner edge that stops about an inch or so above the opening for the trunk. If you want to keep this ridge, you'll have to trim both the old and the new quarter panels so that they blend into the valley at the top of the quarter panel.

So think about these factors *before* cutting either panel. Work in a slow methodical manner to avoid cutting too much. You can always cut some of the panel away but it's a lot harder to weld the metal back on.

Attachment Points—You need to know how the quarter panel is attached before you can remove it. The forward edge of the quarter panel, just inside the door is spot-welded in place. The panel is also spot-welded inside the trunk weatherstripping channel: at the edge of the tail panel at its lower edge behind the wheel (to the quarter panel extension), around the wheel opening (to the outer wheel house) and just forward of the wheel opening (to the rocker panel).

In addition to the spot welds, the quarter panel is blended into the other bodywork with either lead or plastic body filler. On coupes, they're blended near the top of the C-pillar and just inboard of the trunk opening above the deck lid. Ragtops are blended at only one point—just above the deck lid like the coupes.

Initial Cut—You can cut the panel off with a cutting wheel, nibbler, panel cutter or whatever is available. But you'll want to disrupt the surrounding metal as little as possible. This is especially important on

Trim old quarter panel so there's 1/2 in. extra metal beyond cut line. Perform final cut only after replacement quarter panel has been trial-fitted.

Rear of this quarter panel has been rough-trimmed. Up top, trim to outboard of trunk weatherstrip channel. Break remaining spot welds with an air chisel. Note that bumper-bolt reinforcement may have to be cut off old panel and welded on new quarter. Mark its location first before cutting it off.

On convertibles, save lower front edge of old quarter panel. Unlike tail panel seam, this seam cannot be welded due to thickness of rocker panel without damaging the quarter panel. Old quarter panel serves as attaching point for new one, once this edge has been massaged slightly inward.

After major portion of quarter panel is trimmed away, cut off remnants and break spot welds with an air chisel. Don't slip and damage adjacent sheet metal with it, though.

If adjacent sheet metal gets rearranged, straighten it out with a hammer and dolly.

ragtops because you'll be saving part of the original panel.

Trim the quarter panel away so that you're within an inch or so of the spot welds and joints. On convertibles, trim just to the inside of the top of the body curve. This will run from the front of the quarter panel below the quarter-window opening to 2-in. below the top of the opening for the deck lid. And as you're trimming, remember to leave a 1/2-in.-high piece of old quarter panel above the rocker panel, between the front of the quarter panel and the wheel opening. This will provide an attachment point for the new panel.

Refer to *Rough Trimming* and the accompanying photos for more details if your car is a ragtop.

When all the surrounding metal is cut away, you can concentrate on uncovering the bonded joints and removing the spot welds. However, on 1967-68 convertibles, it won't be necessary to uncover the bonded joints on the tulip panel if you're modifying a coupe quarter panel to fit. That's because the joint won't be in the path of the newly modified panel.

Removing Body Solder—To remove solder from the joints, you can use a propane torch and a screwdriver. Carefully heat the joint with the torch while scraping at the surface with the screwdriver. If the joint is leaded, the surface will appear shiny as the lead begins to melt. Remove all lead fom this area because it represents an impurity that will ruin new welds.

If the joint is filled with plastic body filler, the filler will begin to burn. Carefully sand away most of the filler with a coarse sanding disc until the joint and the spot welds that hold it together are exposed. It isn't necessary to completely remove the filler at this point because you'll be breaking the joint shortly.

Remember, you'll find the C-pillar coach joint just a few inches below the top of the roof. And the tulip panel joint is just

inboard of the outside edge of the deck lid opening, running front to back.

Cut-Out Spot Welds—To remove the remaining metal from the car, you have to remove the spot welds. This can be accomplished with a spot-weld cutter or an air chisel. A spot-weld cutter resembles a miniature hole saw with a pilot and cutting ring. Simply line up the pilot with the center of the weld and drill into the panel with the cutter. The cutter removes a nice round piece of steel from the top layer of metal without cutting the second layer underneath. The major disadvantage of this tool is that you have to be able to access the spot weld directly from above. And this isn't possible 100% of the time when you're removing a quarter panel.

An air chisel is just what its name implies: a chisel driven with air. It's simply placed between the two pieces of steel and hammers its way between them and through the spot weld. But you must be careful with this tool as you can cut farther than you'd like in a hurry. Also, sometimes the chisel will skate over the spot weld and make a long cut in the metal. If this happens, position the chisel so that it's 90 deg. to the first cut and finish the job.

Whichever tool you select, take care not to cut through the panels that will remain on the car. If this should happen, it can be remedied by welding a patch into the damaged area.

To break the bond between the upper portion of the tail panel and quarter panel, carefully position the air chisel on the seam between the two panels. Then work the chisel into the seam while working your way down to the area where the bumper would be. Below this point, the flange on the quarter panel lies flat against the tail panel. So work the chisel into the seam toward the center of the car.

When all spot welds are cut, grind off any remaining welds. This will assure a good tight seal between the old panels on the car and the new quarter panel. With the car ready, now you can prepare the quarter panel.

Rough Trimming—When installing a coupe quarter panel onto a convertible, you'll need to trim away the C-pillar on the new panel. And while you're doing this, remember to stay on the conservative side! Always err to the point of not cutting enough of the new panel away rather than cutting too much. Otherwise you'll end up welding patches to replace the metal that you inadvertently cut away.

Start trimming a coupe quarter for a

convertible by measuring for a rough cut. This will get you in the ballpark. Then hang the quarter on the car to see where you have to go.

Begin by scribing a line one inch from the top body curve on the new quarter to the inboard side. Make sure this line is evenly spaced. Marking the quarter every inch or so from front to back will make this easier as will using a tape measure when you have to connect the marks.

Now look at the leading edge of the quarter panel. To avoid any bodywork inside the door aperture, cut the panel about 1-1/2-in. rearward of the aperture. This cut will run vertically to meet the cut line you've laid out from the top body curve.

The final cut line will run on a 45-deg. angle from the corner of the deck-lid opening to the top body curve cut line. Again, this will be the first rough cut to get you in the ballpark.

Trim the metal away, taking care not to distort the metal a significant amount. If you're using aviation snips, many small cuts will distort the metal less than fewer large cuts.

Test Fit—When the quarter panel has been rough-cut, hang it on the car to check your progress. Lower it into the trunk weatherstripping channel first then slide it into position. If everything seems to fit pretty well, use a metal scribe or china marker to transfer the outline of the trimmed area on the new quarter panel onto to the old quarter panel. You'll be using this line to help you flange what remains of the old quarter panel.

The next step is to make sure the rest of the quarter panel fits the car. As you do this, you'll find the lower forward section of the quarter panel won't fit properly because of its flange. One of the easiest ways of remedying this is to trim this flange just above the point where it folds toward the car. By leaving a 1/2-in.-high piece of the old quarter panel in this area, you can weld the new panel to it. To make sure the new panel will be where the old was, lightly tap the piece of old panel in until the new panel is in its proper position. When the new panel fits well, remove it and begin flanging the upper seam.

Make Flange—Position the flanging pliers (available from **The Eastwood Company** and many automotive paint stores) just to your scribed line, making sure they're square to it, and crimp the metal. Do this the entire length of the cut. As you reach the area at about the rear of the quarter window, you'll find that the flang-

ing pliers won't fit in deep enough to reach the bend line. So, just move down an inch or so away from this point and proceed with flanging the panel. This small unflanged area won't present any problems when it comes time to secure the quarter panel.

When the old quarter is flanged, slide the new quarter back into position and check its fit. If the new panel seems to ride over the flange at any point, mark the area so you can trim it later. The new panel should fit into the recess of the flange, allowing about 3/16 in. for welding and filling later on. And while you're working, be sure not to bend what remains of the old quarter panel, otherwise you'll have trouble later joining it to the new panel.

At this point, if your car needs a new outer wheel housing, refer to the story below. If you're not replacing the outer wheel housing, you can begin securing the new quarter panel to the car, page 98.

OUTER WHEEL HOUSING

Now that the old quarter panel is off, take a good long look at the condition of the outer wheel housing. Is it rusted or has the car ever suffered a severe rear-quarter

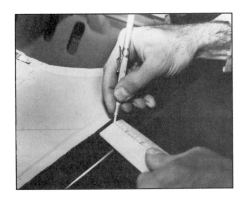
When the rough seam fits fairly well, measure for final trim.

To avoid doing a lot of refinish work inside the door jamb, cut the new quarter panel about 1-in. rearward of the jamb.

With quarter panel slipped onto car, check fit inside wheelhouse. No large gaps should occur between quarter and wheelhouse.

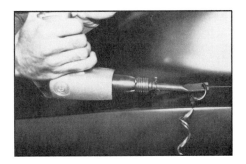
After making measurements, cut new quarter panel with panel cutter, nibbler or shears. Try to avoid distorting the metal as you cut, as this creates more bodywork down the road.

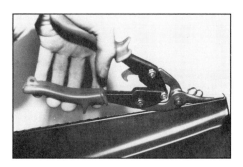
When converting coupe quarter panel for use on a convertible, carefully trim away excess sheet metal.

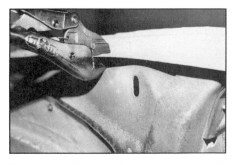
To avoid overlapping joints between new and old quarter panels, one panel must be flanged. Eastwood flanging tool puts a small step in the sheet metal. This allows two panels to form a lap joint without a noticeable seam. A lap joint is also very strong.

shunt? If the old quarter panel had a lot of rust in the area of the wheel arch, chances are good that the outer wheel housing is quite rusted also. Now is the best time to replace it, while the quarter panel is off.

REPLACEMENT

Replacements usually fall into two categories; new and used. New outer wheel housings are available from a variety of sources. And all of them, up to this point in time, are made from painted sheet metal. But there is one exception—1969 models. The aftermarket outer wheel housings for these cars are made from galvanized sheet metal. This has its good points and bad points. The most obvious good point is that it will withstand severe weather much better than uncoated sheet metal.

The drawback is that galvanized metal cannot be successfully gas welded without a lot of extra work. However, if you have access to a spot welder, or are handy with a TIG or MIG welder, you should be able to join the wheel housing without any real difficulty.

Used—Used, rust-free wheel housings can be more difficult to come by, especially in the northern U.S. These pieces can be removed from the donor car the same way you removed the quarter panel. Use a body panel cutter to cut the area around the portion of the wheel housing that you want to save. When that's done, break the welds with a chisel and remove the wheel housing.

Attachment Points—The outer wheel housing has a long flange on its top side that matches up with a similiar flange on the inner wheel housing. And both flanges have registration holes at each end and in the middle. By matching up these holes, you're assured of aligning the outer and inner wheel housings properly.

The coupe outer wheel housing is welded in four areas: to the inner wheel-housing flange, to the wheel opening in the quarter panel, at its front edge to the inside of the front portion of the inner wheel housing, and to the quarter-panel extension in the rear.

Convertibles are a bit more involved due to the extra bracing required. In addition to the areas mentioned for the coupe, three braces, specific to convertibles, are welded to the top side of the outer wheel housing. Starting from the front, this includes the inner quarter-window bracing, the upper quarter-panel support, and the trunk reinforcement in the rear.

Because of the braces and the lack of

rigidity in the body once these are removed, you're better off replacing only the outer portion of the outer wheel housing. Of course, if an entire wheel housing is damaged as the result of an accident or other cause, you'll need to replace the entire outer housing.

REMOVAL

Let's start our discussion with removing and installing the entire outer wheel housing first. A good method for replacing the outer portion of the wheel housing is covered immediately after, since it's a little more involved.

Begin removing the outer wheel housing by straightening the flange on the interior side. A pair of straight-jaw locking pliers will work just fine. Then use a panel cutter or aviation snips to remove the majority of the wheel housing within an inch of the aforementioned spot welds. Now, with the spot welds readily accessible, insert an air chisel between the remaining wheel housing and spot welds then break the welds. Remember, try to avoid damaging any adjacent sheet metal.

When you've broken all the spot welds and removed any remaining traces of the old wheel housing, grind all the spot welds so they're flush with the surrounding metal. This is important because the new outer wheel housing should fit flush against the inner. In turn, this provides a better seal and makes welding easier; especially spot welding.

Trial-Fit—At this point, you can trial-fit the new wheel housing into position. Begin by sliding the flange on the new wheel housing against the old. Then make sure the alignment holes of each flange match up perfectly. When they do, use a couple of locking pliers or C-clamps to hold them in position.

Before tack-welding the wheel housing into position, you have to make sure that it's going to fit into the wheel arch on the fender. Here's where fitting of the fender comes in.

Take the fitted fender over to the car and slide it into position. Match the edge where the trunk weatherstripping fits. Make certain that it's where you would want it to be if you were going to weld it into place right now. Then secure it with locking pliers or C-clamps in at least four places: top, bottom, front and back.

Now stick your head into the wheel well and check the fit of the wheel housing to the wheel arch of the fender. Both pieces should meet with practically no gap be-

tween them at any point. If this isn't the case, shift the wheel housing until it is. Then clamp it back into position. And be sure to clamp the wheel housing to the fender.

Tack-Weld—At this point, you can tack-weld the wheel housing into place anywhere except where it meets the fender. This will allow you to remove the fender so you can have better access for welding the wheel housing.

After tacking the wheel housing in position, remove the fender. Now tack the wheel housing every inch or so on its inside seam. And don't forget to tack it to the rocker panel.

CONVERTIBLE REPLACEMENT

Now that you've got some background on how to install a complete outer wheel housing, let's talk about how you can replace just its outer portion. As mentioned earlier, this is probably the best way to work on convertibles because you don't upset its special reinforcements. This is what we'll be covering here.

When you've got the quarter panel off, take a look at the reinforcement that's welded to the top of the outer wheel housing. As you trim away the old wheel housing, leave a strip of metal about 1/2-in. wide on the outboard side of this reinforcement. This gives a good area to tack the new wheel housing to.

As you look at the forward portion of the wheel housing, you'll see another reinforcement. This is the quarter-window regulator framing. And this piece is also welded to the old wheel housing. As with the upper reinforcement, cut about 1/2 in. to the outboard side of the seam. This will give you an attachment point for the new wheel housing.

Now looking to the rear you'll see the rear portion of the old wheel housing is spot-welded to the inner wheel housing. It's also welded to the quarter-panel extension immediately behind it. You'll have to break these welds to remove the wheel housing.

Plot Coordinates—With the outer portion of the old wheel housing removed, determine how much of the new wheel housing has to be cut away to fit in its place. This can be done by using a simple system of plotting coordinates. It's really quite easy.

Using the center alignment hole on the new wheel housing flange, measure off 1 in. from the center of the hole forward. Then measure another 1 in., and then an-

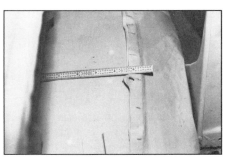

To install a new outer wheelhouse in a convertible, measure (at 1-in. intervals) the distance from the reinforcement that runs along the top of the old wheelhouse to the flange where the inner and outer wheelhouses are joined. Then transfer these measurements onto the new wheelhouse. Don't try to remove the complete outer wheelhouse on convertibles because numerous structural reinforcements tie into it.

other until you reach the front edge of the wheelhouse. Perform this same procedure working back from the center alignment hole to the rear edge of the wheelhouse. Now you'll have forty marks to use as reference points.

Using the old wheelhouse that's on the car, measure the distance at each mark from the upright flange of the wheelhouse to the various reinforcements. Write this measurement at each point on the new wheelhouse. Be sure that you're using the right measurement for each specific point on the wheel housing. If not, when you trim the wheel housing, you might end up cutting too much away. And that means either patching your newly trimmed wheel housing or buying another. You want to avoid doing either of these things.

When you've got all the coordinates marked on the wheel housing, you can play connect the dots. A china marker works pretty good for this. This will show just how much of the upper part of the wheel housing you'll have to remove.

Now simply trim the unwanted portion of the wheel housing away. If you plan on using aviation snips, you'll need two pairs; a left-hand and a right-hand. This way you can make a relief cut by alternating use of the snips. Cut a little with one pair then make a fresh cut about 1/2-in. away from the first cut. Proceed in this manner while using the snips in the same cut you started them in. Also, be sure to stay to the inside of the line.

The only thing left to do is notch the wheel housing about 1/4-in. outboard of the rear alignment tab. This will allow you to slide the new alignment tab into position over the old, thereby disguising your

work. Remember, just work slowly and carefully.

Check Fit—To check the fit of the wheel housing, position the alignment tab so that it will cover the old tab while at the same time sliding the leading edge of the wheel housing into the space between the rocker panel and inner wheel housing. Then rotate the entire wheel housing forward.

Now you'll find that the wheel housing isn't fitting quite like it should, because it isn't overlapping what remains of the old outer wheel housing. Remedy this by making a relief cut just below the point of your previous diagonal cut on the new wheel housing. Then check the fit once more.

The part of the wheel housing where it passes alongside the upper reinforcement should be just to the outside of the flange of the reinforcement. This will give you a small valley to weld the wheel housing. It also will make it easy to cover the work with seam sealer so it will be practically invisible.

Before welding the wheel housing in place, install the quarter panel and check the fit. Look at the alignment between the edge of the wheel housing and the quarter panel. They should be parallel to each other. Also, check the other areas of the wheel housing to make sure that it fits with no gaps. Pay particular attention to the upper portion of the wheel housing as it may tend to move away from the upper reinforcement. And the wheel housing should fit snugly into the area of the rocker panel.

Clamp & Tack-Weld—When you're satisfied with the fit, clamp the wheel housing into place. Then tack-weld it at the top and both bottom edges. Then remove the quarter panel and tack the wheel housing along the top of its entire length. Space these tack welds 1/4-in. apart.

Clean, Primer & Seal—When you've tacked the wheel housing, remove any

loose paint and rust. Then prime the bare metal with a corrosion-resistant primer. After the primer has dried, seal all the seams with seam sealer. Apply it as smoothly as possible in the area of the

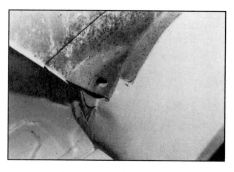

Rear view of trimmed outer wheelhouse shows flange fitting perfectly behind rectangular hole.

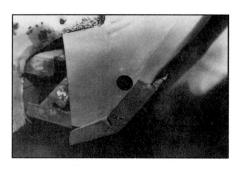

At bottom of rocker panel, new outer wheelhouse fits nearly flush with its rear flange.

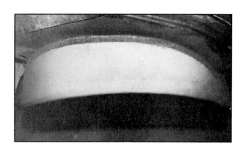

New outer wheelhouse is welded along this upper flange, then sealed with seam sealer.

Basic shape new outer wheelhouse will take after trimming to fit convertible. When installing new outer wheelhouse in a coupe, trimming is not necessary; merely split the old wheelhouse at the seam and remove it.

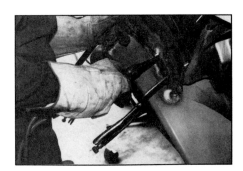

When you're satisfied with the fit of the new wheelhouse, also fit the quarter panel before doing any welding. Hold the wheelhouse in place with clamps or locking pliers.

upper reinforcement to hide any welds. This usually takes two applications of sealer. Apply the first with your finger, pressing the sealer into the seams. After the first application is dry, apply a second coat. Do a small section at a time then smooth the sealer out using the back side of a piece of sandpaper. When this second coat is dry, spray trunk spatter paint on the wheel housing and you're finished.

QUARTER PANEL INSTALLATION

Check Panel Fit—When the new panel is properly trimmed, position it on the car. Then make a last check of every seam on the panel to make sure it fits well. When you're satisfied with the fit, take the panel off. At this point you can prepare the panel for plug welding.

Plug Welding—This welding method is used to weld two pieces of metal together from one side. It consists of drilling a hole in one panel and welding through the hole to join the two. More information on this method can be found in HPBooks' *Metal Fabricator's Handbook*.

Drill the plug-weld holes in the trunk weatherstripping channel, the very leading edge of the quarter panel, on the inside of the wheel lip and on '69 convertible quarter panels, the upper flange where the C-pillar would be. The holes should be 3/16- or 1/4-in. in diameter and spaced about 1-1/2-in. apart. After drilling the holes, use a file to dress them.

Another area that might require the drilling of plug-weld holes is the first fold inboard of the exterior surface of the quarter panel, just inside the door jamb. The galvanized reinforcement for the door striker resides underneath this area and is welded to the quarter panel. But because the reinforcement is galvanized, you can't readily braze it to secure it to a new quarter panel.

Tale of two door jambs. Original convertible door jamb on right has telephone shape to stamping. Coupe jamb at left is decidedly different.

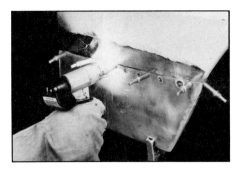

Small spot welder from Eastwood Company welds from A/C arc-welder set to 40-50 amps. Courtesy Eastwood Company.

Brazing Galvanized Reinforcement—If you plan on brazing the quarter panel to it, first grind away the galvanization within an inch of all weld points. Then with plug-weld holes drilled at this outboard edge of the door jamb and around the door striker, you can now braze the panel to the reinforcement.

Spot Welding—Another method you can use for securing the reinforcement to the quarter panel (or any two pieces of sheet metal) is spot welding. Although a production-type auto body spot welder can be used, it's hard to find at tool rental shops. An alternative method is to use the spot welder available from the **Eastwood Company**. This spot welder is used in conjunction with your own arc welder.

To use this particular spot welder, mark the outline of the reinforcement in the door jamb area of the panel, taking measurements as needed. This will guide you as to where the reinforcement is under the panel. If you don't do this, you may end up cutting a nice size hole through the panel that will have to be brazed later. And if you're not sure you possess the skill to use this spot welder, practice on some scrap sheet metal first, not your car.

Then mark the quarter panel as to where you want the spot welds. One weld every 2 in. on the outboard edge of the reinforcement is good. And remember to mark the area around the striker in at least three places.

Clamp In Position—Now install the quarter panel and check its fit one last time. When you're satisfied, clamp the panel in position in at least five places. Be sure the striker reinforcement is securely clamped to the panel. Then grind away the primer from the areas you marked in the door jamb. A clean bare metal surface is important for a good, secure spot weld. Another important point to remember is that the reinforcement and panel must be flush

against each other. Again, this increases your chances for a good spot weld.

Spot Welding—Now slide the two-pronged tip over the end of the welding gun. Then attach the gun to your arc welder and set the output at about 50 amps. After attaching the arc-welder ground cable to your car, position the gun over the area you want welded. Then release the trigger so the electrode rests against the panel and begins to glow. Now slowly pull the trigger back a fraction of an inch so an arc begins. After about 4-5 sec., retract the electrode using the trigger then check the weld. It should appear as a drop of water rippling in a pond. If it isn't, try again. Then repeat the process for the other areas.

And remember to keep the tip of the electrode clean, This will make the process of starting an arc much easier. A fine-tooth flat file works well.

Rivet Upper Flanged Joint—The next step is to turn your attention to the upper flanged joint. This will be at the top of the C-pillar on coupe quarter panels and at the flanged joint you made if you're converting coupe quarter panels to fit a convertible. Before brazing the joint, you must secure it. If you don't, it will warp. Use 1/8-in. rivets with a 1/4-in. grip range to secure the joint.

Use a sharp drill bit to make each hole, positioning them about 3/16-in. away from the edge of the flange. After each hole is drilled, secure the joint with a rivet. If you don't, the old panel will be pushed away from the new, making for two holes that won't line up when you need to install the rivet. And remember to space the rivets about 1-in. apart. If you space them any farther than this, the chances of warpage when welding increase greatly.

Tack-Weld Seam—After riveting, clean the surface with *Prep-Sol* then grind away any paint from the area to be brazed. Hold a wet rag near the area you want to braze (or weld) to minimize panel distortion, then tack the seam with brazing rod, making sure the tacks are in the joint seam. And as you're tacking the seam, remember to avoid working in just one area to avoid excessive heat and warpage. Tack welds spaced about 3/4-in. apart will do the job.

Completing Welding & Brazing—When you've tacked the entire seam, drill the rivets out and grind away any brazing that projects above the surface of the panels. Now turn your attention to welding up the other plug-weld holes you made earlier. Then braze the remaining seams on the panel. Keep in mind that the easiest way to

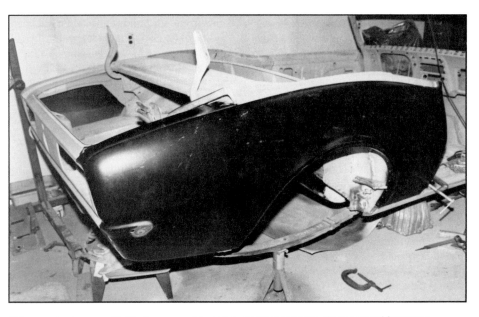

When you're happy with fit of new quarter panel, clamp it to the body at several locations.

Before welding, rivet top of convertible quarter panel. Riveting isn't necessary on a coupe.

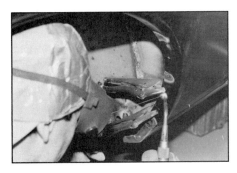

To join quarter panel to body, drill 3/16-in. holes where you want the welds to be. Then weld quarter panel to body through these holes.

Unwanted seams must be completely free of welding flux before applying body filler. Use waterproof filler to minimize chance of seam rusting, then breaking.

Apply good quality primer with built-in metal conditioner for lasting results.

join the quarter panel to the lower half of the tail panel is to braze between the seam. These tack welds will be covered later with seam sealer.

Fill Upper Flanged Joint—With the new panel welded in place, you can concentrate on finishing the upper flanged joint. Depending on your preference, you can use solder or plastic filler to cover the joint. Whichever method you choose, remember to clean off any traces of brazing flux or other impurities from the surface of the joint. And when you've finished soldering, be sure to neutralize the solder. If you don't follow both of these steps, you can almost be guaranteed of having corrosion, paint-blistering and other troubles with the joint later on.

If you're using plastic filler, make sure it's of the waterproof type and apply it according to directions using two or three coats rather than one heavy coat. When each coat is at a semi-soft stage, use a half-round surform file to rough it into shape. Using a surform file on the last coat of filler will also reduce clogging of the sandpaper during the finishing operation.

With the shape roughed-in, use some 80-grit open-coat sandpaper wrapped around a sponge sanding block. This will help you control the rate of the sandpaper's cutting action and allow the sandpaper to follow the contours of the body. If you have any questions about the shape, compare it with the remaining quarter panel.

Then check for any imperfections. If they're large, apply another coat of filler. If they're less than 1/16-in. deep, fill them with glazing putty, after the panel has been primed.

Prime, Sand & Seal—When the surface appears to be blemish-free, apply primer then three or four coats of primer-surfacer. Then use 150-grit paper wrapped around your sanding sponge and sand the surface while looking for imperfections.

Once you're satisfied with the way the panel looks, turn your attention to the other seams on the panel. Bare metal surfaces will be better off if you condition and prime them before applying sealer. And all should be filled with heavy drip-check sealer according to the manufacturer's directions. Don't forget the area where the quarter panel flange fits into the trunk weatherstrip channel. Otherwise you'll end up with a mobile aquarium after the first hard rain. Be sure to work as neatly as possible to avoid applying sealer on the exterior of any panel. Clean up any excess sealer after it dries with adhesive remover.

Now all that remains is to condition and prime any remaining bare metal with primer and apply primer-surfacer to the entire panel. Then block sand the panel with 400-grit paper while keeping an eye out for any remaining imperfections. Wetting the panel with water will help you find them. Any imperfections less than 1/16-in. deep can be filled with glazing putty.

Electrical Systems & Wiring

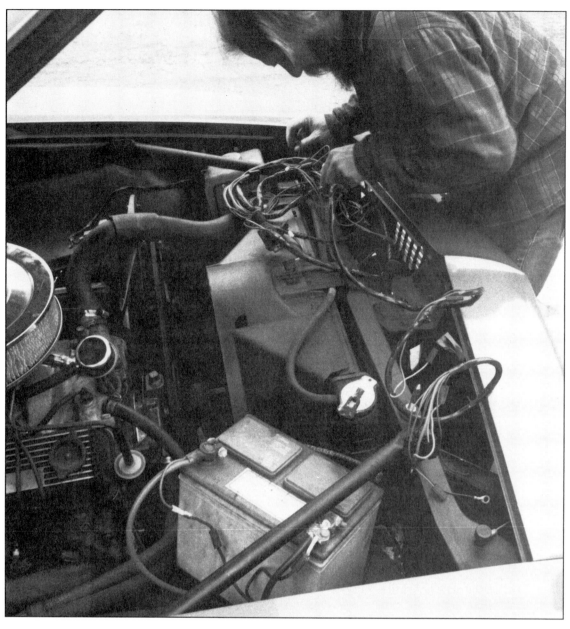

After decades of use and abuse, your Camaro's central nervous system may not work as well as it once did. Cracked insulation, corroded terminals, exposure to the elements and engine heat take their toll. On a restoration, best way to make wiring repairs is by installing a new harness section.

Yard upon yard of wire snakes through your Camaro's innards, connecting the far reaches of that sexy body with the starting, charging, ignition, lighting and accessory circuits. By and large, the GM original electrical system is dead reliable. But after decades of use, and maybe a little abuse too, problems can crop up.

Maybe it's a wire that slowly chafes through its insulation, shorts to ground and poof, your Camaro is a no-go proposition. Or perhaps it's a corroded terminal that won't let even one more electron pass through after years of exposure to salt spray. That trailer-hitch wiring job you let your Uncle Ernie do against your better judgment is now a tangle of electrical tape and frayed ends in your Camaro's

trunk. Packrats, squirrels and mice have been known to take up residence in disabled cars, building nests out of what once was your Camaro's wiring harness. Now how long is it since you moved your car?

An engine compartment is a particularly severe environment for automotive wire, exposed as it is to temperature extremes, vibration, oil and road debris. And if you're bringing a basket-case Camaro back to life—a victim of a serious collision, engine fire or previous race-car modifications—rewiring your car can become labor-intensive.

Sure, frayed or cracked insulation can be temporarily bolstered with electrical tape. And damaged sections of wire can be snipped out and new sections spliced in using soldered butt connectors and covered with shrink-wrap tubing. But really, the right way to repair damaged wiring is to replace it by harness sections.

Pay attention to routing. Keep wires away from hot exhaust manifolds and sharp edges that may chafe away at the insulation. Use cable-ties to keep the wires organized into bunches. Whenever a wire passes through a bulkhead, such as the firewall or radiator support, use a rubber grommet to protect the wire insulation. If you have access to one, follow the wiring diagrams in Chevrolet passenger-car service manuals or reproduction literature such as that available from **Jim Osborn Reproductions, Lawrenceville, GA 30245 (404/962-7556)**.

The wiring harness in your Camaro comes in two halves. The front half distributes electricity to and from all the components forward of the firewall. The rear half handles the same chores for all the components rearward of the firewall including the instrument panel, console, rear window defogger and taillights to name a few.

Both front and rear halves meet at the junction block. They can be found at the lower left area of the firewall.

REMOVAL

Removal of the wiring harness begins at the battery. Disconnect the negative lead from the battery first then the positive lead. This prevents the shorting of wires and possible damage of components if you happen to accidentally ground or short any lead.

If you only want to recondition part of a harness, that can usually be done with the harness still in the car. Just refer to the section below on harness reconditioning.

Label Wires—The next step is to label all the wires. Use hanging tags that are secured with string and that are reinforced in the area of the string hole. As you remove each wire from a particular component, write that component's name on the tag and immediately attach the tag to the wire. Tagging each wire as it's removed helps prevent improper connections later on. You can also make little flags out of masking tape and tag the wires this way if projected downtime is short. Just remember that tape dries out and many types of ink fade quickly on tape.

When all of the wires are tagged, open up any metal hangers that might be holding the wires to the car as you work toward the junction block at the lower left area of the firewall. This is where the front and rear halves of the wiring harness meet.

Fire devastated engine compartment of this late-model Camaro. To replace this harness, you'll need a wiring diagram because old wires that could be used for reference are melted away.

Main Junction—The front and rear harnesses meet at the lower left portion of the firewall. To separate the harnesses, you'll have to remove a long screw that's situated directly between each connector. When the screw is removed, wiggle the connector back and forth while pulling it toward the front of the car.

Sill Plate—To remove the wires from the interior and trunk, you'll have to remove the sill plate from the left side of the car. That's because the wires are normally routed under the carpet alongside the left rocker panel. When you've removed the sill plate, raise the carpet.

When you raise the carpet on first generation cars, you'll find two galvanized metal sheaths (a single, non-galvanized sheath on second-generation cars) that are screwed to the rocker panel. These sheaths protect the wiring harness as it leads to the rear. So remove the screws and sheaths to access the harness.

Rear Seat—Following the harness back, you'll find you may need to remove the rear seat and possibly the left rear-quarter trim panel in order to pull the part of the harness that's in the trunk to the front of the car. Just remember to disconnect the harness from every component in the trunk before pulling the harness forward.

Bag It—If you're just stripping the car at this point, put each half of the harness in a clear plastic bag. Then put both harnesses in a box and store the box in a dry area.

On the other hand, if you're concentrating on reconditioning the harness, proceed to the following section.

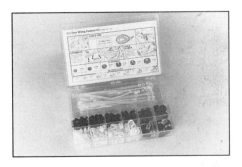

Pay attention to wire routing. Using ties, holddowns and grommets, keep wires away from hot manifolds, fans and pulleys and any sharp edges.

When removing electrical components or wiring, always label wires with tape or paper tags to ease assembly.

COMPLETE RECONDITIONING

Complete reconditioning of the harness entails removing it from the car. Then the elastic wrapping or tubular PVC wrapping is removed and all the wires are checked for damage. Damage to the wire can take the form of the insulation being cut, chafed or burned, or the wire being completely severed. The condition of all the terminals is also checked and any faulty pieces re-

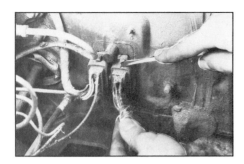

Main junction for front lighting (left) and engine harnesses (right) meet at firewall, below master cylinder. Depress tab with screwdriver to disengage harness junction disconnect.

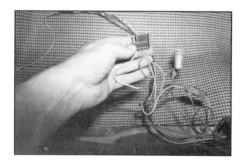

In trunk, look for frayed wires caused by shifting cargo or jury-rigged splice for trailer lights. This rat's nest of cut and spliced wires and electrical tape for trailer lights required a new rear harness.

placed. Finally, the wire wrapping is cleaned and installed if it's in good condition or replaced if it's not.

PARTIAL RECONDITIONING

Partial reconditioning of the harness can be done with the harness in the car. Typically, this level of reconditioning consists of repairing damaged terminals or connectors or repairing the wire wrapping.

For example, let's say that the two wire connectors and terminals leading to the alternator have been damaged. So you're going to need to replace them with new parts. Begin by removing the connector from the alternator. If you look carefully at the connector, you'll see a small tab on the side. This tab locks the connector to the component and is typical of many connectors. To remove the connector, depress this tab while pulling the connector away from the alternator. If the connector won't budge, gently rock it back and forth while pulling on it.

Make Diagram—Before going any further, make a small diagram showing which wire leads to which terminal in the alternator. This is very important. If you end up crossing any wires, chances are you'll damage the alternator's diode rectifiers or even end up burning the wires in the process. And this is true for all wiring that you

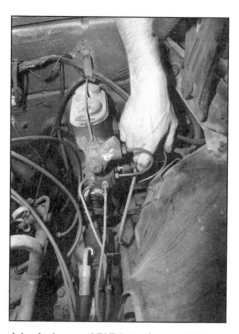

John Anderson of Z&Z Auto, Orange CA begins installing new engine harness in 1967 Sport Coupe by plugging in new bullet connector for brake warning light.

Routing is most important. Here, John loops harness into firewall-mounted hanger.

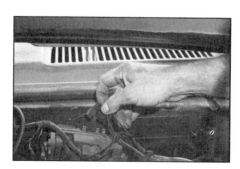

Three sheathed female spade connectors hook up to wiper motor. Label them so you know which goes where.

If you are replacing the wiper motor, undo its mounting hex screws at the firewall, pull it out slightly and loosen the nuts (arrows) at its linkage ball-joint connector.

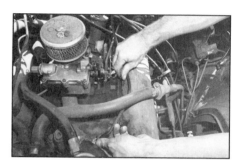

Z&Z repro harness includes heat-resistant sheathing for coolant temperature sending wire and original-type insulated female quick-disconnect terminal. On 1967 model, sheathing clips onto valve cover. Aftermarket air cleaner was replaced later.

Front lighting harness includes leads for battery, alternator, voltage regulator, horn relay, headlamps and front park/turn signals. Note correct ring terminal for alternator.

Horn relay uses spade terminals that slide under screw heads.

Ground wire for horn must have star washer under ring terminal for good metal-to-metal contact.

Don't forget blower motor lead that disappears under right front fender. You may have to loosen and pull the fender away from the body to gain access.

On 1967-69 non-RS models, spring-loaded dual-contact leads for front park/turn signal lamps hide under rubber boots.

Two used parking lamp assemblies for 1970-73 Sport Coupe. Use the one at the right with its original quick-disconnect still intact. If possible, avoid lamps such as unit at left with spliced-in ring connectors.

do. So just repair one connector at a time to avoid any problems later on.

Unlock Quick Disconnects—Next, remove the wire terminals from the connector by looking into the front of it. You should see a small metal tab protruding from each terminal. This tab mates with a step in the connector, effectively locking the terminal into the connector.

To unlock the tab, insert a very small flat-blade screwdriver into the back side of the connector and push the tab down toward the terminal. When the tab is just about flush with the terminal, you'll be able to gently pull the wire and terminal out of the connector. If the wire doesn't almost fall out of the connector, the tab is still locked. So try to unlock it again.

Strip Insulation—When both wires are out of the connector, cut the wire just behind the terminal. Then strip the insulation off the wire using the correct size stripper. If you're not sure what size the wire is that you need to strip, start at the largest opening in the stripper and work down to the smaller sizes. Working in this manner will help prevent cutting some or all of the wire as you strip it.

Crimp Connector—Now slide the stripped end of the wire into the new terminal and crimp it. Be sure that the terminal is crimped tightly to the wire. You can do this by gently trying to pull the terminal off. Once you're satisfied with

the crimp, you can solder the connection for security.

Clean Wrapping—The wrapping on the wiring harness can be cleaned with *Armor-All* cleaner and then treated with *ArmorAll* preservative. The only drawback to *ArmorAll* is that it tends to attract dust. However, some new types of preservatives of the market, such as *Concurs*, claim not to attract dust and dirt. So check around before you spend your money.

If you need to remove the wrapping because it's damaged, you need to replace a wire, or you just want to clean it thoroughly, locate the end of the wrapping. To do this, take a close look at the direction of the wrapping. Decide which end of the harness the wrapping leads to, then look near the end of the harness for the end of the wrapping.

When you find the end, loosen it with a blunt instrument and unwrap it. Contrary to popular belief, the wrapping has no adhesive on its back side. It's held in place simply by the fact that it stretches well and is wrapped tightly around the wires. As you come to the end of the wrapping, take note of the way it was started on the harness. You'll need to use the same method when you install the wrapping.

If you need a more detailed explanation of wiring and the electrical system you can find it in HPBooks' *Automotive Electrical Handbook* by Jim Horner.

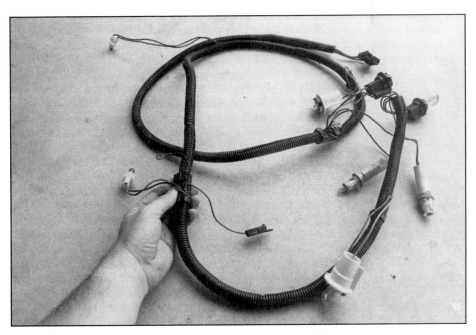

Corrugated, split plastic conduit used on 1970-81 Camaros is a neat way to protect bundles of wires from the elements.

Aside from the dash panel and mechanical components, doors are the most complex sub-assemblies with lots of moving parts that wear out. Beware, second-generation Camaro doors weigh more than 100 lbs apiece.

If the doors on your Camaro are rusted or dented, you can probably bring them back to as-new condition with some hard work and a relatively small amount of money. The results you get depend largely on the present condition of the door and the type of repair it needs. If the door has only minor rust and the window needs to be adjusted, you can get the door back into presentable shape with only a relatively small outlay of time and money. However, if the door you're working on requires a new skin or trim panel, gasket and window, and you want the door up to show quality, plan on spending a considerable amount of time and money getting it into shape.

If you're on a budget, consider used parts in good condition as an alternative to buying new parts. And if your door requires a fair number of replacement parts, buying a complete door in good condition and using the parts from it can save you a considerable amount of money.

The doors for both first- and second-generation Camaros are similar in design and consequently are serviced in a similar manner. Any major differences will be pointed out as we go along. For example's sake, it's going to be understood that you have a door in repairable condition and want to bring it up to show standards. Now let's get started!

SIDE WINDOWS

When the 1967 Camaro was introduced, it featured front vent windows. Then 1968 came along

and so did Camaro with its Astro Ventilation and ventless door glass. Ventless door glass has been used on all Camaros since then. This glass seals to the body without the benefit of a frame. Because of this, the adjustment of the door glass is critical to avoid wind noise and water leaks.

INTERCHANGEABILITY

If your Camaro needs door glass, be aware that 1968 and 1969 glass is interchangeable. The door glass on second-generation Camaros is interchangeable from 1970 through 1981.

However, there is a hitch with first-generation convertibles. The door glass for convertibles is shorter than that on coupes. So, if you need a replacement door glass for your convertible, you'll have to get one from another convertible. The same is true of the windshield and rear quarter windows. Keep this in mind when looking for replacement glass.

POLISHING SCRATCHES

Many Camaros have scratches on the door glass due to the bristles on the window stabilizers being worn away. Sometimes these scratches can be polished out if they're not too deep. If they can't be polished out without distorting the glass, you'll have to get another used or new door glass.

To polish the door glass, you'll need a drill motor, a small (3-in. OD) wool felt buffing pad and a polishing compound such as cerium oxide. This compound can be found in many auto body supply stores. It's also available as part of a glass polishing kit from the **Eastwood Company, Malvern, PA**.

Before you begin, draw a circle around the scratches on the opposite side of the glass with a china marker or a crayon. This will help you remember where the scratches are. Also, keep in mind that door glass doesn't tolerate heat well. Avoiding buffing the glass for more than 30 sec. at a time and only use gentle pressure. If you don't follow these rules, the glass may overheat and distort. And don't think about cooling the glass with water because it may shatter and injure you in the process. Use a drill motor that spins at a maximum speed of 1600 rpm—no faster.

Mix the polishing compound according to the manufacturer's instructions. Then dip the pad into the mixture a few times so that it's well saturated. Position the pad flat against the glass and begin polishing. Start at the center of the scratch and work the pad in a feathering motion to the outside of the circle. Make sure you polish an area slightly larger than the circled area to avoid a bulls-eye in the glass. Let the polishing compound do the work. You don't need to put a lot of pressure on the glass. Always keep the pad wet with compound and keep it moving to avoid excessive heat. Touch the glass frequently while you're polishing it. If it's hot to the touch, it's getting too hot. Stop and let it cool before attempting any more polishing. And remember, don't cool the glass with water.

GLASS REMOVAL

1967 Models—Door glass on 1967 models is unique. The front edge of the glass runs up and down in a channel that's part of the vent-window frame. The bottom edge of the glass is positioned in a rubber sleeve that fits into a channel. The window is retained to this channel by the rubber sleeve and by a nut at the lower rear edge of the glass.

To remove it, lower the window completely. Loosen the nut at the rear of the window and remove it and the stud that fits through the window, along with the washers. Now spray some silicone lubricant over the entire rubber sleeve that the lower edge of the window fits into. This is important because you're going to have to pry the window out of the sleeve. Insert a screwdriver through the access area at the rear of the window and very carefully pry the window upward. When you've raised the window out of the sleeve a slight amount, move the screwdriver to another

area and pry the window up again. Keep doing this until the window is free. When the window is free, remove it through the window opening at the top of the door.

1968-Up Models—The windows for 1968 and-later cars are similar in design and consequently removed in a similar manner. Begin with the window in the up position. Remove the up stops from the rear guide and front lower sash channel. Then loosen the stabilizers at the top of the door frame. Next loosen the upper retaining bolts for the front and rear window guides. Now lower the window completely and remove the lower sash-channel cam

Worn fabric material, or fuzzies, on window stabilizers (arrows) and guide plates can scratch windows when rolled up and down.

Eastwood glass polishing kit removes most scratches too shallow to catch a fingernail. Kit includes cerium-oxide abrasive, polishing wheel, crayon to mark scratch, sponge to wet area and instructions.

Shown here on a scratched windshield, area is masked off to protect paint from abrasive. Powder is mixed with water to form a paste and applied with the wheel at a drill speed of 1600 rpm or less. Use gentle, even pressure.

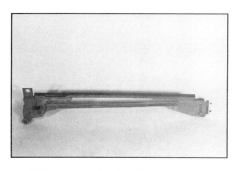

On 1967 models only, this channel and rubber sleeve hold window in position.

If door glass needs to be replaced, begin glass removal by removing front and rear up-stops.

With window lowered, loosen stabilizers and upper retaining bolts for front and rear window guides. Then support weight of glass and remove lower sash-channel cam attaching nuts. Push window away from cam, tilt inward to free from rollers and lift out window. Don't drop it.

Transfer rollers and other glass hardware to new window. This two-prong K-D special tool fits into holes in window cam nut. Don't use brute force.

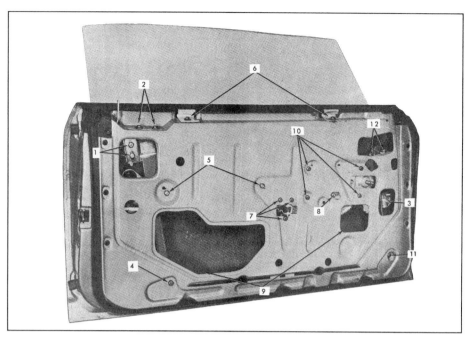

After installing special flat nuts, lube rollers with Lubriplate or equivalent grease for optimal window roller operation.

1. Window Rear Up-Travel Stop
2. Rear Guide Upper Attaching Bolts
3. Window Front Up-Travel Stop
4. Rear Guide Lower Attaching Bolt
5. Inner Panel Cam Attaching Bolts
6. Window Front and Rear Stabilizer Strips
7. Door Lock Remote Control Attaching Bolts
8. Sector Gear Stop Bolt
9. Window Lower Sash Channel Cam Stud Nut Access Holes
10. Window Regulator Attaching Bolts
11. Front Guide Lower Attaching Bolt
12. Front Guide Upper Attaching Bolts

Door and window adjustment details—1968 models; 1967 and 1969 similar. Courtesy Fisher Body.

1. Window Front Up-Travel Stop Bolt
2. Window Rear Up-Travel Stop Bolt
3. Window Stablizer Strip & Adjustable Trim Hanger Plate Bolt
4. Trim Support Hanger Screws
5. Front Guide Upper Bolts
6. Rear Guide Upper Bracket Bolts
7. Rear Guide Upper Bolts
8. Front Guide Lower Bolt
9. Rear Guide Lower Bolt
10. Window Regulator Bolts
11. Inner Panel Cam Bolts
12. Window Bumper Support Bolt
13. Window Lower Sash Channel Cam Nuts Access Holes
14. Door Lock Screws

Door and window adjustment details—1970-81 models. Courtesy Fisher Body.

attaching nuts. Proceed with caution from this point because the window is loose in the door. Push the window away from the cam and lift it straight up through the window opening at the top of the door. Tilt the top of the window toward the interior to free the window rollers from the guides. Then slide the window forward and lift it free of the door.

Sash Channels—At this point you can replace the window if necessary. But you may have to transfer the front and rear lower sash channels because some replacement windows don't include them. To remove the sash, loosen the attaching nuts from each sash with a light hammer and a punch. Insert the punch into one of the holes in the nut and gently tap it free. The nut is only retained with 6 ft-lb. of torque so you don t have to prove you're Charles Atlas and break the window in the process.

Take the sash channels and install them on the new glass using new washers under the nuts. These special washers are available at many automotive glass replacement shops. Tighten the window nuts with your fingers then tighten them an 1/8th of a turn more using a punch. Don't tighten them any further or the glass will break. Now smear some Lubriplate on the rollers and the glass is ready to install. Installation is the reverse order of removal.

ADJUSTMENTS

All the adjustments in the world won't do any good if you don't know what a "perfectly adjusted" window looks like. The top of the window should fit in the middle of the roof-rail weatherstripping and compress it slightly. If the window only touches but doesn't compress the weatherstripping, you'll have a water or air leak. A trick you can use to check the fit of the window to the glass is to thread a dollar bill between the glass and the weatherstrip. As you pull the bill out, you should feel about one pound of tension. And the tension must be equal at all points around the glass. If it isn't, the glass requires adjustment. Also, the window should not bind on any part of the A-pillar or roof when the door is opened or closed.

Before you begin, note that there's five steps for adjusting the glass that must be followed in sequence. They are tilt, fore and aft, up stop, tip-in and down stop. If you don't follow these steps in order, making one adjustment will just throw another off. However, there is one exception to this rule. If the window is square in the opening, you can skip over the first two steps.

Adjust Glass Tilt—Begin by removing the interior trim panel and lowering the glass about 1-1/2 in. The inner panel cam controls the tilt of the window from front to back. Loosen the bolts retaining the cam and shift the window so that the top edge is even with the door opening. Then tighten the bolts.

Adjust Glass Fore & Aft—The fore-and-aft positioning of the glass is controlled by two bolts at the top of the rear window guide. This guide is slotted at the top or has a plate attached to it to allow for adjustment. To perform this adjustment, loosen the bolts slightly at the top of the guide and slide the channel so the glass is equally spaced in the door frame at both its front and rear edges. Tighten the bolts after this adjustment.

Adjust Glass Up Stop—Up stops are used in pairs and should be adjusted when the window compresses the roof-rail weatherstripping too much or not enough as outlined previously. They should also be adjusted if the window is cocked in the door. If adjusting the up stops doesn't cure a cocked window, you may need to adjust the inner panel cam. The up stops (one on each glass channel) are visible through access holes at the front and rear of the door (see nearby illustration). Just loosen their retaining bolts and move each one until the glass presses against the window seal with a slight amount of force. Finally, tighten the bolts.

If your window is still cocked after adjusting the up stops, you may need to adjust the *inner panel cam*. The inner panel cam is a short channel attached to the door that provides a path for the lower regulator roller. The door frame is slotted in the area at the rear of the cam, which provides an adjustment.

Adjust Glass Tip-In—The front and rear window guides are slotted at the top to provide for tip-in adjustment. Moving the guides in moves the window in. Move the guides out and the window moves out. It

sounds simple, and it is, but make sure the glass seals against the weatherstripping evenly in all places.

If you've noticed the glass moving outward when you drive the car, it's probably due to a faulty or missing blow-out clip. This clip is situated near the top of the A-pillar and keeps the glass from moving outward when your car is driven at speed. This clip should be bent so that it prevents outward movement of the glass when you're driving.

Adjust Glass Down Stop—Finish the adjustment procedure by checking how far the window rolls down into the door. If it rolls down too far into the door, evident by extra effort needed to begin rolling up the window, you'll have to adjust the down stop at the bottom of the door until the glass is positioned properly. To do this on 1967 models, shim the area between the rubber down stop at the bottom of the door with metal or plastic shim stock. On second-generation cars, the down stop is adjusted by loosening a bolt at the bottom center of the door and sliding the down stop. There is no down stop on 1968 & '69 models.

VENT WINDOWS

Vent windows were only used on 1967 Camaros. When you were driving along and opened them to just the right angle, they gave you a blast of refreshing air. Unfortunately, sometimes they would begin to open or close by themselves as a result of a lack of spring tension. If your vent windows do this, you need to adjust their spring tension. The tensioner nut and spring are located on the lower vent pivot. Take the door trim panel off and tighten or loosen the nut as required. Unfortunately, you'll have to reach the nut via the access hole below the window regulator, which can be tough. The only alternative is to remove the vent-window frame.

Removal—To replace the vent window, its weatherstripping, or the door glass vertical weatherstripping, you'll need to lower the door glass completely and remove the vent window frame. The frame is attached to the front of the door with a bolt and an adjustable stud. The bolt can be found under the door weatherstripping at the top of the door. The rear part of the vent window frame is secured with a long screw in the window sill and by an adjustable stud near the bottom of the door. Remove all bolts, nuts, screws and the bottom rear stud from the vent-window frame.

Then take the frame out. When you're taking the frame out, have patience. The frame may bind on the glass. To minimize

this trouble, lift the frame 8-10 inches out of the door then rotate it 90 degrees so the front of it is on the outside of the car. Then rock the frame back and forth as you lift it out.

To remove the vent window, remove the tension nut and spring from the bottom window pivot. Next, remove the screw from the inside of the rear frame at the very top. Then remove the two screws from the front portion of the frame, under the bottom of the vent window. The frame is now disassembled; the only thing holding it together is the vent-window weatherstripping. Spray the lower edge of the weatherstripping with silicone and gently pull it out from the frame. However, you may

Down stop prevents window from dropping below window stabilizers and becoming stuck. This type is adjustable.

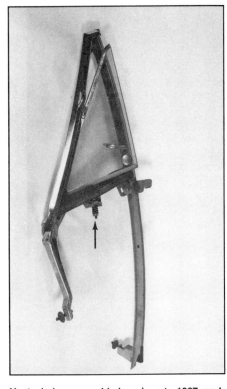

Vent window assembly is unique to 1967 models. Tensioner spring to adjust vent window opening effort is accessible from hole under door window regulator.

107

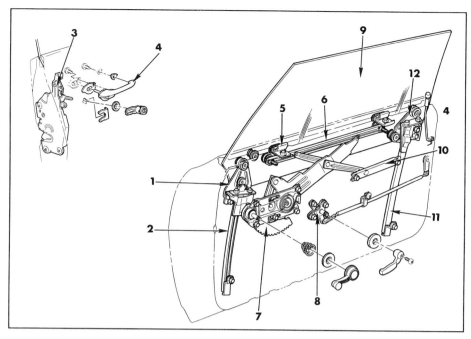

1. Front Lower Sash Channel & Window Roller Cam Assembly
2. Front Guide
3. Door Lock
4. Door Outside Handle
5. Stabilizer Strip
6. Lower Sash Channel Cam
7. Window Regulator
8. Door Lock Remote Control
9. Front Door Window Assembly
10. Inner Panel Cam
11. Rear Guide
12. Rear Lower Sash Channel & Window Roller Assembly

Window regulator and door latch details—1968 models; 1967 and 1969 similar. Courtesy Fisher Body.

When removing vent window frame on 1967 model, don't forget this bolt hidden under weatherstripping.

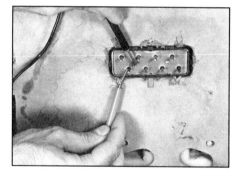

Test power window relay by applying 12 volts and connecting test light between two switch terminals. On 1973-81 models, switch is in console, not door panel.

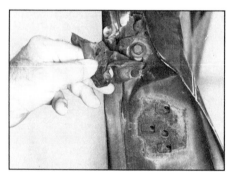

Aside from cleaning and inspection of regulator teeth, only item that may require regulator replacement is broken assist spring. This spring is not sold separately and can uncoil with great force and cause injury if you try to remove it.

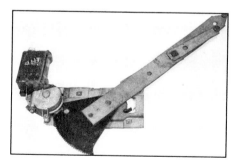

Because many Camaro power window motors are attached to their regulator with hidden bolts, the best way to replace a defective motor is with a good used motor and regulator assembly from a donor car.

have to persuade it with a screwdriver. When the weatherstripping is out, you can remove the window.

Installation—To install the vent window in the frame, spray the bottom edge of its weatherstripping with silicone and feed in into the frame using a large, flat-blade screwdriver for assistance. Then install the tension spring and nut onto the bottom pivot. Finish the job by installing the frame screws you removed earlier. And before you install the assembly in the car, adjust the tensioner.

Now slide the frame into the door, reversing the procedure you used to remove it. Make sure the front stud is installed and protrudes through the door frame. Also make sure the window and guide are positioned in the rear of the vent frame. Proceed to install the rear stud into the vent frame and adjust so that the window doesn't bind when it's rolled up and down.

WINDOW REGULATOR

The window regulator controls the up and down movement of the window. Basically speaking, the regulator is made up of a pinion gear and a rack. The pinion gear is mounted on the end of the window crank on manually operated windows. A rack of teeth, known as a sector, are stamped on one edge of a flat steel plate. This steel plate has arms that are attached to the window via nylon rollers and steel channels. When you roll the window up or down, the pinion meshes with the sector gear. In turn, the arms of the sector gear move the window up and down in the channels. The regulator uses one other component: an assist spring. This spring is attached to the regulator and tightens up when the window is rolled down. When you roll the window up, the spring uncoils and helps lift the window.

Window regulators are pretty durable pieces, but because they live in the door, they're exposed to moisture and the elements. The factory had this in mind when they galvanized them to protect them from rust.

Usually, the factory lubrication of the regulator lasts for the life of the car. But if the window becomes hard to roll up, it might need to be re-lubricated with *Lubriplate* white grease. Lubriplate is available at all GM dealers and some better-stocked hardware stores.

Window regulators seem to have an indefinite life in normal service. The only time they're usually removed is if the assist spring breaks or they need to be cleaned as part of a restoration.

Removal—To remove the regulator (manual or electric), the door trim panel and water deflector must first be removed. You'll also have to remove the window regulator inner-panel cam. To do this, crank the window down three quarters of the way. You will find the channel-cam retaining stud nut(s) by looking through the access holes in the door frame. They're located on the channel at the bottom of the window. When you've removed the nut(s), raise the window with your hands until it's in the full-up position and then prop it there with rubber wedge door stops.

If you're working on a 1967 Camaro, you'll have to spray the bottom of the glass with silicone and then lever it out of the channel. Be careful as the glass may chip. Now remove the window regulator inner-panel cam. This is the small channel bolted to the door that one of the window-regulator rollers rides in. With power windows, disconnect the wiring harness from the window regulator motor at this time. Be sure to disengage the retaining tab on the connector. Then remove the four regulator mounting bolts and remove the regulator through the large access hole. If the regulator is attached to the door frame with rivets, the rivets will have to be drilled out. When installing the regulator, secure it with 1/4-20 bolts and nuts.

Do not attempt to remove the window regulator motor from the regulator without referring to the appropriate *Fisher Body Service Manual*. The regulator lift arms are under high spring pressure and may cause serious injury if the sector gear is not locked in position.

Cleaning & Inspection—Now's the time to clean and inspect the regulator and regulator motor. Remove all dirt and grease with solvent. A toothbrush can help get into the tight places. However, don't let any solvent get into the regulator motor or gear drive because it will ruin them. And you'll have a hard time finding a motor to replace your original. As a matter of fact, you may have to buy a motor and regulator assembly as few motors are available. You're also in the same boat if you need to rebuild your regulator motor because rebuilding parts aren't available. So be careful.

After cleaning, check the regulator teeth to make sure they aren't chipped or mushroomed out from excessive wear. If they are, dress them with a file and clean the regulator again to remove any filings. Now check the nylon rollers. They should move freely on their spindles. With power windows, clean the contacts of the window regulator motor with 400-grit sand-

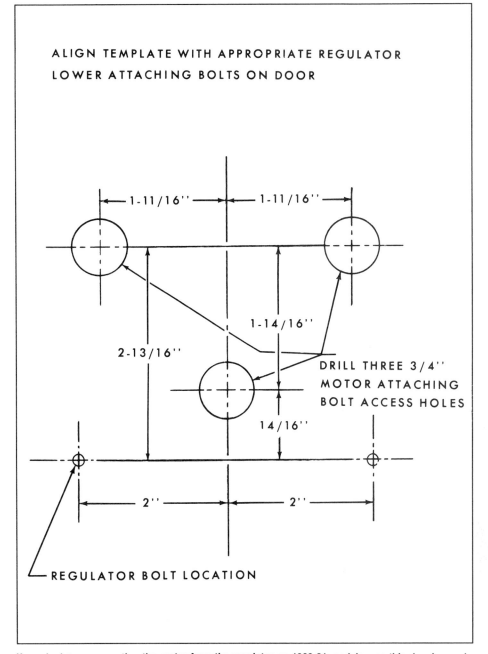

ALIGN TEMPLATE WITH APPROPRIATE REGULATOR LOWER ATTACHING BOLTS ON DOOR

If you insist on separating the motor from the regulator on 1969-81 models, use this drawing and measurements to make a template and a hole saw to find the hidden bolts. Courtesy Fisher Body.

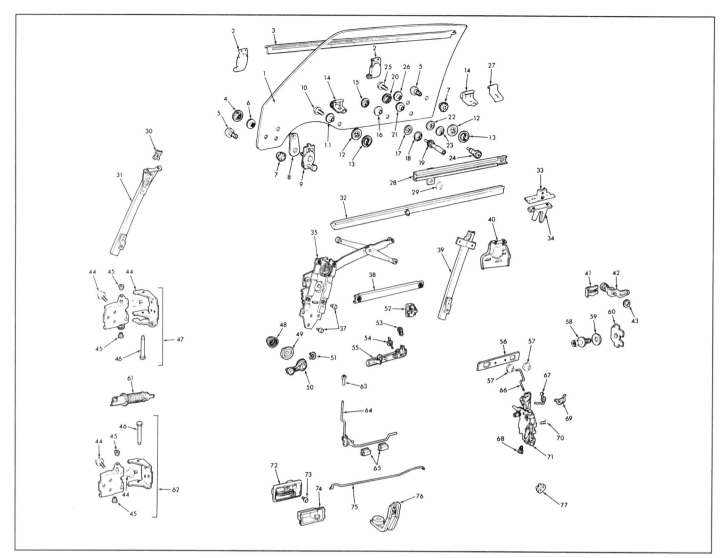

1. Glass, Door Window
2. Plate ASM, Window Guide
3. Sealing Strip ASM, Window
4. Nut, Window Guide Front Roller to Glass
5. Fastener, Window Glass Bearing
6. Bushing, Window Front Roller to Glass
7. Cap, Window Glass Bearing Fastener
8. Washer, Window Guide Front Roller to Glass
9. Roller ASM, Window Guide Front
10. Bolt, Window Cam to Glass Front
11. Bushing, Window Cam to Glass Front
12. Washer, Window Cam to Glass
13. Nut, Window Cam to Glass
14. Plate ASM, Trim Support Hanger
15. Nut, Window Rear Up Stop to Glass
16. Bushing, Window Rear Up Stop to Glass
17. Washer, Window Rear Up Stop to Glass
18. Washer, Window Rear Up Stop to Glass
19. Stop ASM, Window Rear Up/on Glass
20. Nut, Window Guide Rear Roller to Glass
21. Bushing, Window Rear Roller to Glass
22. Washer, Window Guide Rear Roller to Glass
23. Washer, Window Guide Rear Roller to Glass

24. Roller ASM, Window Guide Rear
25. Bolt, Window Cam to Glass Rear
26. Bushing, Window Cam to Glass Rear
27. Plate, Trim Support Hanger
28. Cam ASM, Window Glass Regulator
29. Nut, Window Glass Regulator Cam to Glass
30. Stop, Window Upper on Cam
31. Cam ASM, Window Guide Front
32. Strip, Sealing Window Inner
33. Support, Window Rear Up Stop Guide
34. Guide, Window Rear Up Stop
35. Regulator ASM, Window Elect Less Motor
37. Screw, Window Regulator to Panel
38. Cam ASM, Inner Panel
39. Cam ASM, Window Guide Rear
40. Support, Window Guide Rear Cam Upper
41. Retainer, Lock Cylinder
42. Cylinder ASM, Lock
43. Gasket, Lock Cylinder to Door
44. Screw, Upper & Lower Hinge
45. Bushing, Upper & Lower Hinge
46. Pin, Upper & Lower Hinge
47. Hinge ASM, Upper
48. Spring, Window Regulator Handle
49. Spacer, Window Regulator Handle
50. Handle ASM, Window Regulator

51. Retainer, Handle
52. Clip, Lock Remote Control Rod to Panel
53. Clip, Outside Handle Spring
54. Spring, Outside Handle
55. Handle ASM, Outside
56. Escutcheon, Front Door Outside Handle
57. Nut, Outside Handle to Panel
58. Striker ASM, Lock
59. Washer, Lock Striker
60. Plate, Front Door Lock Striker Anchor
61. Conduit, Electric
62. Hinge ASM, Lower
63. Knob, Inside Locking Rod
64. Rod ASM, Inside Locking to Lock
65. Clip, Inside Locking Rod to Inner Panel
66. Rod, Outside Handle to Lock
67. Spring, Lock Push Button Return
68. Clip, Lock Spring
69. Spring, Lock Over Center
70. Screw, Lock to Panel
71. Lock ASM
72. Handle ASM, Lock Remote Control
73. Screw, Lock Remote Control Handle to Inner Panel
74. Escutcheon, Lock Remote Control Handle
75. Rod, Lock Remote Control to Lock
76. Support ASM, Window Bumper
77. Bumper

Details of door hardware—1970-81 models. Courtesy Chevrolet.

110

paper. These contacts tend to corrode and prevent operation of the window. After cleaning the contacts, spray them with WD-40 to help keep them from corroding. And now is a good time to clean and lubricate the terminals of the motor harness connector. Also check the regulator spring for any signs of fatigue cracking. And be sure to clean the window-channel cam. When everything checks out OK, lubricate all moving parts of the regulator with Lubriplate white grease.

Installation—To install the regulator, reverse the removal procedure. If equipped with power windows, check to make sure electrical connections are secure.

POWER WINDOWS

Except for a brief hiatus in 1970-72 (with a handful of power-window-optioned cars built in 1973), power windows have been offered on Camaro since its introduction in 1967. The system has worked so well that it hasn't changed much at all over the years. This system consists of a wiring harness, 12-volt window regulator motors and switches to control the current feed to the regulator motors. The window regulator motor acts on the window regulator to move the window up and down. The system also uses a relay to handle the relatively high amperage loads and is fused at the fuse block.

Operation—The regulator motor has two terminals. Although both of these terminals can feed current to the motor, only one is used at a time. If current is fed to one of the terminals, the window will open. If current is fed to the other terminal, the window will close. The motor is grounded to the regulator.

The control switch has three terminals. The two terminals that are directly across from each other are used to control the up and down position of the window. The third terminal, which is positioned between the other two and off to the side, provides the voltage feed to the motors. By pushing the switch up or down, you're completing the circuit between the feed terminal and one of the terminals that controls the motor. This is why even the master control switch has only one feed terminal. If you keep this information in mind, testing the operation of the power windows is easy.

Troubleshooting—If all the windows won't operate, check that the fuse in the fuse block is in good condition. If it is, check for a 12-volt feed at the fuse terminals when the ignition switch is in the

ON position. If everything checks out OK, test the relay. Position a test light on the relay terminal where the red/white wire is attached and ground the test light lead. If the test light lights, the relay is in good condition and the trouble is further along the circuit.

Test the operation of the switch by placing a 12-gage jumper wire between the feed terminal and one of the motor terminals in the switch block. If the window operates with the jumper wire in place, but not when the switch is used, the switch is defective. If the window doesn't operate, the feed wire between the relay and the switch block has an open circuit or short circuit.

If everything works OK to this point, check the wiring between the switch and motor. Start by disconnecting the harness from the suspect motor. Be sure to disengage the retaining clip before trying to disconnect the harness. Place a 12-gage jumper wire between the feed terminal of the switch block and one of the motor terminals on the motor. If the window operates now, there's a short or open circuit in the wiring harness between the switch and motor. If the window doesn't operate, the regulator motor isn't properly grounded to the regulator or the motor has failed.

Motor Replacement—If the motor has failed, it must be replaced as a unit, as service parts for the motor are not available. This isn't as easy as it sounds. Some motors are attached to the regulator with bolts that are readily accessible. *But be extra careful.* You must drill a hole through the sector gear and backing plate and insert a bolt or pin to lock the regulator arm in position. If you don't lock the arm and you try to remove the motor, the arm will release with extreme force, possibly crushing a finger in the process. So lock the arm in position! Refer to the appropriate *Fisher Body Service Manual* for details on this proceedure.

Other regulator motors are attached to the regulator with hidden bolts. Locating these bolts on 1969-81 models requires a special template which is shown nearby. With the motor harness disconnected, align the template with the regulator attaching bolts on the door and tape the template in place. Center-punch for the three 3/4-in. holes you'll drill through the door inner panel. Then drill the holes in the panel using a 3/4-in. hole saw and remove the motor attaching bolts.

For regulator removal and replacement, refer to the *Window Regulator* section, page 108.

WINDOW CHANNELS (GUIDES)

Window channels serve as a guide for the window rollers. Each channel is made from galvanized steel and uses two welded mounting brackets that secure it to the door. Two types of window channels are used in the door. The first type is positioned vertically in the door; one guide at the front and one at the rear. These channels (or window guides as they're sometimes called) guide the window when it's rolled up and down. However, you should note that the front window guide on 1967 Camaros also serves as the rear frame for the vent window. Rubber stops are used at the bottom of some channels to cushion the impact of the window rollers when the window is rolled down completely.

Each vertical channel is attached to the door at three points. Two bolts are used at the top and one at the bottom. The lower attachment point sometimes uses an adjustable stud instead of a bolt. This allows for additional adjustment of the window.

The second type of channel runs in a horizontal plane. Three channels of this type are used in each door. One channel is bolted to the lower portion of the window and is attached to the window regulator by nylon rollers. A second channel is bolted to the door and is also attached to the

These window guide plates fit 1969 models. Worn guide at right has scratched the window.

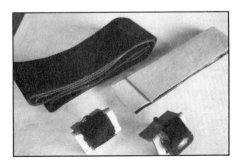

Worn stabilizer at right scratched window and allowed it to rattle. Replace with new stabilizer or recover with fuzzy material epoxied to metal backing.

window regulator via a nylon roller. This channel serves as a cam for the regulator, giving the regulator a stationary plane for lifting the window.

The third channel is used on some windows and is manufactured as part of the front lower sash. This channel allows the front window roller to move forward and backward on the window. This keeps the front of the window from binding as the window is rolled up and down.

Reconditioning of these channels entails cleaning them in solvent, drying them then lubricating them with Lubriplate.

GUIDE PLATES

The window guide plates are found just below the top edge of the door and are attached to the door with two Phillips head screws. The guide plates work together

After disconnecting door latch linkage, remove Phillips-head screws retaining latch to door frame with an impact driver.

Clean latch assembly in solvent, blow-dry and lubricate all pivots with Lubriplate or equivalent. If any parts are worn, replace entire latch assembly.

with the window stabilizers, to prevent the window from moving excessively when it's being rolled up and down. They're covered with cloth to prevent them from scratching the window. But therein lies the, er, "rub." When the cloth on the guide plate becomes worn, the metal that serves as the base for the guide plate becomes exposed. This exposed metal scratches the window every time the window is rolled up or down.

Re-Cover Guide Plates—If the guide plates are worn and you feel industrious, you can re-cover them. Cordura nylon is a good choice for a recovering material since it's very durable and is readily available at many fabric shops. Use a solvent designed to dissolve adhesives to help remove the old covering. Cut the replacement material so it fits the guide plate properly and glue it to the guide plate using a waterproof glue, such as epoxy.

If you lack the ambition to re-cover your worn guide plates, you'll be out of luck trying to order them at the dealer. Some *Parts and Illustration Catalogs* don't have them listed and the catalogs that do list them state they're not required and don't have a part number for them. Your only alternative is finding a used set in good condition from another car—an unlikely proposition.

WINDOW STABILIZERS

The stabilizers are mounted at the top of the door frame by bolts and provide a slight outward tension on the window. When the bristles on the the stabilizers are worn, they allow the window to rattle. And when they are worn through to the metal, they'll scratch the window. One way to remedy this situation is to replace any worn stabilizers with good used or new parts. Another way is to glue a piece of indoor/outdoor carpeting to the metal stabilizer bracket. You can do this with epoxy.

Adjustment—Adjust the stabilizers so they prevent the window from rattling but don't substantially increase the effort required to lower and raise the window.

DOOR FRAME & SKIN
DOOR LATCH

The door latch used on Camaros features a safety interlock. This interlock helps to prevent the door from opening during an accident. Because of the interlock, the latch assembly should be replaced if it's damaged.

The latch is operated by four components: the door handle, lock cylinder, inside locking button and the inside release

handle. And if the car is equipped with power door locks, the electric lock actuator is also connected to the latch. All of these components act on the latch assembly in various ways.

The door-handle push button on first-generation cars acts directly against a release lever on the latch. The lift bar on second-generation cars uses a short piece of linkage to release the latch.

All lock cylinders have a stamped pawl that mates with a rounded lever on the latch to lock and unlock the door. The inside locking button on 1967 models use a rod that acts directly on the latch. All later models use a rod that's supported in the door frame with nylon bushings. The end of the rod fits into a lever on the latch that has an elongated slot.

The inside release handle is connected to the latch by a long metal rod. This rod is connected to a release lever near the bottom of the latch and is secured with a special retaining clip. The actuator for power door locks is connected to a lever on the bottom of the latch by a short piece of linkage. This linkage is retained by another type of clip.

Reconditioning of the latch assembly entails removing all the various retaining clips and linkages, removing the latch from the door frame, and cleaning the latch.

Removal & Installation—Remove the linkage retaining clip from the inside release handle actuating rod with two flat-bladed screwdrivers. The clip has a slot that the actuating rod fits into and a retaining tab. This tab protrudes through a hole in the remote control lever on the latch. To remove this clip, slide the screwdriver under the tab and slide the clip off the end of the lever with another screwdriver. Then remove the actuating rod.

The power door-lock actuator uses a retaining clip with a hole on one end and two "arms" that wrap around the actuating rod. To remove this clip, push the arms off the rod and slide the rod out of the latch assembly.

The remaining linkages will disengage from the latch when the latch is loosened. The latch is secured to the door frame by three countersunk Phillips-head screws. These screws are, in one word, tight. You'll probably have to use an impact driver with a #3 Phillips bit to remove them. Be careful not to round out the screw heads. This is easy to do if you're not careful.

Clean the latch assembly with cleaning solvent or carburetor cleaner. If the latch is really dirty, soak it in solvent until the grime is loose. Then use a toothbrush to remove the grime. Clean the latch a second time with fresh solvent until all traces of old lubricant and dirt have been eliminated. Blow the latch dry with compressed air to remove any remaining solvent.

At this point, lubricate all pivot points and the fork bolt with Lubriplate or motorcycle chain lube. Before you install the latch to the door, coat the side of it that lays against the door frame with an anti-corrosion compound. This prevents moisture from getting between the latch and door frame and starting the rusting process. GM dealers sell a very good anti-corrosion compound in a spray can. However, the job will be neater if you apply the compound to the latch with a brush. Spray some compound into a clean container and apply it to the latch with a small brush. The latch is now ready for installation.

Before installing the latch, position it through one of the access holes in the door and attach the inside locking rod and clip. Now loosely attach the latch to the door with the latch retaining screws. Then attach the remaining linkage and tighten the latch retaining screws.

DOOR HANDLE

Two different types of door handles can be found on Camaro doors: a push-button type on all first-generation cars and a lift-bar type on second-generation Camaros.

Door handles rarely give any trouble although the short-style lift handle used on 1978-81 Camaros is prone to cracking. They usually just become pitted and are an eyesore on an otherwise finely detailed car. Both types of handles are made of die-cast metal (also known as pot metal) which is prone to pitting. If you're looking in the boneyard for a replacement handle, you'll likely find a handle as pitted as the one you're replacing, especially if you have a first-generation car. New handles can be found at swap meets or can still be ordered from aftermarket suppliers or your local Chevrolet or Pontiac dealer. If the dealer has a tough time coming up with a first-generation handle (or two), try ordering one for a medium-duty Chevrolet or GMC truck. They're the same style and fit nicely. Remember to order the handle gaskets as they don't come with the handle.

Replating—The depth of the pits dictate whether the handle can be rechromed or must be replaced. Usually, if the pits are small and shallow, the handle can be replated at a reasonable cost. However, if the pits are deep, it's going to cost more to replate it, if it can be replated at all. The

To remove button, spring and O-ring from 1967-69 door handle, press down retainer, turn it 90 degrees and release. Don't forget to order new gaskets for handles.

deeper the pits are, the more time the chroming shop must spend filling them with copper and buffing them out. Copper is the foundation for the chrome, and in combination with nickel, provides a smooth surface that gives chrome its deep luster. But talk to a replater to be sure.

If you want to get fanatical with this, you can even go one step further. Because chrome plating on many new parts is marginal at best, you could take a pair of new door handles to the plater and have him replate them. This should give you terrific-looking parts. But talk to people who have had excellent replating work on their cars to find out who replated their parts. And then talk to the replater. You'll be much happier with the finished product if you know what to expect.

Removal & Installation—(First-Generation) The push-button style handle is attached to the door by two screws. Rubber gaskets are fitted between the handle and the door. To remove the handle, simply remove the two screws from inside the door. Access is via an access window in the door inner panel after removing the door trim panel.

If you need to remove the button from the original door handle, press the retainer down slightly, rotate it a quarter turn in either direction and remove it. The spring, push-button and push-button O-ring will come off with it. If you have a new design original-type handle and you want to remove the push-button, press the retainer down slightly and rotate it a quarter of a turn counterclockwise.

Removal & Installation—(Second-Generation) The lift bar used on all second-generation Camaros is also made of die-cast metal and is subject to the same pitting problems as the earlier handle. So the process for renewing it is the same. However, the chances of finding a good

Lift bar handles on second-generation cars are prone to cracking. If perusing junk yards for replacements, measure length of yours against donor cars—two different lengths were used.

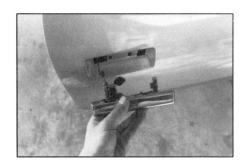

When installing a lift bar, it's easier to assemble latch pushrod on handle before installing handle retaining bolts.

used handle in junkyards are increased because 1970-77 Camaros used the same door handles as 1970-77 Firebirds, 1971-75 Vegas and Impalas and 1975 Monzas.

Beginning in 1978, a shorter, lighter 5-5/8-in. long lift handle replaced the 1970-77 6-1/4-in. long handle. It's prone to cracking (particularly on the driver's side) and does not interchange with earlier handles.

To remove the lift bar, first remove the four upper window-channel retaining bolts. Then loosen the lower window-channel retaining bolt and push the channel to the front of the car. This allows access to the two lift-bar retaining nuts.

New original-style water deflectors are available from Ssnake-Oyl Products to keep moisture away from door trim panels. Make sure strip caulk is pliable and tape deflectors onto door at bottom center and upper corners.

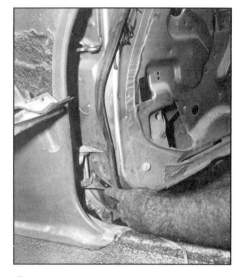

To remove door, support it on floor jack and wood block and unbolt it from hinges, then unbolt hinges from body. Use an S-shaped box-wrench, ratchet-head wrench or swivel-head socket wrench to get at bolt heads.

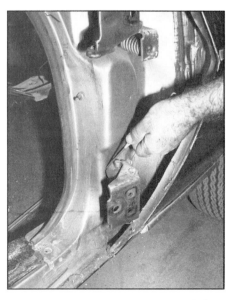

Before removing door or hinges, trace their outline on door and body with a sharp grease pencil.

When removing the lift bar, free the push rod from it by twisting it slightly and pulling it out.

When you install the lift bar, remember to insert the push rod through the hole in the push button lever of the latch assembly. Otherwise, the handle won't unlatch the door.

LOCK CYLINDERS

If the lock cylinder on your Camaro is giving you trouble, it's probably winter. That's when they tend to give the most problems due to smaller operating clearances. Usually a squirt or two of a light penetrating oil, such as WD-40 will free it up. If that doesn't work, open the door and try lubricating the latch assembly. Sometimes the latch assembly becomes stuck and needs a little white grease or Lubriplate to help it out.

If the WD-40 didn't work, you'll probably need to replace the lock cylinder, as the factory does not offer rebuilding kits. You'll also need to replace all the lock cylinders if you have the misfortune to have lost your keys. Don't laugh. The chances of this happening increase with the time you keep your car apart. It seems as though keys grow legs and walk to who knows where.

Before you replace the lock cylinder, it's best to have a new one on hand that's already coded to your key and ready for installation. Your pride and joy will be easy to steal if you have to drive around without a lock cylinder, while shopping for a new one. The process of coding a lock cylinder to a key is relatively easy provided you have the coding information available to you. However, this information isn't available to the general public to prevent theft. The easiest way to code your lock is to ·have a dealership or bonded locksmith do it for you. They can code the lock cylinder to your key for a small charge.

Removal & Installation—The lock cylinder is retained to the door by a simple but effective forked spring clip, which can be found on the inside of the door skin. Before you remove the lock, tape it to the door to prevent it from falling out when you remove the retaining clip. To remove the lock, lever the clip away from it with a large screwdriver, using one of the access holes in the door frame as a fulcrum. Then remove the tape, lock and gasket from the outside.

When installing the new lock, be sure that the pawl fits over the lock extension on the latch assembly. And be sure to use a new gasket underneath it. The gasket pre-

vents water from entering the door through the lock hole in the door skin.

WATER DEFLECTOR

The water deflector is a piece of specially treated, oil-impregnated paper that seals the access holes in the inner door. It acts as a membrane, deflecting water away from the trim panel, keeping the hardboard trim-panel backing and door upholstery dry. The deflector fits between the trim panel and inner door panel, attached to the door by strips of sealing caulk and waterproof tape.

Basically, there are two types of water deflectors. One type has die-cut holes in it for the splined shaft of the window regulator and inside release handle. Some first-generation models with molded door trim panels have a plastic-covered type with a larger hole for the inside release handle. The plastic helps to seal the door release mechanism.

Removal & Installation—To remove the deflector, carefully pull the tape off the door frame. Then lift the edge of the deflector and peel it off the caulking. If it won't separate from the caulking, insert a putty knife into the caulking and cut the caulking with it. If the deflector is brittle, rotted or missing, new custom-cut shields made from the genuine article are available from **Ssnake-Oyl Products. 15775 N. Hillcrest, Suite 508-541, Dallas TX 75248 800/284-7777.** If you rip the deflector, but it is otherwise in good shape, you can repair it with waterproof tape. Or if you're not a stickler about originality, lost or destroyed shields can be replaced with *Visquene* plastic sheeting.

Before installing the deflector, check the condition of the strip caulk at the front and rear edges of the deflector. Although the caulking never really goes "bad," it may lose some of its adhesive properties and not stick to the deflector very well. If it doesn't, replace it, as the water deflector depends on the caulking for a good seal. Position the caulking at the front and rear edges of the inner door panel so that when you install the deflector, the caulking will be positioned near the edges of the deflector. If the bottom edge of the inner door panel has a long, thin slot, insert the bottom edge of the deflector into it, with the black side of the deflector facing the window. If the door panel doesn't have this slot, apply strip caulk to the bottom edge of the deflector. Then press the edges of deflector against the caulking to provide a good seal. Secure the bottom of the deflector with pieces of waterproof tape approximately 3-in. long.

DOOR REMOVAL

Removal of the door is required when you need to perform major work on the door, such as replacing the door skin or repairing or replacing the door hinges. However, before removing the door, get an assistant or at least a hydraulic floor jack and a block of wood, because the door is quite heavy and awkward. Whichever you use, they should help support the rear of the door while you're removing the hinge retaining bolts. And you should remove the trim panel and disconnect the wiring harness from the power window switch and power door lock solenoid (if so equipped) before removing the door. You should also remove the window. This lightens the door and helps prevent damaging the window. Instructions for removing these items can be found under the heading for each individual component.

To help prevent any damage to the door or door opening, it's a good idea to drape a towel or blanket inside the door opening before removal. And while you're at it, run a length of masking tape over the edges of the door to prevent chipping the paint.

The door is attached to the body by two hinges. The lower hinge on 1967 models is cast and the upper hinge is stamped steel. Stamped-steel hinges are used on all 1968-and-later cars. The hinges on '67 cars are also unique in that they use Phillips-head screws for attachment. 1968-and-later models use bolts. Whatever type of fastener is used, each hinge uses a total of six; three attach it to the door and three secure it to the body. Body shops have a special S-shaped 1/2-in. box wrench to remove the bolts, but it can be done with a 1/2-in. socket and a 3/8-in. drive ratchet.

Before laying a wrench on the hinge bolts, mark the position of the hinges with a sharp grease pencil or crayon on the door and body. This will help you reinstall the door in approximately the right position later on. However, if you plan on refinishing the door frame and door opening, you'll be starting from scratch when you begin aligning the door because your reference marks will have been painted over. You could scratch the outline of the hinges on the door and body with an ice pick or scribe, but this only provides a foothold for rust.

You may need to use a universal joint just behind the socket to get a good grip on the bolt head. Start by removing the lower hinge-to-body bolts first; this will minimize the shifting of the door as you remove the bolts. Then move to the upper hinge-to-body bolts.

Jim Altieri of Arizona GM Specialists inspects door sag on 1973 Camaro. Worn-out bushings allow door to sag when it's opened and make it harder to close.

Cast lower hinge is exclusive to 1967 models. Detent roller and pin wear out and can be replaced separately. Installing new roller and pin can be difficult because bottom of original pin is hard. Old pin must be drilled and tapped to accept new pin.

On stamped hinges used on all 1968-81 models (and upper hinges of 1967 models), grind off upset area on end of hinge pin and drive old pin out with a drift.

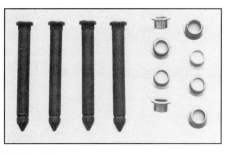

Bushing kits vary with application, but all include new hardened steel pins and bronze bushings.

Tap in new bushings with a wide mallet. Easy does it.

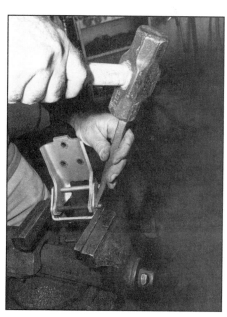

After assembling hinge sections and installing a new pin in the bushings, stake the end of the pin with a chisel to keep it from backing out.

When the door is off the car, you can easily remove the remaining hinge attaching bolts.

DOOR HINGES

The door hinges become worn after time, which can allow the door to rattle against its striker and sag when it's opened. The sintered-bronze bushings in which the hardened-steel hinge pins ride, wear away first on the driver's side, then on the passenger's side. If you can lift the rear edge of an opened door more than 1/4-in., the hinge bushings need to be replaced. Fortunately, these bushings can be

To remove damaged door skin, use a grinder with a 24-grit disc (16-grit if you want to grind it faster). Grind the door edges until you see a line between the skin and door flange. Then drill out the spot welds securing the upper bracing from the door frame.

Corner of this door frame could be patched with fiberglass. But to make a more durable, lasting repair, a hand-fabricated patch panel will be welded in.

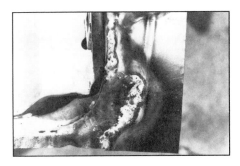

Patch panel welded in place before trimming. Once the weld is ground down, you'd have a hard time finding the patch.

replaced and you can pick them up from some aftermarket suppliers or at your local Chevy dealer.

In extreme cases, signaled by door sag so bad the sill plate is scuffed and you have to lift the door to open and close it, the pins may have worn completely through the hinge bushings and into the hinge itself. This situation will require a new hinge assembly, because new bushings cannot be pressed into a hinge with oblong bushing holes. Also note that the upper hinge of second-generation cars (lower hinge on first-generation cars) incorporates a spring-loaded detent assembly that allows the door to be held open in one of three positions. If these detent ridges are worn down, the hinge assembly should be replaced, otherwise your Camaro's newly restored door may swing out and get chipped paint in tight parking-lot situations.

Bushing Replacement—The bushings can be replaced with the hinges either on or off the car, but to be safe (the doors weigh about 100 lbs. each), avoid chipped paint and get the best working space, the doors should be removed for this operation. This is covered in the *Door Removal* section on page 115.

Take a grinder and grind off the upset area on the lower part of the hinge pins. Then, invert the hinge assembly and use a drift and hammer to knock the pivot pin out of each hinge. The old bushings (two per hinge) can be levered out with a stout flat-blade screwdriver or driven out with a drift the same size as the outer diameter of the bushing.

To install the new bushings, position each one into the bushing hole in the hinge. Carefully drive the bushing into the hinge using light hammer blows. Go slowly to avoid cocking the bushings in the bore.

When the bushings are installed, lubricate the pivot pins with Lubriplate and insert them through the holes in the hinges, joining the two sections of each hinge. With original equipment pins, you'll have to stake the pins in position by supporting the pin head and upsetting metal under the lower bushing with a hardened chisel. Some aftermarket pins are retained with snap-rings, making the staking operation unnecessary.

Reverse the removal procedure to install the doors; that is mount the hinges to the body first, then the doors to the hinges. Note that when installing the hinges on the body, it may be necessary to insert an awl or Phillips-head screwdriver into the sliding nut plate inside the door jamb to align the hinge bolt holes. Align the hinges with the marks you made earlier on the doors

and body. If necessary, realign the door and make a striker adjustment as detailed on page 119.

DOOR SKIN REPLACEMENT

If the outer portion of the door has more than a few minor parking lot dings, you might want to replace the door skin instead of repairing the old one. Replacing the door skin can save you time and you usually come out with a better finished product. If your door is especially rusty, you're better off looking for a used door that's in reasonably good condition to use in its place. Repairing a door that has a badly rusted bottom may not be worth the time and effort it takes to repair it.

Glass & Weatherstripping—Removing the door skin takes a little skill and patience. The door should be removed from the car and positioned on a table. Make sure to remove the door glass from the door to avoid breaking it. If the flange of the door is bent, straighten it with a hammer and dolly before removing the door skin.

Before you begin the skinning operation, remove the door felts from the top of the door. The door felts have barbed retaining clips and are attached to the door with small Phillips-head screws. You'll probably have to spray the screws with a rust penetrant before you remove them. If you're unfortunate enough to round out any of the screw heads, you'll need to drill the remains of the head away. When the screws are removed, carefully bend the top of the seal toward the outside of the door. This will help disengage the barbs on the felt retaining clips. Then pull the felt away from the door.

Attachment Points—The door skin is attached to the upper door frame, or inner door panel, by a brace that's a part of the skin and is spot-welded to the door frame. The sides and bottom of the skin use a hem flange and are also spot-welded. However due to the vent window, 1967 models have two small flanges at the upper front edge of the door instead of having the front portion of the brace spot welded to the door frame. These flanges have to be cut in order to replace the skin.

Required Tools—To remove the door skin, you'll need a portable drill or grinder and a 6-in. OD (or larger) 24-grit sanding disc. The larger the better, because it will cut through the door skin faster. If you're removing the skin from a 1967 door, you'll also need a pair of aviation snips. And be sure to wear safety glasses.

Grind off Flanges—Position the outer edge of the disc against the bottom edge of the door and grind the door skin away until you see the edge of the door frame. This frees the bottom and sides of the door skin from the door frame.

Spot Welds & Braces—You can remove the top edge of the skin by using a chisel, spot-weld remover or drill bit to remove the spot welds from the upper door brace. If you're working on a '67 door, use aviation snips to cut the two flanges at the top of the door. Make the cut closer to the door skin so you'll have more material to work with when it's time to install the new skin. When this is done, you can remove the door skin. Set the skin aside but don't throw it out. The good sections of it can be used as metal patches for other areas of the car if required.

To remove the remains of the hem flange from around the door, you'll need a hammer and a sharp chisel. Place the chisel underneath the remains of the door skin against each spot weld. Carefully tap the chisel against each spot weld until the remaining hem flange is free. But take it easy so you don't distort the edges of the door frame. It's going to serve as a pattern when you install the new skin.

The door frame is the most basic structure for the door. Consequently, it must be in excellent shape if you want the door in the best possible condition. Due to poor drainage, door frames have a habit of rusting at the bottom, especially at the lower front edge.

Door Frame Rust—The first step in restoring the door frame is to remove the rust from this area. This can be done by sandblasting or with a chemical rust remover. If the rust is light and you're on a tight budget, you can usually remove the rust with a concentrated solution of phosphoric acid, such as *Metal Prep,* and water. A mixture of two parts water and one part acid will usually do the trick. If the rust is really stubborn, mix the acid and water at a 1:1 ratio. And always follow the safety precautions on the label. If you can afford it, a derusting shop which uses a chemical solution that won't harm good metal is your best bet.

After you get the door back from the shop, check it carefully for any rust holes that have to be repaired. Pay special attention to the lower front and rear edges of the frame where the metal is bunched. Pinholes usually occur in these areas. When you've located all the areas requiring repair, you can begin repairing them.

If you find pinholes in the bottom corners of the frame, carefully fill them with lead, plastic filler or brazing rod. If you choose to braze, heat the brazing rod first until it softens. Then heat the area of the

pinhole slightly while keeping the brazing rod near the flame. Position the brazing rod over the pinhole and melt it while at the same time, keeping the metal warm. The brazing rod should flow onto the frame and fill the hole. If it doesn't, heat the rod until it does.

If the frame has any large holes, cut the rotted area out until you reach an area that is solid. A Dremel moto-tool with a cutting wheel works well. Cut a patch that is slightly larger than the area you're repairing from the door skin you just removed. Position the patch over the hole from the inside of the door and secure it with C-clamps or locking pliers. Now you're ready to begin welding.

Use a neutral flame and tack weld the patch every inch or so. When you've tack welded the patch, weld it completely around its edge. Avoid heating the panel any more than necessary to minimize distorting it. Repair any remaining large holes in this manner.

When you've finished welding the frame completely, proceed to finish the welded areas. The welds will form a slight ridge where the two pieces of metal meet. Use a pick hammer to slightly depress the welds below the surface of the neighboring metal. Then sand the area so it's smooth with 24-grit paper. At this point, you have to decide if you want to lead the area or use a plastic body filler.

If you're going to use lead, first tin the area. Then use 30/70 lead to fill the depressed area until it's slightly above the surface of the metal. File or sand the lead with 24-grit paper until it's flush with the surrounding metal. Kits for leading are available at many auto body supply stores. If you prefer to use plastic body filler, make sure it's of the waterproof type and follow the manufacturer's instructions.

Prepare Flange—If any spot welds or metal remains from the door-skinning operation, grind them flush to the metal with a coarse sanding disc. And if the flange is bent, straighten it with a flat hammer.

Paint Door Frame—The door frame should be completely finished at this point and ready for painting. It's much easier to paint the inside of the door frame now before you install the door skin. Use some diluted phosphoric acid to etch the metal. Follow this step by priming and painting the inside of the frame. The door frame is now ready for installation of the new skin.

Examine Door Skin—When you purchase the new door skin, check it for damage before accepting it. If the skin is ripped because because of a bad stamping or has any other visible defects, exchange it. If

you walk out the door with the skin and later find out it's damaged, it's going to be difficult if not impossible to exchange it for another.

The new door skin will have the flange partially bent over and ready for installation. Your job is to fold over the remaining metal for the sides and bottom and weld it to the door frame.

Tack Skin—Turn the door frame so that the interior portion of the frame is flat against the table. Now position the new skin onto the door frame. Make sure the upper brace is in the proper position then tack weld it in position. Now carefully turn the door over.

Fold Door Skin Flange—The next step is to fold the door skin flange over the flange of the door frame. Again, make sure the skin is fully seated against the upper door brace. Use a flat body hammer to begin folding over the bottom edge of the skin. Use light hammer blows and do a little at a time. Bend the edge of the skin slightly in the center and work toward the ends. Then bend the edge a little more while working toward the ends. Repeat this until the flange of the skin is seated against the flange of the frame. When you've finished the bottom edge of the skin, use this same technique to fold the side flanges of the skin.

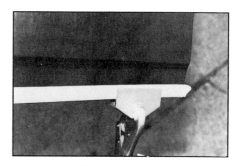

One way to crimp flange of new door skin over frame is with this fixture. Use a hammer and dolly to flatten flange against frame without distorting frame.

With door skin clamped in position, spot weld it in place.

117

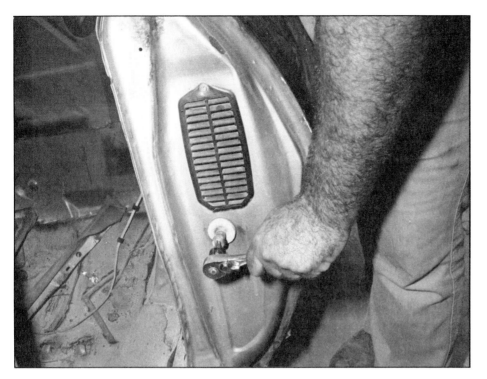

After door hinges are adjusted, center striker in latch with a minimum 3/32 in. between striker head and front and rear of lock bolt. Adjust striker by removing it with an Allen or Torx-head bit and adding or subtracting shims under striker base on quarter panel.

Start installing new door weatherstripping by fastening one formed end with plastic rivets or wide-head, short-shank Phillips-head screws.

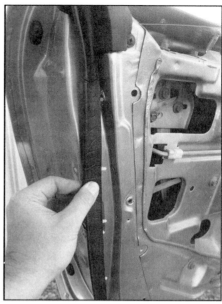

T-shaped, barbed plastic fasteners are built into door weatherstripping and should correspond with holes on sides and bottom of door. Push on weatherstrip and snap clips into door.

Weld & Seal—To keep the skin from shifting and to strengthen the door, tack weld the skin to the frame. Tack the skin every 4 in. while using a damp cloth under the area being welded to avoid distorting the metal. After the welds have cooled, prime the bare metal then seal the joint between the skin and the frame with body sealer. Then condition, prime and paint the entire door except the door skin. You'll do this later when you prepare the body for paint.

DOOR INSTALLATION

Now that you've finished working on the door, you can install it on the car. Be sure to take your time and work carefully, because you don't want to drop it or damage it after spending untold hours getting it into tip-top shape. The first step is to attach the hinges to the door. But before you do, apply a liberal amount of body sealer to the hinge mating surface. This will prevent moisture from getting under the hinge and causing rust. And if you want your car to look its best, paint the heads of the hinge bolts now, before you install them. If the paint chips off them when you install them, touch them up when you're done.

Bolt both hinges to the car and tighten them. Clean up any excess sealer from around the hinges with Prep-Sol or other suitable solvent if the paint is completely dried. If it isn't, you'll have to clean up the sealer later unless you don't mind ruining your fresh paint job. Now coat the remaining mating surface of both hinges with body sealer. Then, with the window down, open the hinges so the door will be fully closed when you position it to the car. This makes it easier to attach the door to the body.

Now if the paint on your car is in good shape, protect it by masking the edges of the door and laying a towel on the door sill. This will help to prevent paint chips. Then grab your friend again and have them hold the door in position while you screw the hinge bolts into the door with your fingers. Don't use a socket to install the bolts at this point because you might cross-thread them and create a bigger job for yourself. If you do cross-thread the bolt hole, you'll have to retap it and use another bolt.

Alignment—When you've installed all the hinge bolts, shift the door so the hinges line up with the marks you made earlier. Tighten the bolts with the door in this position. When everything's tightened up, slowly swing the door until it's almost closed and check its alignment with the rear quarter panel. If the body creases line up, adjust the striker. But don't close the door. If you do and the striker isn't properly adjusted, you may have a real hard time trying to open it. The door must be properly aligned before adjusting the striker. It must be square in the opening, with about 3/16-in.-gap spacing all around. The rear edge of the door should be flush to about 1/16-in. to the outside of the front of the quarter panel. This will reduce the chance of wind noise.

Hinge Bolt Adjustment—Close the door completely and check the fit to the front fender and quarter panel. If the door is out of alignment at either point, or the gap between the door and quarter panel is more than 1/4 in., the door should be adjusted.

This adjustment is accomplished by loosening the hinge bolts as necessary (with suitable support to keep the heavy door from sagging) and shifting the door. Loosen the hinge-to-body attaching bolts if the door needs to be moved forward or backward or up or down. If the front edge of the door is below or above the side of the front fender, loosen the hinge-to-door attaching bolts and move the door until it's flush with the front fender. The adjustment for this condition at the rear of the door is controlled by the striker. Loosen the striker with an Allen or Torx wrench and move it in or out until the rear of the door is flush with the side of the rear quarter panel. And also check the engagement of the striker to the lock bolt once again. After shifting the door all around, the engagement may have changed. At this point you can connect any wiring harness that you disconnected previously and install the trim panel.

Striker Adjustment—When the door is properly aligned, you can adjust the striker. To determine whether the striker has to be adjusted in or out, place a strip of modeling clay on the latch fork bolt. Close the door far enough for the striker to make an impression in the clay. The striker should be centered in the fork bolt, with a minimum of 3/32 in. between the head of the striker and the front and rear of the lock bolt. Make adjustments as necessary by adding or subtracting shims from underneath the striker.

RUBBER COMPONENTS

The door and window weatherstripping and door bumpers make up the rubber components that are used on the door.

DOOR WEATHERSTRIPPING

Door weatherstripping is made of foam rubber and helps to seal out dust, wind noise and water from the interior. This component is often overlooked during the restoration process. A weatherstrip in good condition shouldn't have any tears or signs of rotting. And it should spring back into place when you press it sideways with your fingers. It should also be sealing tightly against the door opening.

Checking For Leaks—A quick way to check for a weatherstripping leak is to apply some powdered chalk to the sealing surface of the entire door opening. You can find this type of chalk at most hardware stores. Now carefully close the door and open it. Then examine the weatherstripping for signs of the chalk. If the door gasket isn't sealing in a particular area, the gasket won't have chalk on it in the area of the leakage.

Another way to check the gasket is to use a sheet of paper. Place a sheet of paper about 6-in. wide in the door opening and close the door. If the weatherstrip is in good shape, the drag on the paper will be moderate to high at all points around the door. If it isn't, the weatherstripping should be replaced.

Weatherstrip Repair—If the weatherstrip has minor cracks or tears, you can use black RTV (room temperature vulcanizing) sealer to repair it. Clean the damaged area with soap and water and let it dry. Then carefully apply a small bead of RTV directly into the crack or tear. If any RTV gets on the outside of the seal, smooth it out using a piece of wax paper. Tape the repaired area without disturbing the repair and let the RTV cure.

Removal—If you need to remove the weatherstripping, pry the plastic button-head pins from each end of the weatherstripping with a pair of cutting pliers or a tack puller. Then remove the Phillips head screws which are positioned on the trailing end of the weatherstrip. Sometimes these fasteners are hidden under the weatherstripping.

As you remove the upper portions of the weatherstripping, you'll notice a dark adhesive is used under the weatherstripping to ensure a good seal. You can usually break the bond of the adhesive by cautiously peeling the weatherstrip away from the door frame. If you can't, use a narrow putty knife to break the bond.

The remainder of the gasket is attached to the door with small nylon retainers shaped like a "T." Probably the safest way to remove these retainers to avoid damaging the weatherstripping is to remove the weatherstripping first. To do this, roll back the end of the weatherstripping until you see a retainer. They're positioned every 4 in. or so. Now carefully stretch the weatherstripping until you can slip it over the "T" portion of each retainer.

These retainers help to seal the piercings that they fit through in the door frame. Because of this, you have to exercise caution during removal to avoid damaging them. If you don't have access to the special tool that removes them, you can remove them with a pair of cutting pliers. Position the pliers directly under the head of the "T." Use the inner edge of the door frame as a fulcrum and lever the retainer out of its hole. To avoid damaging the paint, place a couple layers of masking tape around the inner edge of the door.

Cleaning—If the weatherstripping is being replaced or reconditioned, it's wise to clean it with a rubber compatible detergent such as *ArmorAll Cleaner*. After that, you should treat it with a preservative like *ArmorAll Protectant* or *STP Son Of A Gun*. These products will make the gasket look like new and help to extend its life. And you want to make the door weatherstripping last as long as possible because it's not inexpensive to replace.

Installation—If you were careful removing the retainers from the door, they can usually be reused. If you damaged any by breaking off the head or wiping off the barbs, replace them with others in good condition. They don't have to be new.

Insert the retainers through the holes in the weatherstripping and position the weatherstripping to the door. Starting at the bottom of the door, align each retainer with its hole as you push it into the door frame with your thumb.

Up top, install new outer window felt. On 1970-81 models, it's retained with plastic rivets or wide-head, short-shank Phillips-head screws.

lap by this amount, the door will rattle or bind, depending on whether the fit is too loose or too tight.

Check Fit—To check the fit, close the door almost all the way and measure the clearance from the top of one wedge to the top of the other. If they don't overlap by 1/32 in., remove or add shims under the wedge blocks until you get the right amount of overlap. If you need shims, you can make them from sheet metal using an old shim or wedge plate as a template.

On 1967-69 models, outer felt snaps into top of door with barbed metal clips. Note attached guide plate on this 1967 model.

As you proceed to the sides of the door, apply some black weatherstrip adhesive to the door frame before you install the weatherstripping. Apply the adhesive starting at the top of the front and rear door frame to a point about 9- in. below the top, so it will be directly under the weatherstripping. Finish the job by installing the button- head retainers and screws into their previous positions.

WINDOW FELTS

The exterior and interior window felts seal the window opening from water and debris. Unfortunately, there is no way to recondition these pieces if they are damaged. The exterior window felt uses a rubber-coated metal core and is covered with short bristles or uses a rubber lip for sealing. The metal core has sets of barbs on it that secure the weatherstrip to the door along with Phillips-head screws. Camaros of 1967 vintage also incorporate a window stabilizer into the rear of the window felt. Keep this in mind when shopping for a replacement. For removal of this part, refer to the section entitled *Door Skin Replacement,* page 116. The interior window felt is made of flocked rubber with a steel core and is attached to the top of the door frame or trim panel with barbs or staples. Removal and installation of this component is covered in the *Trim Panel* section, page 137.

DOOR BUMPERS

The door bumpers are positioned at the top and bottom of the rear edge of the door.

They prevent the door from hitting the door opening in the body directly. The bumpers are retained in the door by two barbs. To make their removal easy, spray the area of the barbs with silicone lubricant. Then gently pull and twist them from the door. And if it isn't second nature to you by now, treat them to a dose of rubber cleaner and preservative.

If you need new door bumpers, they're available through the growing number of Camaro parts suppliers and are relatively inexpensive. More often than not, these little treasures are missing from the door. It's little details like this that detract from an otherwise fine car.

DOOR WEDGE PLATES

Door wedges plates are only used on convertibles. As their name implies, they're wedge-shaped plates that help provide additional support for the door so it doesn't rattle when the car is in motion. They're necessary because convertibles tend to flex more than cars with full roofs.

Two plates are used for each door. One plate is made of brass and is positioned at the top rear edge of the door. The other plate is made of nylon and is mounted on the front portion of the rear quarter panel. The wedge plates prevent the door from rattling by keeping it under tension. This tension is produced by an interference fit between the two plates. This means the plates overlap each other when the door is closed. The designed overlap on Camaro ragtops is 1/32 in. If the shims don't over-

Last but not least, install rubber door bumpers (right arrow) and on convertibles, wedge plates (left arrow).

While beyond the scope of this book, now is a good time to rebuild your Camaro's engine. Then, when all the new parts are bolted in and new paint applied, your Camaro will have the stoplight punch and down-the-road performance it deserves.

It's true that a major portion of time is spent during a restoration concentrating on the way a car looks. But just as important, unless you want to relegate your Camaro to museum-piece status, is the way it operates. If you're engaged in a ground-up restoration of your Camaro, chances are the engine is being rebuilt and the transmission and rear axle are getting the once over, too. On the other hand, maybe your Camaro is completely road-worthy and is simply in need of a little freshening after all those years and miles.

The co-author's 1973 Type LT Z/28 is a good example of this latter category. Cosmetically, the car was a rolling basket case—burned-out interior, rusted roof, drooping doors, dented fenders and an El Cheapo paint job with lots of runs and overspray. But mechanically, it gave good account of itself at 82,000 miles. The engine showed good compression, held a tune and passed smog inspection with flying colors. The Turbo Hydra-matic 400 transmission shifted well, suffering only from a few fluid leaks. The Positraction rear axle chattered a bit on sharp turns, but otherwise was quiet and intact. Except for some maintenance items and minor repairs that we'll soon cover, it was kept in running condition, legally registered and smog-checked throughout its 4-year restoration period.

SOURCES

Rebuilding drivetrain pieces is beyond the scope of this book. If you want to rebuild your engine and transmission, there are a few good books on the subject we'd like to call your attention to. Probably 90% of all Camaros ever built were equipped with the small-block Chevy V8. HPBooks' 160-page *How To Rebuild Your Small-Block Chevy* by David Vizard is the best thing around for redoing a 302, 305, 307, 327 or 350-CID V8. Likewise, big-block Camaro owners are in luck because HPBooks has another comprehensive, well-illustrated 160-page book on rebuilding those engines. Pick up *How To Rebuild Your Big-Block Chevy* by Tom Wilson if you have a 396, 402 or COPO 427-CID Mark IV big-block in your Camaro. These books, as well as other fine HPBooks, can be purchased from your local auto parts retailer, or direct from **Price Stern Sloan. 360 N. La Cienega Blvd., Los Angeles, CA 90048. Order toll-free by calling: 800/231-1357 outside CA. Inside CA: 800/523-5579.**

Did you say you were lost in the planetaries? Well, it's not necessary to be intimidated by the Turbo Hydra-matic automatic transmission in your Camaro anymore. The ever-popular Turbo Hydra-matic 350 was installed in millions of six-cylinder and small V-8-equipped Camaros between 1969 and 1981, whereas the high-capacity Turbo Hydra-matic 400 saw duty in 1967-74 behind big-blocks and in Z/28s. Co-author Ron Sessions has put together two in-depth texts on these transmissions: HPBooks' *Turbo Hydra-matic 350 Handbook* and Motorbooks International's *How To Work With and Modify the Turbo Hydra-matic 400 Transmission*.

INSTALLING HARD SEATS AND BRONZE GUIDES

If your Camaro is equipped with a small-block Chevy V8 built before 1971, chances are you'll have valve-seat recession problems using today's low-lead and unleaded fuels. Lead acts as a lubricant between the valve and seat, preventing the exhaust valve heads from pounding their seats into the cylinder head. But lead has its downsides too. Lead contaminates and renders ineffective catalytic converters used on all Camaros since 1975 to reduce hydrocarbon and carbon-monoxide emissions. And recent studies have shown that lead in the environment can cause impaired brain functions in children. Realizing lead's problems, GM began induction, or flame, hardening the exhaust-valve seats on its engines in 1971, which prevents valve recession. If you have a 1971-or-newer Camaro, you can use unleaded or low-lead gas without damage to your engine.

But for the 700,000 or so Camaros built before 1971, the answer is to use anti-wear additives, or have the engine modified so it can use today's fuel. The modification involves machining the head (to accept hardened-steel exhaust-valve seat inserts. Then, most rebuilders go ahead and replace worn integral valve guides with long-wearing bronze inserts.

Follow along with us as Denny Wyckoff of Motor Machine In Tucson, AZ shows us how the insert operation is done on a small-block V8.

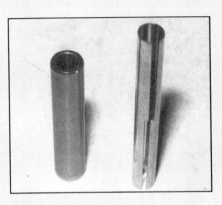

Worn integral valve guides that admit excessive oil into the combustion chamber on startup and deceleration can be restored with silicon-bronze replacement guides (left) or split-sleeve guide liners (right).

Bane of the unleaded fuel era. This 1969 Camaro small-block V8 head has lost compression due to a recessed exhaust valve seat.

Hard chrome steel exhaust valve seat inserts are available from Manley and other sources—a good way to go on 1967-70 Camaros.

After driving liners into bore, Motor Machine's Wyckoff cuts liners flush with top of guide bore, then knurls and reams them to size of valve stems, ready for another 100,000 miles of Camaro service.

After Wyckoff bores cylinder head to accept seat, he drives the seat into position. Seat can then be cut for multi-angle valve job and in essence becomes an integral part of the head that will live happily on a diet of low-lead or unleaded fuel.

If you haven't done so already, steam-clean the engine compartment and undercarriage. Not only will the steam blast away years of gunk but your Camaro will be all the nicer to work on. Remember to cover the carburetor and distributor before hosing down the engine or startup may prove difficult.

Want to keep cast-iron exhaust manifolds looking as good as new? Try using some Eastwood Stainless Steel Coating on them. The coating also acts as a barrier to corrosion. Courtesy Eastwood Company.

As a matter of course, replace all of the hoses in your Camaro, especially cracked, leaking fuel lines. After tidying up the engine compartment, don't forget to replace these hoses near the fuel tank. A leak here could prove tragic!

When going for that original look, check out the post-type hose clamps and GM-stamped radiator and heater hoses available from aftermarket sources.

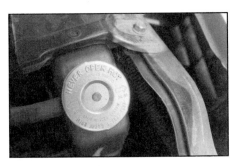

After cleaning, pressure-checking and repainting the radiator, why not top it off with an original-equipment AC-Delco radiator cap. Sometimes, it's the little touches that count.

THE BASICS

But getting back to our original premise, regardless of whether you're doing a frame-up restoration and rebuilding the engine, trans and rear, or just freshening up those pieces to accompany a cosmetic restoration, there are a few things you should do before assembling your Camaro.

Clean Everything—First and foremost, clean the engine, transmission, rear axle and underpan. The best way to do this is at a steam-cleaning facility with a hydraulic lift. Not only will your Camaro be a much more pleasant place to work, but you'll be better able to spot leaks, rust spots, frayed wires and damaged components such as bent cross-members or dented oil pans.

ENGINE

Any high-mileage Camaro engine should be compression-checked and, if available, leak-down tested to determine if it needs a rebuild. If yours is a 1970-or-earlier Camaro and you intend to drive it more than just to and from car shows, the exhaust valve seats and guides should be looked at. Unless you plan to use costly lead or other anti-wear additives to prevent valve recession and rapid valve-guide wear, running an early Camaro on today's unleaded or low-lead gas requires that you install hardened steel exhaust-valve seats and bronze guides (see nearby photo and caption sidebar).

On V8 engines, check for a misaligned front damper, stretched timing chain and worn camshaft lobes. On all engines, replace the valve-cover gasket(s) (a major source of leaks on V8s) and engine mounts, likely dried out and cracked or oil-soaked after all these years. If the rear main seal was leaking, install a new one now while you have the opportunity. And replace any cracked or gummy vacuum hoses (one at a time, so you don't lose track of where they go) that can lead to

troublesome vacuum (air) leaks later.

Oh yes. We almost forgot. Change the oil and filter.

Fuel System—Now is a good time to rebuild the carburetor. Again, HPBooks has some excellent sources for you to service the carburetor on your Camaro: *Rochester Carburetors* by Doug Roe and *Holley Carburetors & Manifolds* by Mike Urich and Bill Fisher. Change the fuel filter and air filter and replace the fuel hoses going to the fuel pump, carburetor, charcoal cannister (1971-up). You don't want a small leak to cause an engine fire and incinerate your restoration project, do you? While you're at it, on 1971-and-later models, install a new filter on the underside of the charcoal cannister. Back by the fuel tank, replace the hoses leading to the engine compartment and those routed up to the vapor separator (1971-74 models). If you experienced fuel-gauge problems before tearing your Camaro down for restoration, now's your chance to check the sending unit inside the fuel tank. See page 73 for details.

Cooling System—Check the radiator for obvious damage such as leaks, bent or broken fins and visible corrosion. If it looks salvageable, take it to a radiator shop and have it cleaned and pressure-checked. If the radiator was very rusty, chances are there's a big sediment buildup of rust inside the water jackets of the engine. Remove the plugs from the lower sides of the cylinder block and drain all coolant out of the engine. On really rusty blocks, sediment buildup will block the coolant from draining until it is poked through with a long screwdriver.

Get all new radiator hoses and heater hoses and a new Delco radiator cap. If you're a stickler for originality, get the hoses with GM markings from your Chevy Dealer or aftermarket sources such as **Classic Camaro Parts & Accessories in Huntington Beach, CA (714/848-9501).** Classic Camaro also has the original style post-type hose clamps used in the late '60s and early '70s.

On the front of the engine, check that the water pump isn't noisy as it spins. Also inspect the viscous-drive fan clutch for a worn-out bearing. And get some new OEM drive belts for the water pump, alternator and accessories.

If your Camaro has a badly stained coolant overflow bottle and a new replacement isn't available, you can clean the old one to like-new condition. Pour some household bleach into the jug so the bleach stands about 2-3 in. deep. Agitate and allow to stand for an hour or so. Then get a baby-bottle cleaning brush from a drug store and scrub the inside of the bottle until it's squeaky clean and white again. Grease stains on the outside of the bottle can be cleaned off with *Gunk* or other engine-cleaning solvents. Paint overspray can be taken off with lacquer thinner. You can give the windshield-washer bottle the same treatment.

Electrical System—Needless to say, if your Camaro was experiencing charging or starting difficulties beforehand, have the alternator and starter serviced and, if necessary, rebuilt now. If originality counts, have your units refurbished—otherwise trade yours in on an exchange basis. Replace the positive and negative battery cables. Clean the battery (and its mounting tray) with baking soda, brush any corrosion off its terminals and check it for sulfated plates and ability to hold a charge. Reproduction top-terminal black-top Delco batteries are available from the **Gurdjian Battery Co., RDl. Box 68. Union City, PA 16438** and other companies. Check *Hemmings Motor News, Cars & Parts* magazine and other old-car magazines for addresses.

CLUTCH & TRANSMISSION

If you can't remember the last time your Camaro's transmission fluid—manual or automatic—was changed, do it now.

Automatic—With an automatic, you have to drop the pan and install a new filter and pan gasket. A teaspoon full or more of filings or aluminum-looking gunk at the bottom of your drain pan indicates that a transmission rebuild may be in order.

Because fluid does most of the work in an automatic and there's so much of it, leaks are common on high-mileage automatics. Areas to check are the seals for the front pump, output shaft, manual-shaft linkage, dipstick, vacuum modulator and speedometer housing. While you're at it,

If the drivetrain stayed in the car during the restoration, at the very least inspect the transmission and rear axle for leaking gaskets and seals. Red droplets on this Turbo Hydra-matic 400 oil pan indicate a transmission fluid leak.

replace the vacuum modulator. These inexpensive canisters are short-lived when exposed to exhaust-pipe heat and a vacuum leak results when a modulator diaphragm goes bad. It can wreak havoc on engine and transmission performance. And it's the source of white tailpipe smoke if the engine is warmed up.

Manual—On manual-transmission cars, the clutch friction disc, pressure plate and throw-out bearing should be inspected. Replace any worn disc, blued or heat-checked pressure plate or noisy bearing. And lubricate the linkage at its pivot points with a high pressure grease.

DRIVESHAFT

Check the universal joints for wear. Place the transmission in **Park** (automatic) or in gear (manual) and jack up the rear end and support the rear axle safely on jackstands. While observing each U-joint closely, grasp one rear tire and attempt to rotate it. If you observe any relative movement at the U-joint yoke and flange, replace it. If your car's U-joints have grease fittings installed in them, lubricate them now with a grease gun.

REAR AXLE

Change the lube in the rear axle by loosening, then removing the differential cover. Allow the old lubricant to drain and then clean out any remaining debris in the bottom of the housing with a clean rag. And if you want to replace an axle shaft seal or bearing, now's the time. Push the axle shaft you want to remove in. This will loosen the "C" clip that secures the shaft to the differential. Simply remove the clip and slide the axle shaft out. When you're installing the clip, make absolutely sure it's fully engaged in the groove of the axle shaft, otherwise THE AXLE WILL COME OUT! Now install a new gasket to the differential cover and install the cover. New 80 or 90W hypoid gear lube can be added through the inspection hole on the side of the differential housing. Fill it up to the bottom of the hole. On Positraction units, fill with the special limited-slip lube and we highly recommend adding the GM special additive, part #1052358 for quieter operation.

TRANSMISSION SWAPPING

So you've got the bug to shift your own gears. Or maybe you're tired of building up your left leg and would rather be shiftless. Whichever you want to do, there are numerous things to consider.

Automatic-To-Manual—If you want to change your Camaro over from an automatic to a manual transmission, all you'll need is a shifter and transmission right? Well, not exactly. You're also going to need shift linkage, a bellhousing, crankshaft pilot bushing, flywheel, clutch disc, pressure plate, release bearing, clutch linkage, clutch pedal and related parts. And you might even need another driveshaft.

Buying all these parts separately could cost a bundle. An alternative to this is buying all the parts from one car, or just buying the whole car. If you go this route, you'll save money and also have the added advantage of seeing how everything fits together instead of guessing.

Manual-To-Automatic—If you're swapping a manual transmission for an automatic, you'll need a transmission, torque converter, flexplate, shifter, shift cable and neutral start switch and associated wiring. And depending on the transmission you select, you may also need a throttle-valve (T.V.) cable and driveshaft.

Engine Considerations—Also keep in mind that the size of the engine may dictate the type of transmission you'll need to buy. For example, if you plan a swap to a Powerglide to fit behind a big-block, it had better be the heavy-duty version otherwise you may be going nowhere—fast. So be sure to consider engine size. Consider that selecting the box on the order blank that said automatic transmission on just about all high-performance Camaros (Z/28s, SS396s) of the late Sixties and early Seventies netted the buyer GM's finest, the Turbo Hydra-matic 400.

Shifter—And don't forget about the shifter. If your car currently is outfitted with a console, it might accomodate the new shifter without any hassles if you merely change the shift plate or surrounding trim.

By thinking a transmission swap through from beginning to end before you actually start it, you'll be money and hassles ahead.

If you plan to drive your Camaro once it's been restored, here's where you'll spend the most time. Imagine slipping behind the wheel of this pristine 1967 convertible and heading out on cruise night.

Now's the time to restore the part of your Camaro you spend the most time in to like-new condition—the interior. If your Camaro has been anything but a zero-mile museum piece, it's likely that the seats, dash pad, carpeting, door trim panels, headliner and rear package shelf—in short, all of the soft interior trim that makes your Camaro so livable—will need cleaning, redying, repair or replacement.

As you haven't painted your Camaro yet, we'll save restoration and replacement of the carpeting, headliner and rear package shelf until *Assembly*, pages 170—172. You wouldn't want to get over-spray all over those hard-to-mask items. Right now, let's concentrate on the parts of the interior you can work on independent of the body (such as the seats and door panels) and those that can be easily masked (such as the gauges and dash pad).

SEATS

VINYL REPAIR

Small tears and holes in Camaro seats can sometimes be repaired with a vinyl repair kit, thus avoiding the cost of recovering the seats. Usually, this type of repair is completed by spraying vinyl dye over the entire piece being repaired. This helps to blend the repair so that it cannot be easily detected.

Various types of kits are available but whichever kit you choose, select a quality kit that allows you to make an invisible, long lasting repair. A good quality kit is available from **The Eastwood Company**. Eastwood's kit includes 16 colors in small contain-

ers and a color-matching guide. If one of the colors doesn't match the color of the vinyl you're repairing, you can mix them to get the desired shade by following the directions on the color-matching guide. The repair kit also includes backing fabric, graining sheets, vinyl cleaner, vinyl sealer and a heat tool.

If you plan on dyeing the seats or other trim afterward, fantastic results can be obtained using Mar-Hyde Vinyl Patching Compound. The only drawback to using this material is that a 500°F heat gun is required to cure it. And you'll have to find some graining sheets that match the texture of your vinyl.

Practice—Before attempting to repair any damage to your interior, practice the procedure outlined below on a piece of scrap vinyl. Become familiar with the particular kit you're using and if you make a mistake, nothing's lost. Practice until you can perform a quality repair.

Clean Surface—To begin the repair, clean the damaged area with the cleaner supplied with the kit. If cleaner isn't supplied with your kit, use a household cleaner such as *Fantastik*. Scrub the vinyl with a toothbrush and liberal amounts of cleaner to make sure you get the dirt out of the grain. Then dry the area with a clean cloth or paper towel. If the vinyl was ever sprayed with a silicone-based protectant such as *ArmorAll* you'll want to remove all traces of it because the repair compound and vinyl dyes won't stick to it.

Small Cracks—If the damage is limited to a small

Eastwood vinyl repair kit works well for small repairs. It contains colored resins, so dyeing repaired sections isn't necessary. For larger jobs, vinyl patching compound from Mar-Hyde is terrific.

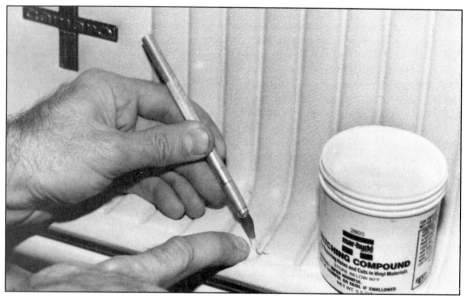

After cleaning surface, work Mar-Hyde Vinyl Patching Compound into small cracks with an X-Acto knife.

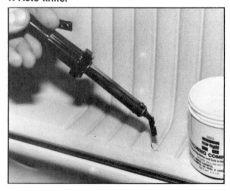

Heat the compound until it turns clear, indicating a proper cure.

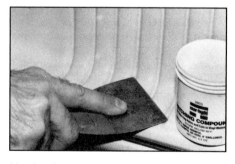

After heating, press a large piece of cold, clean metal against the repair. When enough layers of patching compound have been applied to fill the crack, grain the repair with material supplied in the kit.

crack and none of the vinyl is missing, make a repair by applying the clear vinyl lastomer to the undersides and void between the crack. Choose a graining paper that matches the grain pattern in your seat's vinyl. Then, while the lastomer is still wet, apply graining paper coated with silicone over the crack and tape or hold the graining paper in place. Allow to dry overnight.

Another way to repair small cracks is to use Mar-Hyde vinyl patching compound and a 500°F heat gun. As with the first method, apply the patching compound to the undersides and void between the crack. An Xacto-knife works well for this. Heat the white-colored compound with the heat gun until it turns clear, which indicates it has cured. Then while the patch is very hot, apply graining paper to the repair backed up with a smooth block of metal and press down firmly for about 45 seconds.

Large Holes—If the damaged area is larger than 1/4 in., you must support its underside before making the repair. Cut the backing fabric so that it's about 3/8-in. larger than the damaged area. Then slide the fabric into the damaged area with a small wooden stick. Shift the fabric until it's directly under the damage. In some instances, you may want to put your hand behind the damaged area if at all possible to help you position the backing fabric.

Start the repair by applying a generous amount of clear vinyl lastomer in the hole and allow it to dry for several hours. Then, mix a batch of colored vinyl repair compound with powdered hardener according to the kit's directions. Apply the colored mixture to the hole with a small, pointed artist's spatula, feathering it into the surrounding vinyl about 1/4 in. Select some graining paper and place it on top of the repair area. Heat the wood and metal heat-transfer dowel according to directions by holding it against a home iron for 2-3 min., then pressing against the graining paper. The graining paper must get hot to cure properly. Allow the paper to cool for several minutes before removing it to examine the repair. Repeat as necessary.

REFINISHING & DYEING

If some or all of your seats are faded or off-color, or you're trying to match the color of new and old trim pieces, you can do this with a minimum amount of time and energy by applying a new color coat. Color-coating, or *dyeing* vinyl requires a special cleaner and top coat. The cleaner removes all contaminants from the surface so that the top coat has the best chance of staying put for the duration. Top coats can fall into two categories: ones that stay on the surface of the vinyl and others that actually blend into it. Although the process is pretty much the same for using both types of coatings, we'll be discussing the method that uses a top coat that stays on the surface of the vinyl. Whichever method you choose, be sure to follow the label directions on each product.

Clean & Condition Surface—To refinish the seats, they must first be cleaned with a special vinyl cleaner. These cleaners typically use solvents and water to remove dirt, grease, silicone-based protectants and wax from the vinyl. Wipe the vinyl using a sponge or cloth that's treated with the cleaner, then rinse the surface with water. If water beads on the surface during the rinsing operation, the surface is still contaminated. So apply the cleaner again until the rinse water no longer beads. Use a bristle brush with the cleaner to clean the grain of deeply soiled trim. Then wipe all trim pieces to be painted with vinyl conditioner. The conditioner "primes" the surface, giving the color coat a better tooth to adhere to.

Color Selection—Vinyl top coats are available from automotive paint supply stores. Jot down the interior color code from your Camaro's trim tag on the firewall and take this information with you. And some paint stores may match your vinyl's color if they don't have your particular color formula handy. Just bring in a clean swatch of your vinyl. These paints have a flex agent in them that allows them to flex without breaking. But take note that the more color coats you apply, the more likely that they're going to crack. So only use enough paint to hide the old color and no more.

Before you apply the top coat, mask off any areas that you don't want painted. Then with gun pressure set at about 35-40 psi, spray on the top coat. Using multiple fog coats, keep the gun about a foot away from the spray surface. By the way, some paint stores will put the paint in spray cans for you for a nominal fee if you don't have access to a spray gun. Allow the trim pieces to dry for 2-3 hours at room temperature and humidity before handling them.

RECOVERING

When the seats are out of the car, it's a good time to see whether or not they need to be recovered. A small tear or two may be repairable, but if the tears are long and many, you're better off recovering the seats.

Seat covers for some of the later-model Camaros are still available from GM dealers. Although the cost is high, the covers meet GM's rigid specifications for quality. Very few companies can come close to meeting these specifications.

Many companies offer "exact reproductions" of discontinued seat covers. Some are quite good, others not so good. A good way to check the quality of a reproduction is to compare it side-by-side with your original. You need to check three basic quality-related items: material, color and construction.

Checking Quality—You can begin by checking the material. Feel the thickness and weight of the original vinyl or cloth and compare it to the cover you're thinking of buying. The closer they are in feel, the better. The same thing goes when you compare the texture or grain of the vinyl or pattern of the cloth. And if the seat covers you're replacing have molded vinyl stitching, compare that also.

Color is a little tougher to judge on older seat covers due to exposure to the damaging ultra-violet rays of the sun. Whenever possible, cut off a swatch of material that was folded under the seat, out of direct exposure to the sun, and use it as your color and fabric sample. Just make sure that you compare the color under the same type of light: fluorescent, incandescent or sunlight. Colors can appear to have different shades under different types of light.

Now compare the construction. Look carefully at the thread holding the cover together. The color and stitching should match the factory piece. And if the factory seat cover has two rows of stitching next to each other, the reproduction should have it as well.

Also, check to make sure that any foam used in the covers is the right type and in the right places. For instance, the foam used under the cloth inserts on some Camaros has distinct V-shaped grooves in it that form the main ribs of the seat. If these grooves have rounded shoulders or appear to be "flat" when compared to the original covers, you're better off looking elsewhere.

Location of the padding is another point to check. For example, some factory covers have padding on the underside of the vinyl which runs alongside the cloth inserts. The reproduction covers should have this also.

So as with many other purchases, it's buyer beware. If you can't compare the covers side-by-side, check the return policy of the company you're dealing with before you send any money. That way if

To change or enhance the color of soft interior trim, vinyl dye or paint is available with necessary flex agents that allows trim to be pliable without color flaking off. Rechargeable Air-0-Can from Classic Camaro has enough compressed air to paint many small interior parts.

Cleanliness is critical when painting or dyeing interior trim. Here, Jim Altieri of Arizona GM Classics uses vinyl cleaner to remove all traces of dirt, grease, wax and silicone-based protectants. After applying color coat, allow pieces to dry for 2-3 hours before handling.

you're not entirely satisfied with the quality of the covers, you have the option to return them.

In this section and the next, we'll concentrate on recovering bucket seats. However, the process is almost exactly the same for bench seats, so what you read here will help you no matter what.

Before you begin tearing into this, keep in mind that you want to minimize any more damage to the seat or seat cover. And since the seat will be flipped over many times as you're working on it, it's a good idea to place the seat on something soft like a rug. With that said, let's begin.

Remove Seat Tracks—This is an optional step. To merely recover the seats, it's not necessary to remove the tracks. However, if you want to clean, inspect and relubricate the seat-track mechanism, it's easier to do so with the tracks removed. Turn the seat over so you're looking at the tracks. The side without the release lever is

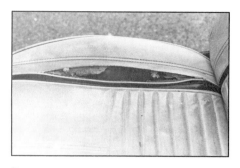

Some seat tears can be patched; this dry-rotted seam tear cannot. New seat covers can restore your Camaro's interior to like-new condition.

If you have new replacements, begin seat recovering by bending or cutting off one-way washers retaining seatback to cushion. If reusing washers, lever them off carefully with a screwdriver.

Spread seatback brackets apart and lift seatback off pivot pins.

Forward edge of seat cushion cover is retained to seat frame by this nylon channel on most models.

On rear seat cushion, hog rings secure cover around circumference of frame. Cut rings off with side cutters. Front cushions on '71-'81 models also have hog rings securing listings to frame from up top. Pull cover back to expose these rings and cut them off.

Remove screws for seatback cover and lift cover off seatback.

Seatback cover hangs on two metal hooks (arrows). Remove all hog rings, then slide cover up and off seatback.

Don't forget to salvage listing wires out of old covers. Many replacement covers don't include them and highly contoured 1971-81 seats won't look right without them.

most easily removed by removing the track spring first. Just unhook it from the track with your hand. Then you can move the track back and forth easily to access the track retaining bolts. If the remaining track doesn't use a return spring, just slide the track back and forth to access the bolts. And when you're removing the tracks, keep in mind that the bottom edge of the track with the slot in the middle is the outboard seat track on first-generation models.

Remove Seatback—The next step is to separate the backrest, or seatback, from the seat bottom. Begin by removing the trim from the hinge bracket. It's retained with barbed clips, which can be seen on the back side of the bracket. These clips can be removed by inserting a flat-blade screwdriver under their heads and levering them out. With the clips removed, pull the trim down and away from the bracket. This will unhook it.

When the trim is removed, you'll see that the backrest pivots on pins welded to the seat frame. And the backrest is retained on these pins with one-way washers and/or E-clips. To remove an E-clip, slide a screwdriver blade into either slot in the clip and lever it away from the pin. If it gets stuck, use a pair of pliers.

Next, remove the one-way washers—one on each side of the seat. But be careful because the washer may break or fly off as you remove it. To prevent this, put a cloth over the end of the pin before removing the washer. You can remove a one-way washer by pushing down on the backrest bracket and sliding a screwdriver under one side of the clip. Pry the washer so that it falls into the groove for the E-clip. Now take a second screwdriver and place it under the washer, directly across from the first one. Proceed to gently lever the washer over the end of the pin. Or if you're replacing the washers with new ones, cut off the old with side cutters.

Follow this procedure for both sides of the seat. With the retaining hardware off, place the seat on its side, spread the seatback brackets apart and lift the seatback off its pivot pins. Finally, pull the plastic washers off the pins.

Remove Cushion Cover—To remove the seat cover from the cushion, you'll need to remove the backrest catch and bumper at the rear of the cushion. On 1971-81 Camaros, these bumpers must be removed with a T-45 Torx-head socket. Then flip the cushion over so that it's upside down.

If you look carefully at the edge of the seat cover, under the seat frame, you'll see that the cover has a flat piece of nylon sewn to it. This nylon fits into a U-shaped channel in the bottom of the seat frame. This is the system that keeps the edges of the cover in place.

To remove the nylon strip from the channel, put one foot on the seat frame and put some weight on it. This will relieve some of the tension on the strip. Then, starting at the rear of the strip, remove the nylon from the channel. To do this, pull the retainer toward the seat springs then fold it toward the edge of the seat. Work in this manner until you've disengaged the retainer from both sides of the seat. Then do the same for the front.

Hog Rings—At this point, the only thing holding the cover to the frame are *hog rings.*

Hog rings are small, metal rings that are wrapped around the tie downs in the seat cover so named, by the way, because of their resemblance to metal rings put through the noses of domesticated pigs.

Broken side spring (arrow) on driver's seat of 1967-69 models is common. Previously unavailable, fix was to adapt spring from passenger's seat. New reproductions are available from Arizona GM Specialists and other sources.

Also available from Camaro aftermarket suppliers are these reproduction foam seat buns for 1967-69 models.

These are what hold the cover to the seat frame. To remove the hog rings, pull the outboard edge of the cover toward the center of the seat. If you look into the groove where the cover is secured, you'll see the hog rings. Cut the rings with a pair of sidecutters taking care not to damage the seat foam. On 1971-81 Camaros, the highly contoured high-back bucket seats have extra hog rings retaining the seat-cover listings, or tie-downs, (rope-like reinforcing strands) to the seat frame between the pleats in the deeply bolstered lateral-support wings. When the hog rings are cut on both sides of the cover and inside the pleats, remove the cover from the seat.

Disassemble Seatback—The rear of the backrest, or seatback, is covered with a metal or plastic panel. This panel has to be removed to access the hog rings securing the seat cover. And to do this you need to remove the screws that attach the panel to the seat.

On 1967-70 models, remove the backrest bumpers. This will require both a medium and a large Phillips-head screwdriver. On first-generation Camaros, you'll also need to remove the backrest release knob. It's retained with a small Allen-head set screw. With the knob removed, you can remove the panel. Gently pull it away from the bottom of the backrest and then slide it upward to disengage it from the retaining hooks at the top.

When the cover is off, take a minute to study the attachment locations for the hog rings. Make a diagram for future reference if you need to. Removal and installation of the backrest covering follows the same general sequence as the cushion (seat bottom).

RECONDITIONING

Reconditioning of the seat cushion can entail repairing part of the seat foam or springs to sandblasting and painting the frame. It just depends on how enthusiastic you are and what kind of shape your seats are in to begin with.

First-Generation—First-generation cars use springs in the seats to give them resiliency, just like an old-style mattress. Broken springs can be repaired with a special type of fastener available from interior trim shops. To repair the spring, the broken ends of the spring are placed in the fastener. Then a new piece of wire is placed alongside the broken area. This helps gives additional support to the spring. Now the ends of the fastener are folded around both pieces and volia, the spring is fixed.

Second-Generation—On second-gener-

ation cars, closed-cell foam is used to give the molded seat its shape and also act as a cushion. Problem is, with exposure to the sun, the foam dries out over the years, eventually decomposing and pouring out onto the floor like so much sand in an hourglass. Without good, pliable molded foam to act as a foundation, it makes no sense to recover your Camaro's seats. As of this writing, new molded-foam seat buns are not available from GM or the aftermarket. You could scour the junkyards for good used foam from a seat that saw little sun or driver miles (tip: get a passenger's front seat—it sees less use). Or you can repair your existing seat's foam

Reproduction wire-reinforced foam seat buns for 1971-81 Camaros are not yet available. You'll have to find a good used bun or repair the originals. To repair damaged cushion, get out that electric carving knife you used last Thanksgiving and cut out the damaged section. Go no deeper than the damage does and cut squarely.

Here, Jay Jett of Jett's Custom Upholstery, Tucson AZ, applies contact trim adhesive to rebond foam patch and cushion.

Once cemented in place, Jay uses carving knife to trim patch to contour of seat cushion. If foam is same density as original, repair is unnoticeable once seat cover is installed.

Ubiquitous hog-ring pliers is best way to install hog rings that retain covers to frame. Notches in jaws hold rings securely as you penetrate fabric, burlap and foam until rings can be clamped onto frame.

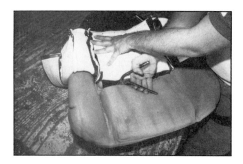

On highly bolstered 1971-81 buckets, install hog rings anchoring listings to frame through grooves in foam.

If cover fits loosely on seat foam, drape some 85/15 cotton filler over buns to take up slack.

Tight-fitting cover is what you want for wrinkle-free appearance. To aid cover installation, spray silicone on buns beforehand. Observe alignment and pull cover snug.

by cutting out the worn or dried-out section and gluing new foam in.

To glue in some new foam, you'll need an electric carving knife (just like the one you use to carve up that tasty Thanksgiving turkey), some 3M aerosol trim adhesive and a chunk of 2-in. Rebond Foam (available from interior trim shops). With the electric knife, cut out the damaged foam and some of the surrounding area, making a square or rectangular depression. Don't cut any deeper than you have to. Cut out a piece of Rebond Foam to match the cutout and glue it in place with the spray adhesive. After allowing the adhesive a few minutes to set up, trim the new foam patch to match the contour of the surrounding foam. Do this with the same care used to carve your Thanksgiving turkey.

COVER INSTALLATION

Begin installing the seat cover by turning it inside out. This will make it easier to access the listings, or tie downs, on the underside of the cover. The tie downs are the pieces of material with a cord running through them. You have to anchor these to the seat frame first. This is accomplished with new hog rings and hog-ring pliers.

Hog-Ring Pliers—Hog-ring pliers are special pliers with notches in their jaws. These notches hold the hog rings securely in place. And be sure use good-quality hog-ring pliers. Many of the pliers that are available are cheaply made out of round steel stock. These aren't worth your time or money because of their poor design. Buy your pliers where the local trim shop buys theirs.

To make installation of the hog rings easier, the pliers should taper to a blunt point. If your pliers don't taper, grind them until they do. You'll also find it easier to use the pliers if the handles are bent at a 45 degree angle in relation to the jaws. This gives you more room to work when you're trying to install the hog rings. If your handles aren't bent, heat them as necessary to achieve the bend.

Anchor Listings To Frame—Now position the inside-out cover onto the seat. Align the piping on the cover so it's on the edge of the seat. You don't need to be perfect here, just close. You'll be able to shift the cover around a bit after you've anchored the tie downs to the frame. The tie downs should fall into the creases in the seat padding.

Anchoring the seat cover to the frame is next. If your's is a 1971-81 high-back seat, first install the hog rings securing the listings for the side wings in the center of the

seat. Then, on all Camaros regardless of seat design, secure any tie downs that run side-to-side on the seat. Then proceed to the tie downs that run front-to-back. This helps minimize wrinkling of the cover.

Place the inside-out cover on top of the seat so that it lines up as it would normally. This is increasingly important as the distance between the tie down and the cover decreases, because there will be less material to allow you to shift the cover. About 3/4 in. from the point where the tie down is attached to the seat cover to the bottom of the tie down cord is the norm.

To secure the seat cover, grab one of the tie downs and hold it in the crease of the padding. And hold it as close to the frame as you can get it. Then place a hog ring into your modified pliers. Now push one edge of the hog ring through the tie down and grab the wire in the frame with the other edge and crimp the hog ring around it. If you don't succeed in anchoring the cover to the frame, try again. You'll probably only hit it every other time. The secret is to keep pushing down as you crimp the hog ring. Install the next hog ring about 2-3 in. from the last hog ring. The closer, the better.

Check Cover Alignment—As you proceed with hog-ringing the cover, take time to check your work. The seams should be straight and square as were the originals. If they're not, try shifting the cover by hand. This can take some muscle, but be sure not to rip the cover. If the cover still won't move, cut some of the hog rings until you can get the seams straight. Then finish anchoring the cover to the seat.

If you're installing new seat covers that just won't snug down for a tight fit no matter how tight the hog rings are and you don't have the time or resolve to return them for a refund or exchange, there is hope. Get some 85/15 cotton filler from an automotive trim shop and drape it over the foam buns in the loose-fitting areas, then reinstall the covers. Nine times out of ten, this will fill out any baggy-looking seat covers. Alternately, you can use some Dacron, which you'll find at your local fabric store.

When you've anchored the tie downs, you can unfold the cover around the edges of the seat. At this point, you want to work from the front of the seat to the back. This allows you to take up any slack in the cover at the rear where it won't be seen.

Engage Nylon Edging—If your seat covers have nylon edging sewn into them, it will have to be folded over and placed in the channel in the seat frame. And this will require pressing down on the front corners, thereby compressing the springs or foam, and stretching the cover as you work your way back.

If your covers require hog ringing, just stretch the material enough so that no wrinkles appear on the sides. Then anchor it with hog rings. Remember to work front-to-back.

Remove Wrinkles—After the front and sides of the cover are secured, check the fit of the cover one last time. If you have any small wrinkles, try stretching them out as you refasten the sides of the covers. Also, some small wrinkles can be caused by insufficient padding. If this is the case, stuff just enough Dacron padding underneath the wrinkles to remove them and no more.

With all wrinkles removed, you can pull the material toward the rear of the seat. Then tuck it underneath and anchor it with hog rings to the seat frame. If a stubborn wrinkle or two pops up at this point, take the seat outside and let it sit in the warm sun for a while. This generally relaxes the cover material and helps remove wrinkles.

Now you're finished with the seat cushion. Recovering the backrest is very similar once you've removed the cover. And the rear seat follows this same procedure as well.

SEAT BELTS

When all is said and done, what can you do with your Camaro's seat belts? After 20 years of buckling and unbuckling, exposure to sun and grease and grime, they're probably somewhat frayed and worn looking. Maybe the retracting mechanism doesn't work all of the time or the female end doesn't latch securely. And for sure, the buckle ends will be scratched up. Unfortunately, you can't buy new replacements from GM. And aftermarket belts would look tacky in a restoration.

Inspect your Camaro's seat belts. If they're just a little faded and the buckles a bit scratched, you can clean the belt webbing and polish the buckles. But if the webbing is frayed, torn or dry-rotted, the retractor jammed or the female latch sprung, the belts will have to be replaced.

You have two options here: find good used belts from a donor car (not impossible) or have your seat belts rebuilt. Rebuilt? **Ssnake-Oyl Products** (not a typo, that's two S's in Snake) in **Dallas, TX (214/233-3047)** takes your old belts, installs new 6000-lb, 3- or 4-panel webbing in the correct color, repairs or replaces the buckle and anchor ends, replates or polishes all metal parts and sews it all together with original-style thread in the original pattern for about $50 per belt set.

And remember, whatever you do, never substitute standard bolts for seat-belt anchor hardware. From the chrome or cad-plating to the tensile strength, these bolts meet the design criteria of the manufacturer to keep you anchored in the car in the event of a collision.

CONSOLE

First-Generation—Two types of consoles were used in first-generation Camaros. The 1967 models used a console with a pot-metal cast top and back with plastic sides. For 1968 and 1969, Camaros utilized a console made entirely of plastic. Also, there were two basic trim levels: with console gauges or without.

The console gauges featured on 1967 models included a fuel gauge, coolant-temperature gauge, clock, oil-pressure gauge and battery-charge indicator (ammeter). The 1968-69 models had the same complement of gauges minus the clock. The oil-pressure gauge is mechanical and is linked to the engine with a small plastic tube.

Unfortunately, on 1967 models, none of the console gauges are numbered so you don't have any real values to work with if something goes wrong. Both 1968 and 1969 models feature numbered fuel, ammeter and oil-pressure gauges.

Moving to the rear, all the consoles included a storage bin, ash tray and illumination light. One problem many consoles had in common was breakage of the hairpin spring that aided in raising of the storage-bin lid. This spring is easy to replace once the console is out of the car.

Second-Generation—Two different consoles were available on second-generation Camaros. A molded, hard-plastic console, similar in design to the 1968-69 unit, was offered from 1970 to 1972. Beginning in 1973, Camaro adopted a padded-vinyl-covered urethane console used in the Firebird. Because second-generation Camaros were designed from the start with full instrumentation capability in the dashboard, no optional console-mounted gauges were available or necessary.

Like its 1968-69 counterpart, the 1970-72 Camaro console is constructed of multiple pieces of hard ABS plastic, with provision for a manual 3-speed or 4-speed shifter or a stirrup-style automatic shifter and lighted quadrant. The console also contains a storage bin with lid, ashtray for rear-seat passengers and inboard seat-belt receptacles.

The 1973-81 console deletes the rear ashtray and seat-belt receptacles of the previous design. But there's a new storage shelf formed at the forward end of the console where it fits beneath the dash and the storage-bin lid is now foam-padded vinyl, providing a comfortable right-hand armrest. When power windows are ordered, the switches are in the console. With automatic transmission, a Firebird-style shifter is used and the lighted shift quadrant is moved to the center of the instrument cluster, as with column-shift cars. Originally, all consoles were black, but beginning in the late 1970s, the console color matched the seats and other interior trim.

Most seat cushion covers have nylon edging sewn into their front and sides that engages channel in frame. Depress cushion by hand and snap edging into channel.

After joining seatback and cushion, install one-way washers with deep-well socket and hammer. Install remaining seat trim and voila, it's 1971 again!

Give your Camaro's seat belts a good cleaning and inspection. Polish buckles and anchor ends on a wheel. Replace belts with frayed webbing, broken latches or stuck retractors.

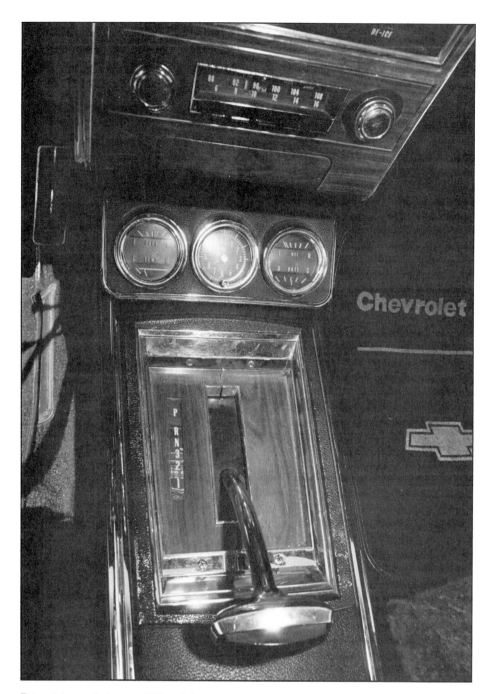

Pot-metal console top on 1967 models can be detailed with semi-gloss enamel. Circular-dial console gauges were optional in 1967 only.

REMOVAL

If for some reason you haven't removed the console yet, do so now.

1967-69 Models—Before you begin to remove the console, here's one item to keep in mind. Some of the console retaining screws fit into rivet nuts in the floor pan. If the screws are rusted tight to the rivet nuts, the nuts may spin if enough torque is applied. So you may have to end up drilling off the heads of the screws to remove the console. This will leave you with rivet nuts that are useless.

This situation can be remedied by driving out the old rivet nuts from the underside of the car with a long drift and then installing new rivet nuts. And as bad as this may seem, new rivet nuts require a special crimping tool for installation, so this can get expensive too.

Another way of remedying this situation is to weld nuts in position as required or simply use nuts on the underside of the floorpan and hold them as the screws are being tightened. The disadvantage to this is the nuts will spin if you need to loosen the screws.

First, remove the shifter knob on manual-transmission cars. The shift knob is held in position by a large locknut, located just underneath the knob. The shifter handle used with automatic-transmission cars doesn't need to be removed to remove the console. As with manual-transmission cars, all you need to do is remove the shifter trim plate. And if your console is equipped with gauges, disconnect the wiring harness (found by following the har-

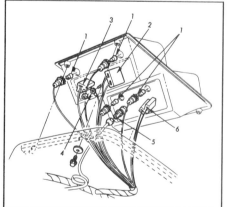

1. Cluster Lamps
2. Ammeter Conn.
3. Temperature Gauge Conn.
4. Ground Wire
5. Oil Gauge Conn.
6. Fuel Gauge Conn.

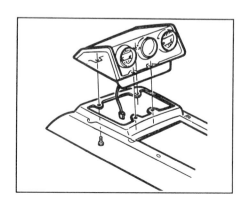

Wiring details—1967 console gauges. Courtesy Chevrolet.

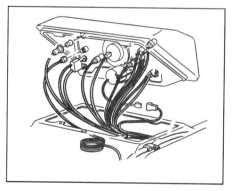

Wiring details—1968-69 console gauges. Courtesy Chevrolet.

ness from the front of the console to the underside of the dash) and oil-pressure line.

With these items out of the way, all that's left to do is remove the fasteners securing the console to the car. In 1967 Camaros, two screws, just forward of the shifter plate, on the side of the console, secure the front of the console. You'll also find a retaining screw inside the console storage bin and underneath the ashtray. These last two screws are the ones that fit into the rivet nuts, so keep this in mind if the screws don't seem to loosen.

The console in 1968-69 Camaros uses a bolt underneath the rear edge of the shifter plate and underneath the ashtray. The easiest way to unfasten the front of the console is to slide underneath the car and remove the two nuts that hold the console retaining bracket. These are found in the transmission tunnel, above the transmission.

1970-72 Models—Remove this console in a similar manner to the plastic 1968-69 console. It bolts to the transmission tunnel with six screws in three brackets. Access to these mounts is gained by removing the shifter and trim plate and through the storage bin. Removing and installing this console is a lot easier with one or both of the front seats removed.

1973-81 Models—The Firebird-type console is retained to the transmission tunnel with four screws and instrument-panel support with two screws. Access to the forward mounts is underneath the radio. The transmission-tunnel mounting screws can be reached after removing the storage bin tray. First remove the snap-ring retaining the shifter knob, and slide the knob off the shaft. Then remove four Phillips-head screws retaining the shifter trim plate and lift off the plate. If your Camaro has power windows, unbolt the storage-bin tray and disconnect the wiring harness before sliding out the console. As with the 1970-72 console, removal and installation is facilitated by first removing the front seats.

CONSOLE GAUGES
(1967-69 models only)

As stated earlier, all of the gauges used in the console are electric except the oil-pressure gauge which is mechanical.

Without access to some specialized testing equipment, the only practical way of testing the fuel gauge, ammeter or temperature gauge is to try eliminating other possible causes of failure.

Electrical Gauges—For instance, if the fuel gauge isn't working, check the "GAUGES" fuse first. If the fuse is in good condition, check the cleanliness and

tightness of all connectors. If everything's OK, take a careful look at the wiring itself. It shouldn't be cut or grounded against another part of the car and its insulation should be intact.

If all these items check out, that only leaves the sending unit and fuel gauge as the possible culprits. If you or a friend have another fuel gauge in good working order, substitute it for yours. If the substitute gauge works, you know yours is faulty. Various electronic repair shops around the country can repair your old fuel gauge. On the other hand, if the second gauge doesn't work, it's time to replace the in-tank sending unit.

If you follow these steps, you'll be able to narrow down the cause of an inoperative electrically powered gauge, whether it's a fuel gauge, ammeter or temperature gauge.

Mechanical Gauges—The oil-pressure gauge is another story, because it's a mechanical unit. A small-diameter plastic tube runs from the rear of the block, next to the distributor, to the gauge. Oil is fed through this tube to the gauge. Con-

sequently, if the gauge isn't working, it means one of three things; oil pressure is low, the plastic tube is kinked or damaged, or the gauge is faulty.

Start your diagnosis with the engine first. Make sure the oil is of the right viscosity and is up to the FULL line on the dipstick. Then start the engine briefly. If you hear a ticking noise in the area of the valve covers, something is definitely wrong. *Shut the engine off immediately!*

Now disconnect the oil line leading to the gauge in the car. You'll find it on the top of the block, just to the left of the distributor. Then connect a known good oil pressure gauge in place of the old one and start the engine again. If oil pressure is low or non-existent, have a professional look at the car. And don't drive it to his shop unless you want him to rebuild the entire engine. Have it towed.

Now if the engine doesn't make any ticking sounds, but sounds healthy, check the oil-pressure-gauge feed line along its entire length. It shouldn't be kinked or chafed. If both the engine and the oil-pressure-gauge feed line check out OK, the gauge is faulty and must be replaced.

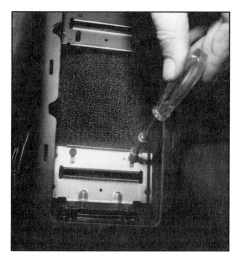

When removing 1973-81 console, look for these hidden hex screws under storage bin.

Reproduction plastic 1968-69 consoles from Classic Camaro include woodgrain kit.

Console gauge lenses scratched or clouded over? Get new ones or polish to see-through brilliance with Meguiar's #10 plastic polish.

On 1973-81 console, padded vinyl storage bin lid often cracks. New lid, spring and latch are available separately.

Hastily removed trim panel often results in metal clip (arrow at right) tearing through hardboard backing, rendering that attachment point useless. Straighten bent tabs (arrow at left) to remove trim panel garnish moldings.

Christmas tree plastic clips often break or lose their barbs when trim panel is removed. Use new clips at installation time.

Armrest screws on 1970-71 models are removed from back of trim panel.

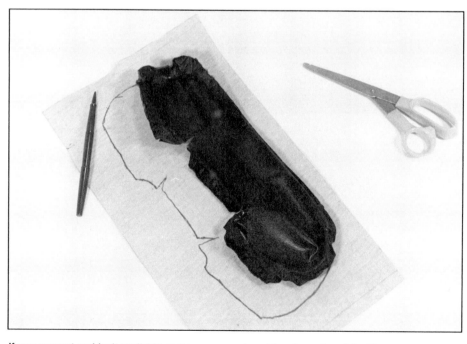

If new armrest pad isn't available, make a cover using old pad as a template. Then cement new cover in place.

CONSOLE RECONDITIONING

With the console out of the car, you'll be able to disassemble it entirely for cleaning and inspection.

1967-72 Models—If your 1967-72 console has any small cracks or holes in the plastic or metal body, they can usually be repaired with an epoxy-based adhesive. Just remember, the crack must be properly prepared beforehand. Do this by grinding a groove in the crack, along its entire length, then fill it with adhesive.

When it comes time to repaint your console, use enamel, not lacquer. Enamel paint is far more durable than lacquer, and for some reason, consoles are the first items to get damaged when something bad happens in the interior. *Ultra* by Plastikote semi-gloss black is a good choice. It's a little shinier than the original paint, but it's long lasting. If you want that original-look semi-gloss, have some made up specially by your local automotive paint store. It's also available from some of the aftermarket suppliers.

If yours is a deluxe console with wood grain trim, repairing cracked, faded or cigarette-burned wood grain insets is as easy as buying reproduction precut replacements from aftermarket Camaro parts specialists. Or you can pop for a new reproduction console such as those available from **Classic Camaro Parts & Accessories in Huntington Beach, CA.**

On 1967-69 models, if any of the gauge lenses are lightly scratched, try polishing the scratches out with a good plastic polish like *Meguiars*. Removal of deeper scratches will require sanding with 3M 1500-grit *Microfine* sandpaper followed by application of the polish.

If you're in the market for new gauge lenses, and you want a true restoration, make sure the new lenses are exactly like the originals in every way by comparing them with your old lenses. Some aftermarket lenses aren't like the originals in that they use white type instead of green. This doesn't make them bad, just different. You can check the color of the type by shining a flashlight through the lens as you look at the type.

1973-81 Models—With age and exposure to sun, this padded-vinyl console tends to crack and the foam padding leaks out like so much sand in an hourglass. If the console itself is cracked, your best move is to buy a new one from your GM dealer. Replacement consoles come only in black, so if yours is interior color, you'll have to dye it to match. For power-window control switches, you'll have to cut out new holes. If the console lid is cracked, or its closing spring or latch broken or missing, these items can be ordered separately.

DOOR TRIM PANELS

Camaro door trim panels have been produced in a wide variety of styles and treatments. Many are interchangeable. For example, the 1968 and 1969 molded door trim panels are identical except for the location of the door lock button. On 1968 models, the button is located at the top rear edge of the panel. In 1969, Chevrolet moved the button closer to the center of the door to help deter theft.

On second-generation Camaros, trim panels are manufactured in one, two, or three pieces depending on model year and trim level. Some are made of injection-molded plastic while others are vinyl or cloth-and-vinyl covered hardboard. A hybrid door panel with an ABS lower half and vinyl-covered hardboard upper half was widely used from 1972 to 1976.

Cleaning—If you're fortunate, the trim panels will only require cleaning. The plastic trim panels can be cleaned with a mild solution of soap and water. Use of a soft bristle brush will help lift the dirt out of the grain. Vinyl-covered and molded vinyl trim panels can be cleaned with a vinyl cleaner such as *ArmorAll Cleaner* and a soft bristle brush. If a stain is hard to remove, try using *Bestine*. which is a solvent that evaporates very quickly. Use of a quickly evaporating solvent is important because it avoids damaging the surface. Bestine can be found in art supply stores. When the panel is clean, treat it with a preservative to prevent it from drying out and cracking.

Cloth trim panels can be cleaned with upholstery or carpet shampoo. Mix the shampoo with lukewarm water and agitate the mixture until you have plenty of suds. Apply the suds to the cloth with a soft bristle brush and scrub lightly. Avoid soaking the cloth to prevent future rotting of the fabric and thread.

Fasteners—All trim panels use a variety of fasteners to secure them to the door. All trim panels use Phillips-head screws in the armrest area to provide a solid anchor point for closing the door. The perimeter of the panel can use nylon fasteners which are shaped like Christmas trees, a metal retaining clip which slides into a plastic sleeve in the door frame, or Phillips-head screws which are usually positioned at the bottom of the panel.

Hockey-Stick Armrest—Many second-generation cars (1975-up) use an armrest that's shaped like a hockey stick. These armrests use a hard plastic inner core with a soft urethane exterior coating. Unfortunately, they have a bad habit of tearing at the front retaining screw after a period of time. And sometimes the handle portion of the armrest breaks. A repair kit that's offered can repair either type of damage. To begin the repair, saw off most of the handle with a hacksaw. Then slip the replacement handle provided in the kit over cut portion of the armrest and secure it with screws. This repair kit works fairly well but is quite noticeable (see adjacent sidebar). The only way to avoid this situation is to purchase a new armrest from the dealer and dye or paint it to match the original. See the sidebar, page 136.

Vinyl Repair—Cracked or torn trim panels can be repaired satisfactorily using a vinyl repair kit available from auto body supply stores and restoration supply houses. These kits typically include an adhesive, a solution to color the repair and graining sheets to provide the repair area with a grain. The quality of the kit is usually proportional to its cost; the best kits being the most expensive. Remember, you get what you pay for. Refer to page 125 for more details.

Lock Button & Ferrule—One area that takes quite a beating in normal service and is immediately apparent upon walking up to the car door is the inside door lock button and its surrounding trim molding—called a ferrule. The chrome-plated plastic buttons get chafed and scraped and bleached by the sun. Replacing them is as easy as unscrewing what's left of the old buttons and screwing on new ones.

But the ferrule suffers two ways. First, its star-shaped inside gasket wears out, allowing the knob to flop around and chafe. Second, and especially on second-generation Camaros with deluxe interior trim and padded upper door trim panels, the ferrule's chrome rings tear through the padded vinyl. Here the fix is to buy slightly oversized aftermarket ferrules, such as those offered by **Classic Camaro, Huntington Beach, CA,** to cover the torn area.

With the trim panel removed from the door, bend out the old ferrule's retaining tabs and install the new oversized ferrule. Then, using a good fabric cement, such as *Goodyear Pliobond*. cement the padded vinyl to its hard backing and secure it in place by bending flat the ferrule tabs. Check the outside of the trim panel to make sure no cement leaked out on the vinyl, then put the trim panel in the sun to dry. If necessary, use two small C-clamps to hold the ferrule in position as it is drying.

RESKINNING

Molded door skins for first-generation cars are available as complete trim panels or vinyl replacement skins. The complete trim panels usually are ABS-backed and have carpeting at the lower edge. The backing may have depressions indicating locations for manual window cranks or power window switches. To mount either piece, carefully cut a slit or hole through the back side of the door skin in the appropriate area, using your old door panel as a guide.

A vinyl replacement skin fits over the existing door panel. This process is more involved than simply buying a new door panel but can save you as much as 60% if you want to do the work yourself.

Removal—Before reskinning the trim panel, you must remove it from the door. To do this, first unscrew the lock button and remove it. Then remove the Phillips-head screws that retain the armrest. These screws are positioned below or inside the armrest. Now remove the window cranks. They're attached to the window regulator shaft with an open-ended wire clip. The open end will be pointing toward the knob on the window crank. The easiest way to remove these retainers is with special flat-bladed pliers designed for the job. Many auto supply stores have these tools in stock. Slide the tool between the crank and the scuff plate from the end opposite the knob. Grab the bent ends of the clip with the pliers and pull the clip out. If you can't get the special tool, you can use a small flat-bladed screwdriver to remove the clip although your chances of damaging the trim panel are increased. Push the trim panel away from the crank and insert the screwdriver into the looped end of the clip. Then pry the clip away from the handle.

Now remove the Phillips-head screws from the bottom of the trim panel. These screws are used on all first-generation and some second-generation cars. If screws aren't used, a plastic retainer that resembles a Christmas tree is. These retainers are used on both the sides and bottom of many second-generation cars. Although a special tool is available to remove them, you can remove them from the door frame by locating each retainer and prying them out of the door frame with a flat-blade screwdriver. The sides of the trim panels on first-generation cars use a metal clip that will slide out of the door if you gently pull on the trim panel.

The top of the trim panel is attached to the door by merely sliding under a metal trim rail on 1967 models only or "hangs" over the top of the door on all other cars.

HOCKEY-STICK ARMREST REPAIR

Hockey stick armrests used on many 1975-and-later cars often break in forward handle area. Effective but not-so-attractive fix involves sawing off broken section and bolting on hard plastic handle to front of existing armrest.

If you have a 1975-or-later Camaro with armrests that look like hockey sticks, chances are the armrests are broken. And by far the most common area of breakage is the screwhole at the front of the handle. This is partly because the handle is used to close the door. And partly because the plastic beam inside the handle fatigues prematurely.

There are at least three ways to repair the handle. The first repair consists of cutting off most of the handle and replacing it with a hard-plastic clone. But the problem with this method is that you can see the repair. And the clone looks ugly.

The second method can be used if the only damage consists of the front armrest screw pulling through the handle. And it involves using a metal reinforcement available from J.C. Whitney.

The reinforcement is fitted into the scalloped area for the front armrest screw. Then a screw is inserted through the reinforcement into the good part of the handle. All that's left is installing the front armrest screw through the reinforcement into the door. Because the reinforcement fits into the scalloped area, the repair isn't obvious—

especially if you paint it the same color as the armrest. So this is a good fix if you're on a budget.

The third method is a home-grown cure for armrest handles that are broken more than 1-in. away from the front retaining screw. It entails cutting the back side of the soft vinyl with a utility knife just about down to the armrest. Then cut off the hard-plastic core of the handle near the armrest.

When the core is removed, carefully wrap the exterior of the soft-vinyl handle with duct tape. Be sure that the cut sides of the handle line up with each other. Otherwise you'll end up with something akin to a pretzel when you've finished the repair.

Now mix 2-3 oz. of fiberglass resin and hardener. When this solution is thoroughly mixed, stand the armrest on end and pour the solution into it. If you want to increase the strength of the handle even more, grind a washer to fit into the end of the handle. After pouring the resin into the handle, cap the end of the handle off with the washer. Then let it harden.

When the resin is hard, simply drill a hole where the handle screw originally was and you're finished.

After all trim-panel hardware and clips are removed, simply remove the trim panel off the door frame. When you've removed the molded trim panel from some Camaros, you'll find a large spring in the area of the window crank. This spring keeps tension against the door panel and the window crank. Be sure to install it when replacing the trim panel.

To remove the molded door skin, you first have to remove the metal trim from the door. The Camaro insignias are retained by metal push-on clips. Remove these clips by cutting them with sharp cutting pliers. The metal moldings are attached to the door with metal tabs. Before you remove the moldings, take a careful look at the way the tabs are bent. The

If vinyl is tearing out around lock ferrule hole, cement vinyl to backing before installing new ferrule.

It doesn't look like much now, but once this vinyl door skin is installed over trim panel backing, it'll look as good as new and save you money in the process.

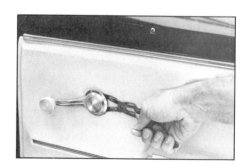

You'll need a lock-clip tool or equivalent to remove the horseshoe clip retaining the window winder.

tabs at the ends of the molding are bent in the opposite direction of the inner tabs. This keeps the molding from shifting and backing out. Keep this in mind when you're installing the molding. To remove the molding, simply straighten the tabs and remove the molding from the door. If the molding at the top edge of the carpet binds, it's probably hanging up in the carpet. Work slowly to avoid bending the molding.

Staples and glue are used to secure the skin and weatherstrip to the cardboard backing. The staples can be removed with a flat-bladed screwdriver or a pair of cutting pliers. Be careful not to damage any component as you remove the staples. You can peel away the vinyl skin that's glued to the cardboard backing with your fingers. Try not to take any cardboard with it.

At this point, carefully peel the vinyl skin away from the molded-foam backing. Work slowly to avoid tearing the foam because the foam serves as the base for the skin. If you do pull a chunk of foam out of the backing, carefully cut it away from the skin and glue it back into place with trim cement.

Installing New Skin—Warm the new skin by laying it out in the sun for awhile or carefully heat it with a hair dryer. This makes it more pliable and much easier to work with when covering the trim panel. Just don't melt it.

When the skin is warm and flexible, try a dry run of the installation process. Position the skin onto the foam in the area of the arm rest first. Work from the armrest toward the outside of the skin, making sure the peaks and valleys of the skin match those on the foam backing. When you are satisfied with the fit of the skin to the foam, begin installation. Get the skin warm again, then spray the back side of it (except for 1/2 in. around the edge) and the molded foam with trim cement. Position the skin to the foam starting with the arm rest. Then carefully work toward the outside of the skin, making sure there are no air bubbles and the skin matches the foam backing. Spray the remaining edge of the skin and the edge of the backing with trim cement. Wrap the edge of the skin around the backing until it's well secured. The lower edge of the skin where it meets the carpet can either be tucked in behind the carpet or carefully trimmed so it's flush with the top edge of the carpet. Staple the skin to the perimeter of the backing from the hidden side.

Inner Window Felt—You can install the inner window felt with screws, rivets or staples. If you're trying to duplicate the "factory" look, use staples. You can use 0.030-in. wire to make your own "staples." First, position the felt to the panel (if your panel is so equipped) and secure it with small C-clamps. Carefully drill pairs of 1/32-in. holes every 4 in. through the felt and backing. Each pair of holes should be 1/2-in. apart. Insert the wire through each set of holes, then cut and crimp the wire as necessary. Whichever method you use, paint the heads of the fasteners with flat black paint so they won't be noticed.

Insignias & Hardware—Now cut slits through the skin from the back side of the panel for the window regulator shaft and other hardware that protrudes through the skin. Install the trim molding, remembering to bend the two tabs at each end toward each other to prevent the trim from shifting. Finally, install the Camaro insignias using new press-on retaining clips.

REAR QUARTER TRIM PANELS

Except on convertibles, rear quarter trim panels for first-generation Camaros are padded vinyl over hardboard. These can be repaired in a similar manner to door trim panels, as long as the hardboard hasn't decomposed due to exposure to moisture. Camaro convertibles are different in that they use a metal backing. However, door trim panel repair procedures will work for these pieces as well. And whatever the reason, reproduction replacements are available.

Second-generation Camaros switched to molded ABS rear quarter trim which cannot be repaired if scratched or dried out from sun exposure. If yours are damaged, comb the junkyards because new replacements are not available from GM or aftermarket sources, as of this writing. If you find a good used set that's the wrong interior color, they can be dyed along with your other interior trim.

REAR SAIL PANELS

The panels finishing off the inside of the rear roof (C) pillars, called *sail panels,* are vinyl-covered hardboard on 1967-69 models and hard ABS plastic on 1970-81 models. On second-generation Camaros, the rearmost portion of the sail panels is exposed to intense sunlight directly under the rear window. Therefore, unless the car spent most of its life in a cloudy area like Seattle, WA, the plastic in this area will be bleached and brittle. As of this writing, new replacements are not available. The co-author's 1973 Type LT was an Arizona car with sail panels turning to dust. Luckily, **Camaro Country in Marshall, MI (616/781-2906),** came up with a good set

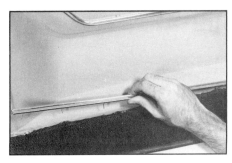

Transfer all garnish moldings and bright trim from old panel to new.

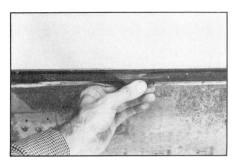

Stout staples retain door inner weatherstrip to trim panel. Remove these with needle-nose pliers, straighten and reuse with new weatherstrip.

To ease installation of staples through new weatherstrip, drill 1/32-in. holes for staples, tap into position with hammer and drift or 1/4-in. short extension. Then cinch staple ends in place by compressing them with small C-clamp.

Cement new skin to trim panel backing using clothespins to hold edges in place until cement dries. After installation of insignias and garnish moldings, reskinned trim panel is ready for 20 more years of service.

from a snowbelt car that could be dyed to match the interior color.

REAR PACKAGE SHELF

This hardboard and vinyl item gets destroyed by both water leaks in the rear window area and intense baking in the sun. Therefore, replacing it is a necessary part of every restoration. If necessary, dye the replacement shelf to match the interior color and make cutouts for options such as a rear speaker or rear window defogger. Then, the vinyl lip that forms the front edge of the shelf is glued in place with 3M contact cement. It's best to wait to install the shelf until after you've painted the car.

HEADLINER

As with the carpeting and rear package shelf, don't install the new headliner until after you've finished the body and paint work on the car. Concentrate on cleaning and if necessary painting or dying the parts that accompany the headliner: sun visors, headliner trim moldings, coat hooks and shoulder-belt trim pieces.

INSTRUMENT PANEL

Much of the reconditioning of the instrument panel and gauges can be done in the car. Briefly, this includes bezel cleaning, gauge testing, and replacement of the

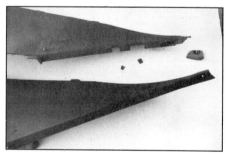

ABS plastic rear sail panels on 1970-81 cars decompose with exposure to sun. If shopping for good used panels, try junkyards in areas with cold, cloudy weather.

Using old package shelf as template, transfer cutouts for rear speaker and defogger, if so equipped.

switches and gauges. Procedures for these operations are covered below. However, if you're trying to bring your car into A-1 shape, or you just want better access to the components, you're better off removing the instrument cluster from the car. The instrument cluster houses the gauges and speedometer in the instrument panel.

DIAGNOSIS

You're better off diagnosing the gauges when they're in the car. You should only remove the gauges for service after checking them in the car. That is, unless, you plan to take them to a professional service for testing.

Before testing any gauges, make sure the wiring between them and their sending units is in good condition with no visible breaks. Repair any wires that are in poor condition before testing the gauges. Also clean all connections with some fine steel wool or sandpaper. This will assure good contact in the circuit.

You can test the operation of the coolant-temperature gauge and sending unit by removing the sending unit and placing its business end in a pan of boiling water. With the ignition on and the harness attached to the sending unit, the gauge needle should be about two thirds of the way toward the "H" marking. If it isn't, perform this test again using a known good sender. If this still doesn't cure the problem, chances are very good that the gauge is faulty.

If the fuel gauge, ammeter, oil-pressure gauge or clock isn't working, you can diagnose them following the procedure outlined under *Console,* page 134.

If you plan on doing any bench testing, you may need a signal generator for testing depending on the gauge(s) you're working on. A signal generator simulates the signals that would be sent to a particular gauge. This allows you to test them through their working range.

If you don't have access to this equipment or the knowledge to use it, you're better off having the gauges checked by a professional provided the testing fee and possible repair is less than a replacement gauge in good working order.

INSTRUMENT REMOVAL

1967-68 Models—Instrument clusters on 1967 and '68 cars are attached in the same manner. Screws secure the cluster at the top and bottom. However, before removing the cluster, disconnect the battery negative cable. And reduce the chances of scratching the top of the steering column by covering it with tape. When you've

taped the column up, remove the cluster retaining screws and slide the cluster toward the steering wheel. You'll only be able to pull the cluster so far before the speedometer cable and wiring harness stop your progress.

Disconnect the speedometer cable by pulling the cable retaining clip toward the steering wheel while you pull the cable toward the front of the car. To remove the wiring harness connector, simply push the two tabs on the connector toward each other and pull the connector free. Also, grab the socket of the high-beam indicator lamp at the top of the cluster, and while wiggling it back and forth, pull it out. Now take the cluster over to the bench to work on it.

Looking at the back of a 1967 or '68 instrument panel, you'll see they're quite simple, featuring a printed circuit, individual light receptacles and one wiring-harness connection.

If the printed circuit board has a broken circuit, you can try soldering an 18-gauge wire across the broken section. Sometimes this works and sometimes it doesn't, but it's worth a try.

To remove the gauges from the cluster, simply remove the screws that hold the gauge bracket to the cluster. However, before you do this, take note that the ground strap for the high-beam indicator light fits between one of the retaining screws and the metal gauge bracket. If this strap is not in this position, the high-beam indicator won't work.

With the gauge bracket removed from the cluster, you can remove the individual gauges. For details go to page 140.

1969 Models—Before you can begin removal of the instrument cluster retaining screws, and eventually the cluster itself, you'll need to remove the battery negative cable, the radio control knobs and disconnect the wiring at the headlight, wiper and power-top switches.

Begin instrument cluster removal by gently tugging on the radio control knobs. When the knobs are off, remove the radio-shaft lock nuts with a deep-well socket. You can use a pair of needle-nose pliers in a pinch if the lock nuts aren't too tight.

The wiper switch and power-top switch (convertible models) are both retained by screws which are accessed from the back of the instrument panel. When removing the wiper switch, be sure to disconnect the ground wire from the dash.

Removal of the headlight switch is accomplished by pushing in the switch shaft release button on the switch body while pulling the shaft out. Then remove the lock ring from the front of the dash with a pair of needle-nose pliers. Insert the tips of the pliers into the depressions in the ring and rotate the ring counterclockwise. When the switch is out of the dash, tag and remove the washer hoses from it, if your car has headlight washers. Then disconnect the wiring harness from it.

The cluster on 1969 models is held to the dash panel by two nuts on its right side. These nuts can be reached from the back side of the dash. The cluster also uses retaining screws at its bottom, left side and top. The bottom and side screws can be easily accessed. However, the top screws are covered by the dash pad. So you're going to need to remove the dash pad, (see page 140), before attempting to remove the cluster.

Before you start removing the instrument cluster from the dash, apply some tape to the top of the steering column in the area just forward of the cluster. This will prevent scratching the column when you remove the cluster. With the tape in place, gently pull the panel forward.

Now remove the speedometer cable from the back of the speedometer. Do this by pushing the cable retainer on the back of the speedometer toward you while gently pulling on the cable. If the cable won t come off, chances are you're not pushing the retainer far enough. Then tag the wires leading to the electrical gauges and remove them.

The 1969 instrument clusters are almost as simple as the earlier models, but they're different in that they don't use a printed-circuit board. Each gauge and warning light is wired individually. So take notes of wiring colors and sizes as you disassemble them.

1970-81 Models—There were actually two different instrument clusters used in these years; one from 1970 to 1978 and another in 1979-81 models. But for the purposes of removal and repair, the procedures are essentially the same.

Access to the gauges is greatly enhanced if the steering wheel isn't in the way. If you want to make the job much easier and avoid possible scratches on the gauge lens bezels, remove the steering wheel, page 142.

Disconnect the battery negative cable. Remove the six screws (two above the ashtray) retaining the lower trim cover and heater/air-conditioning control panel beneath the instrument cluster. This will allow you to sneak a hand up behind the dash to disconnect a few things.

Next, find the small-diameter cable running from the transmission shift indicator

(quadrant) to the steering column. Mark its position with a felt-tip marker and disconnect it from the column. Then, with a large flat-blade screwdriver, compress the retaining clip connecting the cable to the speedometer and pull the cable housing free of the gauge cluster.

The headlamp switch is next. To remove it, first locate a small, spring-loaded button on the side of the headlamp switch. Depress this button while pulling on the headlamp switch knob to remove the headlamp switch shaft. With the shaft removed, use two small screwdrivers or slightly opened needle-nose pliers or snap-ring pliers to unscrew the notched split ring on the headlamp switch bezel nut.

Reach up behind the cigarette lighter and disconnect its single wire lead. Pull out the lighter. Now insert two fingers inside the lighter socket (make double sure the battery is disconnected!), holding tension against the sides while you unscrew the outer portion of the barrel from the rear.

Remove the four (1970-78) or six (1979-81) screws retaining the cluster face plate to the gauge cluster assembly. Also remove two screws straddling the steering column from underneath. Tilt the face plate outward and disconnect the electrical connector for the rear-window defogger (some models) and wiper switch (and its fiber-optic tube if so equipped). Remove the face plate.

Remove the four screws retaining the gauge cluster assembly to the dash pad and pull the assembly out partially to gain access to final electrical connections. Disconnect the connectors for the printed circuit, clock and ammeter/voltmeter (if so equipped). Remove the cluster assembly to a bench for further disassembly.

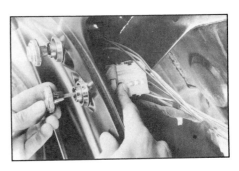

To remove headlamp switch knob, push button on side of switch and pull knob firmly out.

Headlamp knob retaining ring is slotted. If careful, you can remove it by pushing on slot in counterclockwise direction with screwdriver.

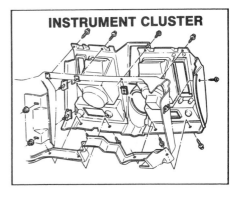

INSTRUMENT CLUSTER

Instrument cluster mounting details—1969 models. Courtesy Chevrolet.

Rear of 1967-68 instrument cluster shows delicate printed circuit and black bulb sockets. Now's a good time to replace any burned-out bulbs.

On 1973-81 models with automatic transmission and floor shifter, disconnect this clip for shift quadrant from steering column.

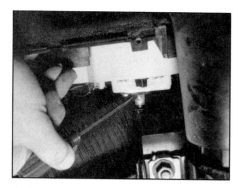

To disconnect speedometer cable, depress spring clip with screwdriver and pull cable housing away from cluster.

Undo cluster face plate screws, pull out plate and disconnect wires from cigarette lighter, wipers and rear defogger. As we said, it's much easier if the steering wheel is removed first. It's best to cover top of column with masking tape to prevent scratches to lenses or paint on column.

Melted spot on printed circuit indicates location of short circuit. Determine cause of short, then replace circuit.

Clear plastic gauge cluster lenses are no longer available for 1970-78 Camaros. To remove minor scratches, polish with Meguiar's #10 plastic polish or equivalent.

GAUGE REMOVAL & INSTALLATION

The gauges gracing the first- and second-generation Camaros range from the standard speedometer and fuel gauge to the optional clock, ammeter, oil-pressure gauge, coolant-temperature gauge and tachometer.

When you've removed the instrument gauge panel or console gauge panel from the car, you can remove the gauges. On 1967 and '68 Camaros, the fuel gauge and speedometer can be removed by unscrewing the five 1/4-in. screws around the perimeter of the metal gauge housing on the back of the cluster. With the housing off, remove the three 5/16-in. nuts from the printed-circuit board behind the gauge. And remove the lamp holders from the printed-circuit board as well. Now carefully peel back the circuit board and remove the two 5/16-in. stud nuts from the back of the gauge as well as the 5/16-in. screw beneath them. While making a mental note that the resistor board is positioned above the bakelite insulator, pull the gauge from the cluster. To remove the speedometer, simply remove the two rubber-mounted screws from its backside. On 1969 Camaros, the speedometer and fuel gauge can be removed in a manner similar to that of the '67 and '68 models. However, gauge removal for this year requires removal of the gauge bezels. Do this by first removing the bezel retaining screws located at the top of the bezel. With the screws out, gently pull the top of the bezel out so it rotates toward the ground. This will unhook the bottom of the bezel.

On second-generation Camaros, the plastic bezels for the gauges are incorporated into a one-piece surround molding which mounts to the front of the gauge cluster with eight screws. You'll probably want to remove this bezel assembly for cleaning and polishing with *Meguiars #10* plastic polish or equivalent. All gauges on 1970-81 models come out the back of the cluster assembly. Take extreme care not to bend or scratch the gauge cluster's printed circuit.

Remove the clock mounted in the instrument panel by first unscrewing the small screw in the middle of the adjusting button, and removing the button. Then remove the retaining screws from the back of the clock.

The fuel gauge can be removed by first disconnecting the harness from it. If you can see the back of the gauge, you'll find that three 5/16-in. nuts with lock washers retain it. You'll also see that a small resistor board is directly behind the left and

right nuts. What you can't see is that the terminals and bakelite insulation board is also retained by the washers and nuts.

So after you remove the nuts and washers, carefully slide the resistor board, left and right terminals, bakelite board and center terminal off the gauge lugs. Then you can remove the gauge.

Unlike many other components, when the time comes to install the gauges, all you really need to do is reverse the removal procedures. Just be sure that you have the wiring in the right places as per the factory wiring diagram or your own. And make sure all of the instrument-panel bulbs light.

DASH PAD

All first- and second-generation Camaros were equipped with a soft dash pad, or crash pad, on the upper part of the dash panel. The crash pads are made of a vinyl-clad foam (first-generation and 1979-81 cars) or rubberized material (second-generation through 1978) which helps to reduce the chance of injury in an accident.

The pads used in 1967 and '68 Camaros used a steel frame with a foam and vinyl overlay. This steel frame held up well under the repeated exposure to the sun. Unfortunately, this wasn't the case in 1969, when fiberglass was used as the framing material. The heat from the sun would cause the plastic to expand. However, when it cooled, it didn't contract evenly. This produced warpage of the pad. Consequently, it's considerably harder to find an unwarped dash pad for a 1969 model.

For 1970, Camaro switched to a fully integrated plastic instrument panel consisting of a hard ABS multi-piece lower section and a one-piece padded fiberglass upper section and rubberized dash pad. Problems with these instrument panels included poorly fitting lower sections and warping and "blackening" of the rubberized crash pad upon extended exposure to sun. Dash-pad cracking, however, was not a problem.

In 1979, the Camaro got a slightly revised dash design, with a new foam-padded vinyl upper section. These dash pads tend to crack with age.

REMOVAL

1967-69 Models—Removal of the dash pad on 1967-68 models is quite similar. The ends of the pad are covered with the A-pillar interior trim. So to remove the pad, you must remove this trim. And in order to remove this trim on convertibles, you must also remove the latch assemblies for the convertible top.

The dash pad has studs on its back side that fit through the dash panel. Nuts are used on the ends of the studs to keep the pad in place. Also, on '68 models, screws are used at the ends of the dash pad on either side. The nuts are accessible from the back of the dash panel. However, it can get crowded in a hurry back there, making it hard to get to some of the nuts. Removing the glove box makes it easier.

When all the nuts and screws are removed from the pad, carefully pull it away from the dash. If the pad won't come, you probably haven't removed all the retaining hardware. So check again before you damage the pad.

On 1969 models, the dash pad is also retained by clips which slide into slots in the top of the dash panel. When you've removed all the nuts and screws, lift the bottom of the pad until the bends on the clips are visible. Then pull the pad toward you and out.

1970-81 Models—Rather than a padded piece added on to the top of the existing instrument panel, the 1970-81 dash pad is designed as an integral part of the dash, forming the foundation to which the instrument cluster, headlamp switch, windshield-wiper switch, cigarette lighter, rear window defogger switch and A/C center vents attach to. So replacing the dash pad on these model Camaros involves disassembling the complete dashboard.

Start dash pad removal by performing the steps listed in this chapter under *Instrument Removal*, page 138. Next, remove the windshield garnish moldings inside the A-pillars (2 or 3 screws each), with the lower screw partially obscured by the door weatherstripping. Then remove the screws for the vents and A/C outlets underneath the lower section of the dash. Remove the glove box strap and slide out the glove box insert. This allows you to remove some of the heater and air-conditioning ducting above the glove box and gain access to two of the nuts for the clips retaining the forward edge of the dash pad to the metal section adjacent to the windshield.

Next, disconnect the power lead, speaker lead(s) and antenna from the radio. Pull off the radio knobs and remove the radio control-shaft nuts with needle-nose pliers or a deep-well socket. Remove the radio. This gives you access to a third dash pad-to-dashboard clip. Remove the nut for it directly in front of the radio speaker.

Finally, remove two hex screws retaining the dash pad to the dashboard, directly above and in front of the recess for the instrument cluster. Lift the dashboard

assembly free of its metal support. Angle it up and over the steering column, taking care to unsnag any wires or hoses that may have been left connected.

Remove the dashboard assembly to a workbench. Separate the dash pad from the lower panel assembly (4 screws). Dye the new dash pad to match your interior (unless it's black). Transfer any hardware not supplied with the new dash pad from the old one.

RECONDITIONING

Reconditioning of the dash pad can take several forms. If the pad is only dirty, clean it with something like *ArmorAll*

Cleaner. Then coat it with a preservative like *ArmorAll Protectant*. Be careful not to use a harsh cleaner or you may end up going through the paint on the pad, which brings us to our next point.

Painting or Dyeing —If the pad in your car is interior color, and black is showing through it after you clean it, the only way to bring it back around is to repaint or redye it. The same is true if you are replacing the dash pad with a new one as service replacements are available only in black. This requires that you use a paint with a flex agent or special interior dye available from your local automotive paint supply store. Make sure that you first clean the

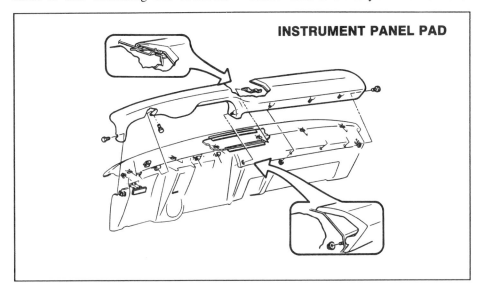

Dash pad mounting details—1969 models. **Courtesy Chevrolet.**

On 1970-81 models, dash pad replacement involves removing entire dashboard. As stated earlier, removing steering wheel first gives you more maneuvering space.

dash pad with *Prep-Sol* to remove any traces of dirt, wax and especially silicone-based preservatives such as *ArmorAll*. Otherwise the paint or dye won't stick. Follow the directions on the label.

Frame Cracks—If the mounting bosses are cracked on pads with plastic framing, they can be repaired using an an adhesive such as epoxy. Be sure to give the epoxy plenty of time to set up, especially in cooler temperatures.

Pad Cracks—Cracks in the pad itself are a little more involved. If the pad only has one or two cracks, you can repair them by following the procedure found in the *Seats* section of this chapter. However, if the pad has a lot of cracks, you might be better off buying a new or good used pad.

INSTALLATION

If you've had your car totally apart for a restoration, you're better off installing the dash pad before installing any other components. With one exception: those items under the dash that are near impossible to access once the dash is installed. These include the front radio speaker, heater ducting and the various vacuum hoses for the heater and air-conditioning controls.

Installation is basically the reverse of removal. Just be sure not to over-tighten the nuts or screws as you'll surely damage them or the component they're attached to. And when you're installing the pad on a 1969 model, tape the clips into position on the underside of the pad. This will prevent them from falling out when you install the pad. Also, be sure all the clips are fully engaged in their respective slots in the dash panel. If you don't, the loose clips will rattle and may allow the pad to warp, since it isn't retained at that point.

STEERING WHEEL

There are several reasons why you may need to remove the steering wheel. First, if it's the original wheel, it's possible it has developed hairline cracks during the 20 years of use and abuse it has been subjected to. Or, the previous owner may have installed an aftermarket wheel that will detract points at a car show.

REMOVAL

Horn Button or Bar—Removal of the steering wheel begins with the horn button or bar. The button can be levered off with a screwdriver or pulled off with your fingers. The bar is retained to the wheel with Phillips-head screws at the rear of the spokes. After removing the button or bar, you can remove the screws that hold the horn button retainer and steering-wheel trim ring, if used.

Steering Wheel—The actual wheel is next. Now unless you're the lucky type who's toast always falls jelly side up, you'll need a steering wheel puller to remove it. To remove the wheel, first remove the nut and washer from the column center shaft. Before mounting the puller, screw one of the bolts from the puller kit you think will fit into the steering wheel hub. Only use your fingers. Do not use wrenches on the bolts, otherwise you may strip out the threads with the wrong-size bolt. If the bolt's the proper size, you'll be able to screw it in at least three full turns.

When you've located the right bolts,

To remove steering wheel, first undo horn button or pad, then this horn switch retaining ring and plate.

Make sure steering wheel puller is square to wheel before tightening center screw. If necessary, tap end of center screw to pop wheel free.

142

remove them from the wheel hub and position the puller over the center shaft. Then screw the bolts you selected earlier through the wings of the puller into the holes. Screw the bolts in an equal amount to avoid cocking the puller and make sure they're in at least six revolutions.

With the puller in position, tighten the center screw until the steering wheel pops free of the column. However, if you have to rotate the center screw more than, say, three quarters of a turn after it contacts the steering shaft, and the steering wheel doesn't come off, gently tap the center screw with a hammer while you keep tension on it. This will dislodge the wheel.

ASSEMBLY

Next, slide the steering wheel over the steering shaft, making sure the wheel spokes are in the proper position with the front wheels pointing straight ahead. You'll probably want to install the column in the car for this. Then install the lock washer and nut on the steering shaft and tighten the nut securely. You'll know the nut is too tight if you feel binding when the steering wheel is turned. With the nut properly torqued, stake the end of the steering shaft with a center punch or chisel. This will help prevent the nut from backing off.

Finally, install the steering-wheel trim ring, horn-button retaining ring and horn button or horn bar.

HEATER & A/C CONTROLS

Camaro's heating and cooling system is conventional in design and shares many common features with other GM cars of the same respective model years.

Before you begin to work in and around your air conditioning system, read the warning and safety tips in the nearby sidebar. Air conditioners and their components can be hazardous to your health. Once you've armed yourself with the proper protective gear, you can perform the following checks to diagnose any problems your A/C or heater may have.

Check Refrigerant Charge Level—One of the leading causes of poor A/C performance is low refrigerant level. You can check refrigerant level with an inexpensive gauge, available at many auto and discount stores. This gauge is connected to a valve on the low-pressure side of the system, which looks just like a tire valve stem. If indicated pressure is low, remove the gauge and add a can of refrigerant to the system through the same

fitting you connected the gauge to, with the engine running and the A/C on. Then check the pressure again.

Leak Check—The problem with low refrigerant level is that it indicates a leak somewhere in the system. And if refrigerant can leak out, air, moisture and dirt can get into the system. These items can destroy the delicate system components, particularly the compressor and dessicant in the receiver/dryer.

A general rule of thumb to follow if you're wondering whether or not to have the system checked is this: If the leak is small, meaning you haven't had to add refrigerant to the system in at least two years, you can probably not trouble yourself with finding it. However, if your system looses refrigerant more often than this, you should have it checked by a qualified mechanic.

In addition to performing a visual inspection, the mechanic will usually use an electronic "sniffer" to check for refrigerant leaks beneath the various components and lines. Because refrigerant is heavier than air, it will be most noticeable below the components.

If you'd like to try this yourself, electronic "sniffers" are available from spe-

cialized tool manufacturers such as Kent-Moore in Roseville, Michigan, but their cost is about equal to having a carburetor rebuilt, so be prepared.

Stuck Compressor—If after storing your car for a month or more and the A/C compressor drive belt breaks when you turn the A/C on, you might think the compressor is seized. You might be right, but then it might just be that the oil in the compressor has run off some of the moving components. So before you replace the compressor, slip a socket or wrench over the compressor shaft lock nut and rotate the shaft in the opposite direction of normal rotation. If you can "break" the shaft loose, rock it back and forth a few times. Then rotate the compressor shaft by hand at least three complete revolutions. This will help the oil to flow where it's needed in the compressor. Finally, run the compressor for at least a minute to run things in once again.

Stuck Temperature Doors—If you encounter a problem with the heater not producing enough heat, or you can't shut the heat off, a misadjusted temperature door may be the culprit. To check this, slide the temperature control lever back and forth completely, at a fast rate. If the tempera-

Heater blower motor and housing are easily accessed from inside engine compartment.

143

Black rectangle at left is temperature door on this 1967 model. Foam sealing this door deteriorates, leading to poor heater output.

ture door is properly adjusted, you'll hear the door "hitting" when the lever is at the extreme ends of its travel. If you don't hear the door hitting, check to make sure the cable is properly attached at both ends. If this checks out OK, you'll have to remove the cable from its retaining bracket, slide it in one direction, reinstall it, then check it again. If it still isn't hitting, try sliding the cable the other way.

Vacuum Leaks—Another leading cause of poor heater and A/C performance is a vacuum or air leak either between the engine and control panel or buried under the dash. If you suspect a vacuum leak or hear a hissing sound under the dash, start the engine and connect a vacuum gauge to the intake manifold Tee. Jot down the reading. Then disconnect the vacuum line leading to the control panel, plug the open end of the Tee and recheck the reading. If it's low, you know there's a leak somewhere in the system. Look for old, cracked vacuum hoses or hoses that have become disconnected. Be prepared to spend a lot of time on your head peering up into the bowels of the instrument panel with a flashlight and a very cramped arm.

More information on operation and troubleshooting can be found in factory service manuals and in *Harrison Air Conditioning Manuals*.

144

With all major bodywork completed and stripped of all trim, this Camaro awaits a rendezvous with destiny in the paint booth.

The time is drawing near for you to perform what is probably the single most satisfying phase of your Camaro's restoration—painting it. After all those hours of hard work, you'll finally be seeing what the car will look like when it is finished, glistening in your driveway. But before that gorgeous color coat is applied, there's a bit more work to do to make sure its shine is long-lasting.

CAULKS & SEALERS

In order to assure a weatherproof car, different types of body caulks and sealers are used. Caulks and sealers serve the same basic function; keeping water, dust and noise out of the car. Basically, two different types of sealers are used. When the gap to be sealed is quite small, such as a lap joint between two pieces of sheet metal, you'll usually find a heavy-bodied drip-check sealer. If the gap is larger, weatherstripping caulk (also known as strip caulk) is used. These sealers can be found at your local automotive paint supplier.

APPLICATION

The surface that you'll be applying the sealer to should be cleaned with *Prep-Sol* then allowed to dry. This gives the sealer the best chance of adhering to the surface.

Use a small, clean acid brush to spread the drip-check sealer after you've applied it to the surface. And when you work with this type of sealer, it can get pretty messy if you're not careful. So keep some adhesive cleaner or lacquer thinner on hand for cleaning.

If you accidentally get some sealer on an area that you didn't intend to, and the area has old paint on it, or is metal or glass, clean off the sealer with adhesive cleaner before it dries. However, if the paint is fresh, allow the sealer to dry before removing it. This will avoid dissolving the paint and save you a lot of anguish.

Strip caulk is pressure sensitive and sticks when it's pressed into place. If the gap you're filling is wider than the width of the weatherstripping, fold the weatherstripping until it's wider than the gap then roll it between your hands to blend it together. Place it in position then simply press it into the gap being sure to seal the edges.

The various types of sealers are used where they'll do the most good. The following list covers the different types of sealers and where they're used on the car:

Body (Strip) Caulk
- Rear quarter panel and tulip panel pinch welds. Found under the stainless-steel C-pillar molding on 1967-69 convertibles.
- Quarter window access panels on 1967-69 convertibles.
- Cowl-to-plenum pinch-weld intersections.
- The inside of the A-pillar trim.
- Door hinges where they meet the body and door.
- Around the vents on the backside of the kick panel.

When it comes to paint, cleanliness is absolutely essential. These metal cleaning and conditioning chemicals help ensure that the surface will accept new paint.

Strip caulk is pressure-sensitive and sticks when pressed in place. In addition to sealing body seams, it's used to retain water shields to doors as shown.

Brushable seam sealer seals seams where two panels overlap. Applying sealer with your finger, then texturing it with a stiff tooth brush gives your work that factory original look.

Seam sealer can also be applied with a pump gun. With a screwdriver, check for loose, dried-up seam sealer that would allow leaks into the interior, pry it out and apply new sealer.

Drip Check Sealer

- Rear seatback support sides and bottom where it meets the body.
- Floorpan-to-rocker panel joints in the interior.
- Front seat platforms to floorpan in the interior.
- Driveshaft tunnel where it meets the firewall under the instrument panel.
- Rear seatback bottom hooks where they meet the floorpan.
- Cowl-to-firewall seam.
- Cowl-to-dash seam.

SHEET METAL ALIGNMENT

After installing all the sheet metal on the car, you'd be wise to align it before you paint the car. This will avoid scratching any paint as you loosen the hinges or whatever and slide the sheet metal into position. Sure, the scratches may be small and in an unnoticeable area, but scratches can allow moisture to get under the paint and give a home to that age-old enemy: rust.

Three items indicate misaligned sheet metal: incorrect or uneven gap spacing between adjacent panels, high closing effort and poor sealing. Correcting each of these items in this order will minimize the time and effort you'll spend aligning the sheet metal. Specifications for sheet-metal alignment, other than the ones discussed here, can be found in the appropriate factory assembly manual or *Fisher Body Shop Manual*.

Before beginning, you should know that there's a proper sequence for aligning each panel on the car. Begin with the quarter panels, then move to the doors, front fenders, hood and deck lid. Let's take a look at these one at a time.

REAR QUARTER PANELS

Proper sheet-metal alignment begins with the rear quarter panels because these are the only stationary panels on the car. Because their position is fixed, the alignment of the moveable panels on the car such as the deck lid, doors and front fenders are predicated on the position of the quarter panels.

For example, let's say you align the trailing edge of the front fender with the leading edge of the door, spending a considerable amount of time to get it right. Then you realize that the trailing edge of the door doesn't mate with the quarter panel evenly from top to bottom. Because the quarter panel is welded into position, you can adjust the door only. So you adjust the door for a good fit to the quarter panel, but in the process you throw the alignment of the front fender to the door way off. Sort of like borrowing from Peter to pay Paul. Now you have to readjust the fender, wasting valuable time.

If your car hasn't required new quarter panels, you can begin aligning the trailing edge of the door to the quarter panel as discussed below.

However, if you're replacing the quarter panels, you need something to align them to, otherwise every panel on the car may be misaligned. To prevent this, the doors and deck lid must be aligned properly before removing the quarter panels. When the doors and deck lid are properly aligned, you can use them to align the quarter panels. And be sure that the quarter panels are properly aligned after securing them into position and before any welding is attempted. For more information on replacing quarter panels, refer to *Quarter Panel Repair*, page 92.

DOORS

The doors on Camaros are inherently tougher to keep adjusted than those on a four-door car. In order to provide access to the rear seats, doors on two-door cars must be longer. The added length adds weight which, in turn, makes the door more prone to sagging. Second-generation Camaro doors weigh more than 100 lbs. each with the side glass installed.

Before aligning the door, assess the gap spacing, closing effort and sealing. Use the body lines in the sheet metal to judge alignment, not the body side moldings which can be readily moved.

The gap spacing should be even between the door and the quarter panel and should be within specifications; approximately 1/8-3/16 in. depending on the year Camaro you're working on. An uneven gap can indicate worn-out door hinge bushings or misalignment due to improper hinge adjustment or accident damage.

Check door closing effort by standing approximately 2 ft. away from an unlatched door. Then try to close it by putting your hands on the trailing edge of the door and putting your weight against it. If the door doesn't close, chances are the striker is in too far. And the door shouldn't spring open when it's unlatched. This also indicates striker misadjustment. We'll talk about striker adjustment shortly.

A popular way of checking the quality of the seal between the door and its mating surfaces is by using a dollar bill. Place the bill against the seal and close the door. Now try pulling the bill out with moderate force. If it slips out easily, the seal is poor and requires attention. If it's difficult to pull the bill out, the seal is also poor, requiring attention. Work your way around the door and be on the lookout for a poor seal. Remember, sealing is not a case of "if some is good, more is better." We'll cover any problems you might have a little later in this book.

For demonstration purposes, we'll assume your door needs attention to all three areas: alignment (also known as gap spacing), closing effort and sealing. Let's take a look at these one at a time.

Gap Spacing & Striker Alignment—If the door is misaligned but you're not sure if it's moving up or down, apply a piece of masking tape across the gap between the quarter panel and the door when the door is closed. Then cut it in the area of the gap. Now open and close the door while comparing the piece of tape on the quarter panel to the tape on the door to see which way the door is moving.

If the door seems to move when it contacts the striker, mark the position of the striker with a pencil and then remove it. An Allen wrench can be used to remove both the Allen-head and torx-head style strikers in many applications. However, stubborn torx-head strikers will require the use of a torx bit.

Now recheck the panel alignment. If the body lines of the door and quarter panel align with each other when the striker is removed, all you need to do is center the striker in the latch. If the alignment is still off, you'll have to adjust the door as outlined below.

Just keep in mind that any adjustments that you make to the door will affect the alignment and sealing between door glass and weatherstripping, so make door-glass adjustments later. For this same reason, roll the window down to avoid damaging it or the weatherstripping while adjusting the door.

Door Alignment—The door should be adjusted following a three step sequence: up and down, fore and aft, then in and out. First, adjust the door up or down by loosening the hinge-to-door-retaining bolts. When the door is properly adjusted for height, tighten the hinge bolts. Next, adjust for fore-and-aft gap spacing by loosening the bolts on the body side of the hinge. Then when the door has the proper fore-and-aft gap spacing and the hinge bolts are tightened, you can adjust for any discrepancies of the door edges being too far in or out. This is accomplished by loosening hinge-to-door-retaining hinge bolts and shifting the door as needed.

The bolts that secure the hinges to the body pillar provide fore and aft as well as gap alignment adjustment. So if the spacing between the rear edge of the door and quarter panel is incorrect or uneven, loosen the bolts attaching the hinge to the body pillar and shift the door to correct the misalignment problem. Be sure to use the body lines to check the alignment, not the body side moldings. Then tighten the bolts and close the door so you can recheck the alignment. Like the rear of the fender, the rear edge of the door should be flush to 1/16-in. above the front edge of the quarter panel to minimize wind noise. This adjustment is handled by moving the door striker in or out as needed.

Now check the alignment of the leading edge of the door to the front fender. The gap spacing should be even and the edge surface of the door should be flush with or slightly below the trailing edge of the front fender. If the surface of the door protrudes farther out than the fender, you can almost be certain you'll have wind noise. If the gap spacing is uneven, loosen the fender retaining bolts and adjust the front fender to alleviate the problem. To move the edge of the door below the fender, loosen the hinge-to-door retaining bolts and push the door inward. Then tighten the bolts and check your work.

Check Closing Effort—With the door properly aligned to the quarter panel and

Dead giveaway for misaligned door is this worn striker. Door on this Camaro had to be lifted onto striker to close.

Door sagged at rear, so door hinge-to-body bolts were loosened slightly, rear of door lifted ever so slightly and hinge bolts retightened.

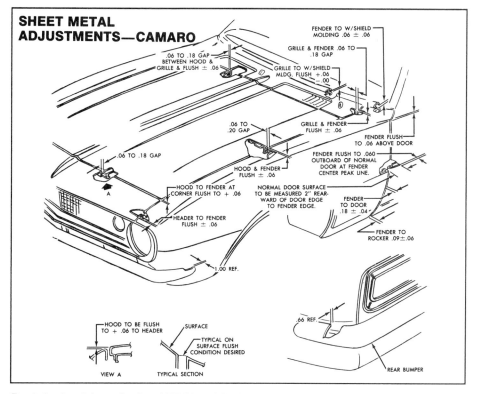

SHEET METAL ADJUSTMENTS—CAMARO

FENDER TO W/SHIELD MOLDING .06 ± .06

.06 TO .18 GAP BETWEEN HOOD & GRILLE & FLUSH ± .06

GRILLE & FENDER .06 TO .18 GAP

GRILLE TO W/SHIELD MLDG. FLUSH +.06 −.00

.06 TO .20 GAP

GRILLE & FENDER FLUSH ± .06

.06 TO .18 GAP

HOOD & FENDER FLUSH ± .06

FENDER FLUSH TO .06 ABOVE DOOR

FENDER FLUSH TO .060 OUTBOARD OF NORMAL DOOR AT FENDER CENTER PEAK LINE

HOOD TO FENDER AT CORNER FLUSH TO + .06

NORMAL DOOR SURFACE TO BE MEASURED 2" REARWARD OF DOOR EDGE TO FENDER EDGE

FENDER TO DOOR .18 ± .04

HEADER TO FENDER FLUSH ± .06

FENDER TO ROCKER .09±.06

1.00 REF.

.66 REF.

HOOD TO BE FLUSH TO + .06 TO HEADER

SURFACE

TYPICAL ON SURFACE FLUSH CONDITION DESIRED

VIEW A

TYPICAL SECTION

REAR BUMPER

Front sheet-metal gap details—1967-69 models. Courtesy Chevrolet.

front fender, you can install the striker. The striker should be centered in the latch—never use the striker to adjust the positioning of the door. The door should now close and latch securely with moderate effort. If you have to slam the door to get it latched, something is amiss with door alignment.

Check Sealing—After the door is aligned, evaluate the alignment of the door glass to the seal. It should be square to the opening and exert some pressure against the seal when the door is closed. You can use the dollar-bill technique explained earlier to test the seal. If the window isn't square to the opening or isn't sealing properly, adjust the glass using the procedure covered in Chapter 10, *Door Reconditioning*, page 106.

When the door and glass are adjusted, move on to the adjustment of the front sheet metal, beginning with the fenders.

FRONT FENDERS & HOOD

Next on the agenda is aligning the front fenders. Start by taking a look at the rear edge of each fender. It should be evenly gapped 3/16 in. from the front edge of the door. It should also be about 1/16-1/8-in. outboard of the door. This helps prevent wind noise. If the fender-to-door gap is uneven, add or subtract shims from either the top or bottom of the rear of the fender. These shims are made from sheet metal with a "U" shape cut into them. To install or remove them, loosen the bolt they're under and slide them in or out. Do this until the gap is even.

If the gap spacing is even with the door but not within specs, it can be adjusted by adding or subtracting shims as needed. These shims are found under the bolt at the rear of the fender which screws into the firewall in a horizontal position.

If the rear of the fender isn't approximately 1/16-1/8-in. outboard of the door, loosen the bolts at the rear of the fender. Then slide the fender until it's within specs and tighten the bolts.

When you've installed and adjusted both fenders, bolt the hood into position. Final adjustment of the fenders is easier with the hood in place.

Slowly close the hood and check its alignment and gap spacing with the fenders. The gap should be even from front to back and should be about 1/8-in. wide, give or take 1/16-in. You can either shift the hood or the fenders to correct most gap irregularities. But you may have to shift both if there's a big difference in spacing.

To adjust the hood, loosen the hinge-to-hood retaining bolts. You can also use this adjustment to correct gap spacing problems between the rear edge of the hood and cowl and between the front edge of the hood and header panel. You can also shift the header panel slightly if you need to.

Hood Height—Also make sure the top of the hood is even with the cowl, fenders and header panel. If the rear of the hood needs to be raised or lowered, this can be accomplished by loosening the hinge-to-fender retaining bolts and moving the hinge as necessary. On second-generation cars, you can use three threaded adjusters on the cowl if hood height needs to be tweaked.

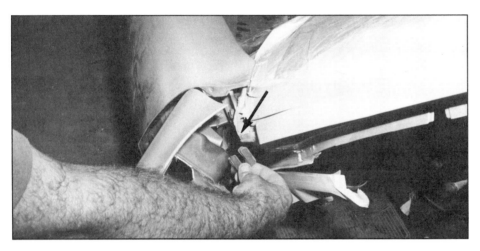

In and out adjustment for tops of fenders is made by adding or subtracting shims (arrow) under tab and bolt here at front of door openings.

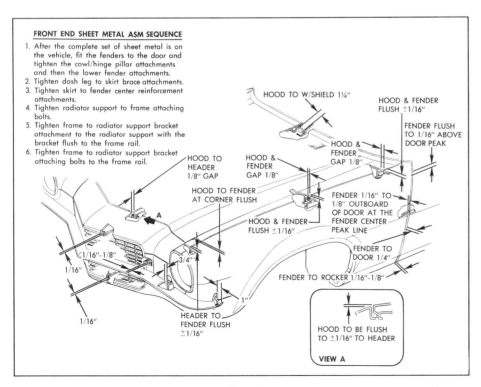

FRONT END SHEET METAL ASM SEQUENCE

1. After the complete set of sheet metal is on the vehicle, fit the fenders to the door and tighten the cowl/hinge pillar attachments and then the lower fender attachments.
2. Tighten dash leg to skirt brace attachments.
3. Tighten skirt to fender center reinforcement attachments.
4. Tighten radiator support to frame attaching bolts.
5. Tighten frame to radiator support bracket attachment to the radiator support with the bracket flush to the frame rail.
6. Tighten frame to radiator support bracket attaching bolts to the frame rail.

HOOD TO W/SHIELD 1¼"

HOOD & FENDER FLUSH ±1/16"

FENDER FLUSH TO 1/16" ABOVE DOOR PEAK

HOOD & FENDER GAP 1/8"

HOOD TO HEADER 1/8" GAP

HOOD & FENDER GAP 1/8"

HOOD TO FENDER AT CORNER FLUSH

FENDER 1/16" TO 1/8" OUTBOARD OF DOOR AT THE FENDER CENTER PEAK LINE

HOOD & FENDER FLUSH ±1/16"

A

FENDER TO DOOR 1/4"

3/4"

1/16"-1/8"

1/16"

FENDER TO ROCKER 1/16"-1/8"

1"

1/16"

HEADER TO FENDER FLUSH ±1/16"

HOOD TO BE FLUSH TO ±1/16" TO HEADER

VIEW A

Front sheet-metal gap details—1970-73 non-RS models; 1974-81 models similar. Courtesy Chevrolet.

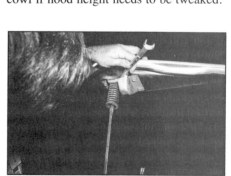

If hood doesn't latch securely, loosen lock nut and adjust height of latch pin with flat-blade screwdriver. A little grease on the pin helps minimize closing effort.

The height of the front of the hood can be adjusted with the rubber-covered threaded adjusters that screw into both front corners of the core support. Simply pry off the rubber cap, loosen the locknut and raise or lower the adjusters until the hood is properly positioned.

Striker & Latch—If the alignment of the hood changes when you close it, it indicates the striker or latch plate has to be moved. As you do this make sure that the latch catches the latch pin securely. You'll hear a prominent click if it does. And also make sure that the safety latch works properly. It should catch the hood when the hood is first released from a closed position. As with the doors, the hood should latch securely.

DECK LID

The deck lid must be aligned to its opening in the body first. So to prevent the latch from throwing the deck lid off, remove it. Then close the deck lid and hold it closed while you assess its alignment to the quarter panels and tulip panel. The tulip panel is the panel which separates the rear of the roof from the trunk. Deck lid gap spacing should be even at all points. If not, loosen the hinge bolts and shift the deck lid in a direction that will solve the alignment problem. Close the deck lid and check your work. When the lid is properly aligned, carefully open it and tighten the hinge bolts. Then close the lid once again and check its alignment to be sure it hasn't shifted.

Install Latch—Now install the latch. The latch retaining bolts should be just loose enough for you to move the latch by hand. At this point, you can temporarily install the trunk weatherstripping and adjust the latch. Hold the weatherstripping in place with two to three pieces of masking tape. The weatherstripping has to be in place to adjust the latch properly. Then latch the deck lid in place if you can. If you can't, loosen the latch adjusting bolts and move the latch. The latch is properly adjusted when it latches almost immediately after it contacts the weatherstripping.

Check Seal—Finish the job by checking the seal between the deck lid and weatherstripping. You can use the dollar-bill test mentioned earlier to do this. If the seal isn't adequate, raise or lower the latch or the rubber bumpers (1967-69 only) at the rear of the deck lid until it is. And if you really want to be assured that the seal isn't going to leak, you can perform a water leakage test. This test is covered under *Wind Noise and Water Leaks,* page 178.

PAINT PREPARATION

After many hours of tedious work, you've got the body together. And now you want to paint it. That's great. But don't get over anxious now and do a second-rate job after all your first-rate work. Any paint job is only as good as the preparation that comes before it. Nothing can make up for poor preparation. The surface underneath the top coat has to be flawless to get a prize-winning paint job. It's that simple. And you can do a great job if you discipline yourself. But if you don't, the end result will be less than spectacular, or worse.

Basically, there are three reasons for poor preparation: laziness, lack of time and lack of knowledge. Nothing we can say here can change the first reason. Either you're willing to avoid taking shortcuts that detract from the job or you're not.

Lack of time is dependent on you. If making it to that next show and getting 3rd or 4th place is more important than waiting another month to get Best of Show at the next meet, go right ahead. But you'll be wasting your time. You'll eventually wish you had taken the time the first time around to do the job right. Then it'll take longer because you'll have to repeat some of the operations performed the first time. So do it right the first time.

The third item, lack of knowledge, can also be controlled. Take time to read reference books on body preparation and painting. An excellent book on this subject is HPBooks' *Paint and Body Handbook.* Also, take time to go to local paint suppliers and ask their recommendations about various painting systems offered by major paint manufacturers. And if you don't

have enough information to satisfy your needs, give the paint manufacturers a call. They know their paints best and will be able to tell you what's compatible with what.

In the preceding paragraph, you'll notice we said paint systems. The major paint manufacturers; PPG, DuPont, Rinshed-Mason (Inmont), Sherwin-Williams and Martin Senour to name a few, spend a great deal of time and money developing and testing products that are compatible with each other. Mixing paint products from different manufacturers can cause compatibility problems such as cracking and lifting. The best way to avoid these types of problems is to stay with a particular manufacturer's products from start to finish. This includes everything from metal conditioner and primer to color coatings. If you follow this rule, your chances of getting a good, durable paint job are greatly improved.

Now about surface preparation. Before we begin, you should know that the factory-applied paint is the best you're gonna get. The sheet metal that comprises the body is chemically cleaned, electromagnetically coated with primer, then baked. Then the top coats are applied and baked to help them flow out for a high gloss.

Because of these and many other steps, it stands to reason that if you can avoid breaking through this finish, your car will be better off for it. And if you have to remove the color coat, try to avoid damaging the primer. There really isn't any practical way of duplicating the electromagnetic coating (ELPO dip) process that the factory primer undergoes. You can remove the color coat without upsetting the

Sheet metal aligned. Your Camaro is now ready for final surface preparation and paint.

SOFT FASCIA REPAIR

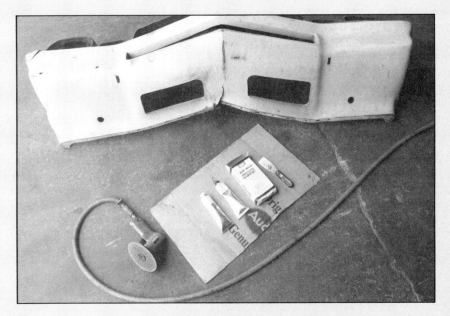

Small nicks and scratches in soft-fascia urethane bumpers on 1978-81 models and 1970-73 RS grille surround can be repaired with this SEM Weld flexible repair kit.

One of the victories of GM stylists over government bumper regulations was the adoption of soft-fascia, flexible front and rear ends on 1978-81 Camaros. Compared to the cobbled-looking, battering-ram aluminum bumpers of 1974-77, the flexible bumpers present an integrated, almost bumperless look that complements the svelte lines of the second-generation cars. Vanity does have its price, and beneath the soft urethane skin of these flexible bumpers are labyrinthine supports and a structure that adds considerable cost, weight and repair complexity. Routine fender benders in excess of 10 mph closing speed may require complete removal and replacement of the front nose because it "sacrifices" itself, deforming, crushing and absorbing energy in a collision.

However, minor parking-lot dings as well as small tears, holes and scratches in the soft urethane skin can be repaired on the car. A popular repair method involves using 3M's *Flexible Parts Repair Material* (FPRM), which is applied and sanded much like plastic body filler—your basic Bondo.

First, clean the repair area with soap and water, followed by a wax and grease remover such as Pre-Kleano. Then, with a 36-grit rotary grinding disc, rough up the repair area. Bevel the edges out 1-2 in. around the repair area to aid adhesion of the FPRM. Be careful not to grind away too much urethane—it's much softer than steel! Next, feather-edge the repair area with a dual-action (DA) sander and 180-grit sandpaper. Wipe or blow the area clean and inspect.

If the skin is perforated, clean the underside of the repair area and apply auto body repair tape. This provides a firm base for the filler.

As with plastic body filler, mix equal portions of filler and catalyst on a palette or piece of cardboard according to the manufacturer's directions. This may vary depending on temperature and humidity. Stir in one direction to avoid getting bubbles in the mixture. Once mixed, you'll have about 10 min. to apply the FPRM before it becomes to hard to spread.

Apply the filler with a plastic squeegee in uniform strokes, building it up higher than the surrounding panel. Allow it to dry for 20 to 30 min. if air temperature is 60 to 80 degrees F. and reapply if any low spots are evident.

Then, sand the hardened FPRM with a dual-action sander and 180-grit sandpaper, followed by 240-grit paper. Or you can block sand wet with 320 then 400-grit sandpaper on a rubber sanding block.

Once you're satisfied with the repair, apply primer with a flex-agent added to it. Scuff-sand to see if any high spots show through and if not, you're ready to prime and color-coat the panel. Remember, though, that the color coat must also have a flex-agent added to it. This may require masking off the body, painting the bumper(s), then masking off the bumper(s) and painting the body. Finish the job by cleaning the entire bumper with a solvent such as Prep-Sol, scuff-sanding it with 400-grit sandpaper, and applying two coats of clear that contain a flex agent. This will keep the repaired area from peeling and help blend the repair.

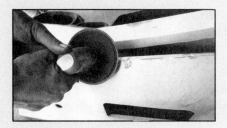

Grind a V-shaped groove and bevel edges of repair area before applying flexible repair material.

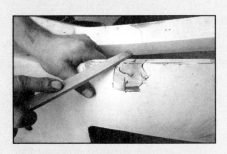

Mix equal parts of filler and catalyst, apply with a squeegee and allow to dry for a half-hour or so. Knock off high spots with a file, then sand with 180-grit paper followed by 240-grit. Take care to maintain character lines of bumper edges.

Finish by painting as you would the rest of the car. But make sure paint contains flex agents or it will peel and flake off within weeks.

FIBERGLASS SPOILER REPAIR

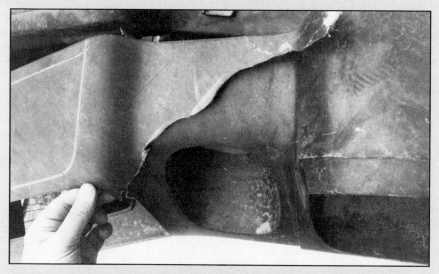

All Camaro rear spoilers and header panel of 1974-77 models are constructed of plastic. If replacements aren't available, these can be successfully repaired using a fiberglass cloth and resin patch.

All front and rear spoilers are manufactured from plastic. Because of this, they can be repaired using either plastic body filler or fiberglass. Structural repair adhesive adheres well to a properly prepared surface and offers outstanding strength. If damaged, textured chin spoilers, such as those used on 1967-77 Camaros, are best replaced with inexpensive ABS reproductions. But if you insist on originality, they can also be repaired using the following method. However, the grain will have to be cut into the repair using a grinder equipped with a fine grinding point.

If you need to repair a crack, begin by drilling a hole at the end of it. This keeps the crack from spreading. Then grind the crack with a small, round grinding bit, forming a V-shaped depression the length of the crack. This gives a surface which body filler or fiberglass cloth and resin can adhere to. Now mix the filler or fiberglass resin according to the label directions. Apply any filler to the crack with a plastic squeegee. If using fiberglass, cut the cloth slightly larger than the crack area, soak it in resin and apply it by hand to the spoiler.

When the filler or fiberglass is dry, block sand it with 240-grit sandpaper until it's level with the surrounding area. Then spray two or three coats of primer-surfacer on it and let it dry.

When the primer-surfacer is dry, block sand it with 400-grit or finer sandpaper. This assures that no sanding marks will appear later. Complete the repair by shooting it with four to six coats of color.

Cut cloth patch slightly larger than repair area and soak with resin.

Grind V-shaped groove along length of crack and apply resin-soaked cloth patch. After resin dries 1-2 hours, sand, prime and paint it along with rest of car.

primer by using a special type of paint stripper. But we'll talk more about that in a little bit.

If your Camaro still has exterior trim on it, carefully mask it off using a high-quality masking tape and masking paper available through auto-body supply stores. Simply put, there's tape and then there's tape. If you use the 5- and 10-cent store variety, don't cry when the bargain-basement masking tape lifts off some of your Camaro's pristine paint or that impossible-to-get factory decal (such as the co-author did) when you pull it off after painting. Try to avoid using newspaper as the newsprint can bleed. Also, make sure that the interior trim, if not already removed, is completely sealed off. Cover the tires and engine compartment and mask off the fuel tank, exhaust tips and front anti-roll bar (if these are still installed) to avoid time-consuming cleaning and detailing later.

REFINISHING

For our purposes, we'll discuss the refinishing process from paint removal to application of the primer-surfacer. We'll assume your Camaro is covered only with the factory-applied acrylic lacquer paint and that you're going to refinish it with the same. We'll also assume that you're going to have to remove rust from underneath the paint. And keep in mind that spot repairing is a little more difficult than repairing the full panel. That's because you have to blend the repaired surface with the surrounding area. And you also have to match the paint, which can be difficult with light-colored metallic paints.

Custom Or Original?—Just because this is a restoration book doesn't mean that you

Important stickers and trim that cannot be removed must be carefully masked. Don't use cheap tape over decals; the tape may destroy the decals when removed after painting.

absolutely have to restore the car to just-off-the-assembly-line condition. It really depends on what you want to do with the car. If you plan to show your Camaro and enter it in stock/restored classes, or if you want to get top dollar for your Camaro if and when you sell it, it's wise to paint the

Just because it's a new genuine GM replacement, don't assume all panels are ready for paint right out of the box. Stress marks in metal from old parts dies must be hammer and dollied, sanded and filled or your Camaro will have that genuine "ripply" look.

When using rotary sander, use even pressure and keep wheel parallel to work to prevent grinding off character lines. Here, Jim Altieri of Arizona GM Specialists finishes repair of once-rusty C-pillar that was sandblasted and filled with USC All-Metal filler.

CAUTION!

Above everything, your safety comes first. Always wear safety glasses or goggles when sanding. And be sure to wear a respirator so you don't inhale paint and metal dust into your lungs. While we're on this subject, be aware that catalyzed paints contain *isocyanates*. Isocyanates contain cyanide, a deadly poison. You *must* wear the *proper* respirator when mixing and spraying these paints. Wearing a respirator doesn't mean you're a sissy. It means you're smart and you'll probably live longer than someone else who isn't using one. So wear the proper respirator and live.

car in its original color. Nine times out of ten, original cars fetch higher prices than modified ones, and changing exterior color counts as a modification. On the other hand, if you plan on making the car an everyday driver, or want to modify it with aftermarket seats, shifter, steering wheel and gauges, a roll bar, aftermarket wheels, chrome engine goodies, fender flares, hood scoop and the like, color isn't going to matter very much—paint quality, however, will.

Now let's suppose that the Camaro you're refurbishing is Lime Green (which in your opinion is about as desirable as pond slime) and you had your heart set on Fire Engine Red. If you must have a red Camaro, go ahead, but try painting it an original red offered from Chevrolet in the same year as yours. This compromise can help minimize the dollar penalty when—perish the thought—you sell your Camaro.

Reading Codes—To determine what exterior color (or colors) your Camaro was painted at the assembly plant, check the Fisher Body trim plate on the driver's side of the firewall in the engine compartment. At the upper right hand corner of the plate, below the sequential body unit number (not to be confused with the VIN number) and directly to the left of the letters **PNT**, are two 2-digit numbers (2-tone paint cars), a 2-digit number and a single letter (vinyl-top cars) or a single 2-digit number (single-color cars). The first (or only) 2-digit number is the code for the lower body color (or complete body). For example, a code 44 for a 1973 Camaro means the car was originally painted Medium Green Metallic. The second 2-digit number or letter is the code for the upper body color or vinyl roof covering. A letter code G indicates a green vinyl top. Refer to the Appendix at the back of the book for a complete list of exterior color codes.

To help you decipher the codes, take this information to your local Chevy dealer or automotive paint shop. There, they can match up the codes with a paint chip chart and show you what the color should look like.

Removing Paint—If you just want to repaint your car the same color, or even a different color for that matter, and no rust is apparent, you don't need to remove the old paint, provided it's the original paint. If your Camaro suffers from an El Cheapo $99.95 One-Day Wonder enamel paint job a previous owner commissioned, that stuff will need to come off because it will be incompatible with the new acrylic lacquer you'll be applying.

Scuff-Sanding—If your Camaro is still showing its original coat, all you have to do is clean, then scuff-sand the old paint and spray on the new color coats, although best paint durability is ensured by removing the old paint down to the factory primer. This reduces the chances of paint checking and can be achieved by using a special paint stripper, covered under *Paint Stripping,* below.

If you plan on painting over the original paint, just remember to do the sanding with 400- or 500-grit sandpaper. If you use sandpaper any coarser than this, the paint won't be able to hide the sand scratches. Mount the sandpaper on a rubber sanding block and soak it in water for at least five minutes. This makes the sandpaper more pliable. Then flood the panel you intend to sand with a liberal amount of water and begin sanding.

As you sand, make sure that the surface remains wet. After a few minutes of sanding, wipe off the panel and take a look at it. It should look equally dull in all places for the new paint to adhere properly. When it does, you can move on to applying the color coat, which is covered later in this section.

Paint Stripping—The quickest and best way to remove old paint is with paint stripper. If you don't have to go down to bare metal, the paint stripper you'll want to use is Ditzler DX-525. This type of stripper will remove color coats without removing the factory primer. If you don't need to remove the factory primer, don't. You'll never be able to do a better job of priming. Just apply the stripper according to the label directions. If you need to get down to bare metal, as in the case of removing rust, use a heavy-bodied stripper as described on page 87.

REMOVING RUST

The best way to remove all traces of rust from sheet metal is to use a sandblaster. A sandblaster can get into all the nooks and crannies that a sanding disc would just pass over. Just remember to keep the pressure high enough to remove the rust but low enough to avoid warping the metal. For more information, see *Sandblasting,* page 88.

CONDITIONING METAL

When rust is removed, the bare metal must be chemically cleaned. This is accomplished with metal conditioner. This product cleans the surface and coats it with a phosphorus agent which helps guard against further rusting. Most metal conditioners have to be diluted with water before they're applied. Again, be sure to follow the label directions. Note that some conditioners are only suitable for galvanized or aluminum surfaces.

You can apply this type of conditioner with almost any type of small brush. An acid brush, 3M Scotchbrite pad or a clean rag will work fine. Just be sure not to spray any of the conditioner as it contains acid. After you've applied the conditioner, wipe it off before it dries. If you don't the surface will begin rusting shortly. However, if the conditioner dries before you have a chance to wipe it off, simply re-wet the bare metal then wipe it dry. Then allow the metal to air-dry for 10 minutes.

To help save time conditioning metal, you may want to consider using Dupont Vari-Prime. This two-part primer contains a metal conditioner, so separate metal conditioning isn't required.

PRIMER

At this point, the surface is clean and ready for primer. Notice we said *primer* and not *primer-surfacer*. In addition to priming the surface, primer-surfacer contains solids which help fill minor surface irregularities. It's these solids which prevent primer-surfacer from adhering to bare metal surfaces as well as a straight primer. So you're better off spraying the surface with straight primer. Then you can proceed to spray the primer-surfacer over the primer.

Primers are available in many different variations. These include zinc-phosphate primers, chromate primers and zinc-phosphate-chromate primers. Although many of these don't require a catalyst to harden, some do. And some require that their application be followed by primer-surfacer. Because of all these variations, your best bet is to find out as much as you can about the primer you intend to use. As a general rule, the primers that require a catalyst are the most durable. But they're also usually the most expensive. And they contain isocyanates just like the catalyzed color coats. So read and follow the label directions exactly.

PRIMER-SURFACER

As the primer is drying, you can mix the primer-surfacer. Primer-surfacers require a thinner that evaporates rather quickly. So be sure to use a fast or medium-drying thinner for mixing as temperature conditions allow. Apply two to three medium-dry coats, allowing sufficient *flash-off* time between coats. Flash-off time is the time it takes for the surface of the primer or paint to change from a high gloss to a

normal gloss. Under most temperature conditions, this will work out to a maximum of ten to fifteen minutes. This allows enough time for the solvent to evaporate out of the primer-surfacer.

Wet Sand—After applying the last coat of primer-surfacer, allow it to dry. When it's dry, you can begin sanding it with 600-grit wet-or-dry sandpaper mounted on a rubber sanding block. The sanding block will help you find the low spots that need to be filled. When you're sanding, be sure to keep the sandpaper and surface wet with water. This helps keep the dust down.

Tight, enclosed areas, such as door jambs are another matter. Ordinary sandpaper tears under these conditions and is difficult to use. For these instances you can use 3M Scotch Brite or Scuff and Clean pads instead of sandpaper. These pads are made of a fibrous material which is abrasive.

Surface Inspection—When the primer feels smooth to the touch, stop sanding and dry the surface. The primer will appear to have a satin finish when it's properly sanded. Then give the primer a close look. Bare metal indicates high spots. If this happens, clean the surface with Prep-Sol and give it another couple coats of primer. Dull spots in the primer indicate low areas. If these areas are less than about a sixteenth of an inch deep, you can fill them with spot putty. Otherwise, you'll have to use body filler.

Just remember, extra time spent block sanding and making sure there aren't any defects will pay off handsomely when the paint is applied. The "secret" to a great paint job is plain 'ol hard work and great preparation.

SPOT PUTTIES

Spot putty is a filling material which fills minor surface imperfections. It's usually applied straight from the tube with a small squeegee. Use the squeegee to spread the putty as you would with body filler. Just be sure to work quickly, otherwise the putty may begin to dry as you're working it. This will cause it to roll up into little balls, requiring you to remove the putty and start over. After the putty has dried, give the repair another couple coats of primer-surfacer. Then sand the area and check it for flaws.

COLOR COATING

When all flaws are removed, you can prepare for the color coat. If you plan on color coating the car in the garage or driveway, the best paint to use is acrylic lacquer. Acrylic lacquer paints have the

endearing quality of drying quickly. This helps prevent airborne dust, dirt and insects from ruining your paint job. And acrylic lacquers can be easily repaired if you make a mistake. All you have to do is sand the mistake away and repaint.

Before mixing the paint, be sure that you're using a high quality thinner that's right for the temperature conditions you're spraying in. Use a thinner that will allow the paint to flow out properly but not attack

After final scuff-sanding with 400-grit paper, you're ready to apply primer. Look for scratches and imperfections and then apply primer/surfacer.

Spot putty can be applied over primer/surfacer to fill minor surface imperfections.

Cut-in door jambs, deck lid underside, wheelwells and other hard-to-reach areas first.

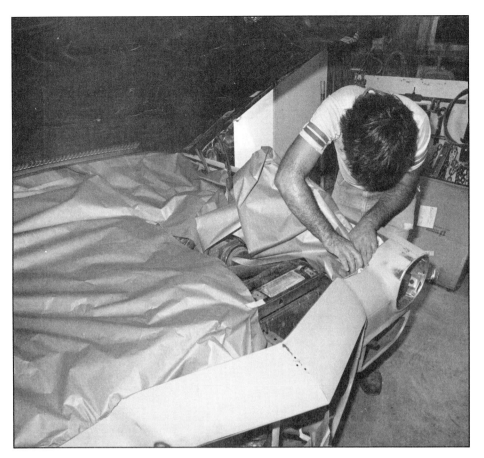

Paint gets everywhere and we mean everywhere! So mask off all areas you don't want overspray on. That includes the engine compartment.

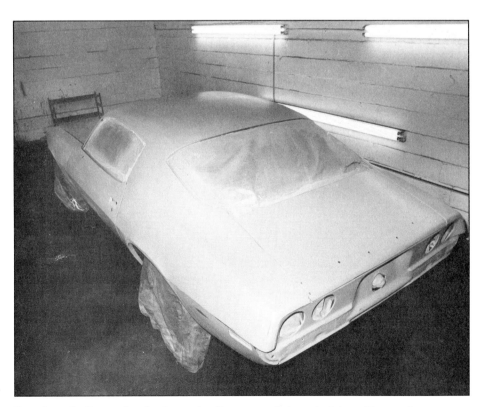

Hose down the floor and walls of spray booth or garage to keep dust down. Then push your Camaro in and cover the wheels and tires with sheets. Apply your color coats (or color coat and clear coat) now as per paint manufacturer's instructions.

the primer you've meticulously applied. A paint is said to flow out well when it appears to be a smooth film.

If you need to paint or "cut-in" door jambs, the insides of the doors and other areas, paint them first. This helps avoid getting overspray on the exterior paint, if the exterior is painted first. If you get overspray on the primer, wait until it's dry then sand it off. Then carefully mask and paper these areas to prevent getting overspray on them.

Just before you're ready to shoot the color coat, clean the surface with R-M's Pre-Kleano and dry it. Then wipe the surface with a tack rag. This ensures that the surface is free of dust and dirt. A clean surface is essential if you expect the paint to adhere properly. Don't use Prep-Sol at this point because it isn't formulated for use under color coats.

The first coat of paint on the car should be misted on. After this first coat, take time to sight down the side of the body looking for any imperfections that didn't show up earlier. If you find any, repair them as necessary. Then apply four double wet coats of color, allowing sufficient flash time between each coat.

If you're spraying a light-colored metallic paint, you may need to apply five or six double wet coats to achieve proper hiding of the primer. And if you plan on color sanding the paint, which is described next, apply an extra coat of paint. This extra paint will be sacrificed during the color-sanding operation. However, don't be tempted to put on ten or more coats of paint. All this does is promote cracking. After you've applied the paint, wait at least a day before attempting to sand or buff the surface.

After paint has cured, you can color-sand out small imperfections with 1000-grit paper, then polish to a mirror-like finish. Liquid Ebony (right) removes swirl marks from polishing wheel.

COLOR SANDING

Some people want a surface that goes beyond the factory finish. This is accomplished by color sanding. Color sanding the paint removes the high spots on the paint and when properly buffed, will give the paint a mirror-like appearance. However, it is not recommended for metallic colors, due to the possibility of cutting through the top layer of paint and creating bulls-eyes.

Color sanding is performed with an ultra-fine wet-or-dry sandpaper such as 1000-grit mounted on a rubber sanding block. Soak the sandpaper in a bucket of water for about five minutes. This helps soften the paper. Then flood the panel you're going to sand with lots of water. Now lightly sand the surface using back and forth strokes. And be sure to keep the surface wet. Every minute or so, stop sanding and dry the surface off. The paint should appear equally dull at all points when you've removed enough paint. When all the panels are sanded, you can buff the surface with polishing compound followed by a glazing compound.

RUBBING, POLISHING & GLAZING

If you want your Camaro to look like it just left the factory, buffing the paint with a polishing compound is all that's required. Use a lamb's wool bonnet with a nap about an inch thick. Some bonnets are radiused at their edges. This helps prevent you from accidentally cutting through the paint with the side of the pad. Use this type of bonnet if at all possible.

Polishing compounds have very fine abrasives which act like sandpaper to remove very minor surface imperfections. The 3M Company markets a polishing compound that works great. Apply a small quantity of the compound to the paint and then work it into the paint with the wheel rotating slowly. When you've spread the compound, increase the wheel speed and keep it moving to avoid damaging the paint.

The edges of panels are particularly prone to damage. This damage occurs when you're trying to buff the edge of one panel where it meets another. Because the edge of the adjacent panel is exposed, it's easy to rub right through the paint. You can

prevent this by inserting a piece of cardboard into the seam separating the panels. With the cardboard in place, you can rub to the very edge of the panel without disturbing the paint on the adjacent panel. Or you can apply some masking tape to the leading edge of the adjacent panel.

As you're rubbing and polishing, remind yourself not to bear down on the buffer. Extra pressure isn't necessary to achieve a good finish if you're using the right grade of polishing or rubbing compound.

When you've finished polishing the car, it may exhibit swirl marks when viewed down the side. These marks can be removed with a machine glazing compound such as *Liquid Ebony*. Liquid Ebony is applied in a manner similar to polishing compound and will make your car look terrific. Just follow the directions on the bottle. And be sure to use a clean polishing bonnet. A dirty polishing bonnet will be putting in the swirl marks you're trying to get out.

After all your hard work, stand back, admire your car and give yourself a pat on the back. You deserve it.

Voila! Out of the paint booth, our 1973 Z/28 sees the first light of day!

14

Aside from the fun of top-down motoring, rag top Camaros of the 1967-69 vintage are fast appreciating in value. But be prepared to make the investment in time to keep the folding top mechanism and fabric in tip-top shape.

CONVERTIBLE TOP

To satisfy fresh-air aficionados, a convertible-top version of the Camaro was offered in 1967, 1968 and 1969. And when you made your choice a convertible, you could stay with a manually operated top or opt for a power top. The choice was yours.

TOP MAINTENANCE

Proper maintenance is critical to the life of the top. To help your top live longer (and the rest of your car for that matter), keep your car garaged when you aren't using it. Or at least shield it with a car cover. And when cleaning the top, use mild detergent and a soft bristle brush. Clean a section at a time then rinse each section thoroughly with clean water.

To help the top stay soft and pliable, you may want to "dress" it with preservative after washing it. Choose a product that doesn't add any color to the top, or attract a lot of dust. Dust tends to chafe the top if the top is lowered and raised often.

Clouded Rear Window—A plastic rear window that's dirty or clouded with a milky haze requires special attention. First, rinse off any excess dirt with clear water. This helps prevent any further scratching of the soft plastic. Then wipe the window with a wet cloth to remove any remaining dirt.

Plastic rear windows that are lightly to moderately clouded can be cleaned with a product such as *Meguiars Plastic Polish*. Windows that are heavily clouded can be cleaned with *Meguiars Professional Plastic Cleaner* followed by Meguiars Professional Plastic Polish. Meguiars products have the added advantage of an anti-static coating to help reduce dust buildup.

POWER TOP OPERATION

Camaro's power top uses two hydraulic cylinders to raise and lower the top. An electrical motor mated to a hydraulic pump provides pressure for this system.

The top is controlled by a small assortment of components including a dash-mounted control switch, wiring, an electric motor/pump/reservoir assembly and feed hoses to connect the motor and pump to the top cylinders. Hydraulic oil in the form of Dexron II or Type A transmission fluid is the lifeblood of the system.

The power top is activated by a dash-mounted switch. This switch is responsible for directing battery voltage to the electric motor positioned in the bulkhead behind the rear seat backrest. In turn, the motor is connected to a pump. This pump contains a rotor which circulates hydraulic fluid in the system. Two hoses, an upper and a lower, connect the rotor to the double-acting hydraulic cylinders that lower and raise the top.

Top Down—When the power-top switch is held in the top-down position, battery voltage is fed to the dark green wire on the motor, causing it, and in turn, the pump to run. At this point, the pump pushes hydraulic fluid through the upper hose to the top side of the cylinders. Fluid entering the cylinders forces the pistons downward, lowering the top. The fluid that's in the bottom of the cylinders returns to the pump via the lower hose. Here it's stored for recirculation to the tops of the cylinders when the top needs to be raised.

Top Up—When the dash-mounted switch is held in the top-up position, battery voltage is fed to the red wire on the motor. This causes the motor and pump to run, feeding fluid through the lower hoses to the bottom side of the cylinders. In turn, the fluid pushes the piston rods upward, raising the top. As when the top was lowered, fluid that was originally in the other side of the cylinders is returned to the pump where it is stored for later use.

DIAGNOSIS & SERVICE

To properly diagnose a problem with the top, you have to determine if the failure is in the hydro-electric portion of the system or the folding top itself.

You can isolate the problem (if it isn't readily apparent, such as a blown fuse) by disconnecting the tops of each cylinder rod from the top linkage. Then manually raise and lower the top while feeling and listening for any signs of binding of the top. If the top binds, try lubricating its pivot points with a couple drops of engine oil. However, if the top binds only when it's being latched to the header, chances are the door or quarter windows are out of alignment with the side roof rail weatherstrips. You can align them following the procedure outlined for window adjustment, page 106. If the top works freely, the problem is in the hydraulic or electric portion of the system.

If you don't hear the motor working when you throw the top-actuating switch, first check the condition of the battery and the fuse in the fuse block. If these check out OK, check the green and red wires at the motor and pump assembly to be sure they're getting voltage. You can do this with a test light. If you don't have a professional test light, an old taillight socket which has two wires running from it and a good bulb in it will work just fine.

Connect the green wire that runs from the switch to the motor, to one of the leads from the test light. Connect the the other test-light lead to a good ground. Now have someone push the top switch to the down position. If you're getting current to the motor, the test light will light. If it's not lit, check for voltage at the top switch.

Check the red wire in the same manner when the switch top is in the up position. If the light doesn't light, check for current at the top switch. Performing these tests will help you isolate the problem to either the current-feed portion or the hydraulic portion of the system.

If you can hear the motor running but the top won't raise or lower, check the plastic hydraulic lines leading to each top cylinder. If they're kinked, they won't allow fluid to flow, which can be the cause of the problem. However, if the lines appear to be in good condition, you'll want to check the fluid level in the pump reservoir.

Now before you jump right into this, be

careful to prevent spilling hydraulic fluid in the car. To catch any accidental spills, lay some rags next to and around the pump. Also, the top should be up to allow working room.

Because the filler plug is on the end of the motor/pump, you'll also need to remove the pump so you can access it. To do this, remove the clips that secure the top cylinder lines to the rear bulkhead, and pull the pump attaching grommets out of the floor pan.

With the pump out in the open, you'll find the filler plug on the end opposite the electrical leads. Remove it with a flat-blade screwdriver. Now hold the pump in the same position as when it's in the car and check the fluid level. You can do this by eye or with a clean pipe cleaner bent 90 degrees. The fluid level should be 1/4-in. below the lower edge of the filler-plug hole, when the top is in the down position.

If the fluid level is lower then this, add some Dexron II or Type A transmission fluid until it's at the proper level. Just make sure not to overfill the reservoir as you'll only put bubbles in the system. Bubbles will prevent the top from being lowered or raised.

Now if the motor is getting power, and its connections are clean, but you don't hear it working, you'll need to replace it.

TOP REPLACEMENT

After a period of time, the vinyl portion of the top eventually dries out and begins cracking. Or it may shrink so much you can't close the top. Or it may fade. Whatever the reason, the time will come when the top must be replaced. So you must choose a top that best suits your needs.

Replacement tops are made from various types and weights of vinyl, although some are made of canvas material. The better quality tops have longer warranties against shrinking and cracking. And when you select a replacement top, you usually have the choice of a plastic or glass rear window.

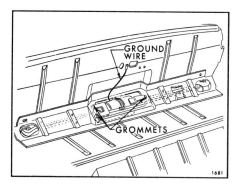

Location of Hydro-Lectric motor and pump assembly. Courtesy Fisher Body.

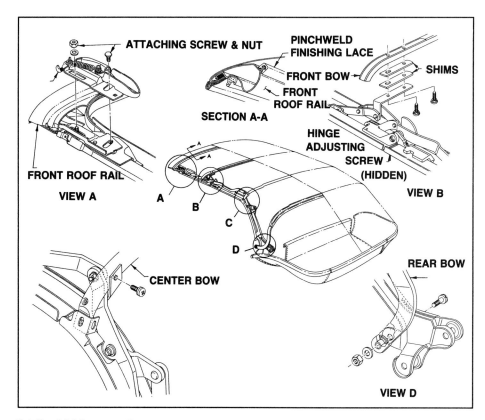

Details of folding-top adjustments—1967-69 models. Courtesy Fisher Body.

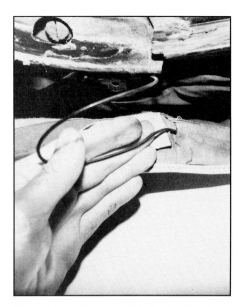

This cable fits in a pocket in the top. The tension it creates keeps the sides of the top from flapping.

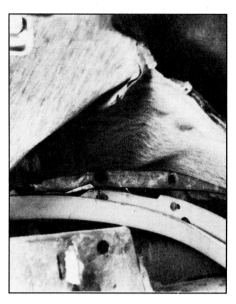

Rear of convertible top is stapled to tacking strips. In turn, these strips fasten to body with hardened bolts.

Tops with plastic windows are less expensive than those with glass rear windows. However, a glass rear window offers better visibility and doesn't cloud with age like plastic does.

To prevent chafing and also give a smooth, neat appearance, cloth pads are positioned between the outboard edges of the top frame and vinyl. So, when replacing your top, you're better off installing new top pads.

To keep the sides of the top tight, spring-loaded cables run along the outboard horizontal edge of the top. These cables are fitted one per side and are housed in a pocket sewn into the top. It's a good idea to replace these cables whenever the top is replaced. However, you can use the old cables if they aren't frayed and the tension springs keep the cables taut.

Material Removal—Replacement of the top isn't all that difficult, but it can be quite

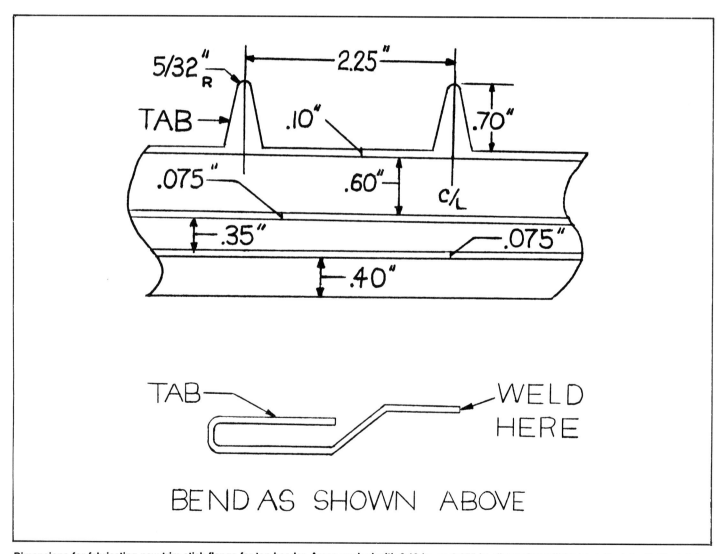

Dimensions for fabricating new trim stick flange for top header. Areas marked with 0.10-in. or 0.075-in. dimension will be bent. Each tooth-like tab is bent at its root to help retain trim stick. Double number of tabs last 6 in. of each outboard end to secure edges of trim sticks.

time-consuming when it's done properly. In addition to a new top and pads, and a normal complement of hand tools, you'll need a tack hammer, a good supply of galvanized tacks, some waterproof tape, weatherstrip adhesive and a good deal of determination. In addition, you'll need to fabricate four wedge blocks from pine. These blocks will help you fit the top. Measurements for these blocks are shown on page 160.

If you don't have the ambition to install a new top, take your car to a seasoned professional. Although the bill will probably be more for the labor than you'll pay for the top, it will pay off in a top with no signs of drawing or looseness. Whether or not you choose to install the top yourself, it pays to know how the top is attached to the car. This way you can properly critique your work or somebody else's.

The top is retained at three different points with staples. The lower rear of the top is stapled to long strips of compressed paper held in a metal channel called a *trim stick*. On 1967 models, there are three trim sticks: one long one and two short ones. The 1968-69 models use three trim sticks but these are nearly equal in length. All trim sticks are anchored to a metal lip that runs along the inside of the body opening.

Tacking strips are also used to secure the top and top pads to the top frame. Like the trim sticks, the top is tacked or stapled to these tacking strips. One of these tacking strips runs the length of the top header—the part of the top frame positioned directly above the windshield. This tacking strip is on the underside of the header and the front edge of the top is stapled to this strip.

Two smaller tacking strips are positioned on the top side of the header. The front part of the top pads is attached to these strips. The remaining tacking strip is found in the top of the rearmost frame bow. The rear window, convertible top, and the rear of the tops pads are stapled to this strip. The pads are also attached to the center bow with screws to keep them from shifting.

As you remove the old top, pay attention to these details. Take notes, photos or draw diagrams, but make a point of removing the top slowly, so you'll see first-hand how the top was installed.

To protect the car during top removal and installation, apply liberal amounts of masking tape to the top of the windshield header moldings as well as the moldings at the rear of the top. You'll also want to cover the car body with clean blankets to avoid scratching the finish.

Begin top removal by marking the left, center and right rear trim sticks, then unbolt them from the car. Pry the staples out of the trim stick so both the lower edge of the top and rear window are free.

Then remove the rear welt and the tacks securing the top to the rear bow. Moving to the front, unlatch the top and carefully remove the windshield header weatherstripping. This weatherstripping is held in place with the same retainers used for the door weatherstripping. It also uses one Phillips-head screw positioned at each outboard edge. The weatherstrip may also be glued in position, so work carefully. When the weatherstripping is out of the way, remove the tacks from the front of the top.

The sides of the top are held to the top frame with spring-loaded cables. The screws for these cables are found on the center bow and just to the outside of the top latches. With these screws removed, you can remove the top and cables from the car.

Moving to the rear bow, remove the tacks retaining the rear window, and remove the window. Now only the top pads remain. Remove the four screws that hold each pad to the center bow and the staples that secure the pads to the header.

After the entire top is removed, carefully set it aside as you may want to refer to the screw-hole positioning (among other things) as you're installing the new top. You'll also need to take the old front welt to a trim shop to get it re-covered. Be sure to check its length before and after it's re-covered so you're confident it's long enough for installation.

Restoring Frame—After the top material is removed, take a good look at the top frame. Are the tacking strips in good condition or are they missing chunks? Does the top frame have any signs of rust? Is the top header rotted out? If you can answer yes to any one of these questions, now's the time to do something about it, not after the new top is installed.

If the top frame is rusty, you may want to sandblast it and paint it. Sandblasting also tests the mettle (pun intended) of the frame. If any part of the top is weak, but it's not quite visible yet, it will be after you've finished sandblasting. Just be sure to avoid blasting any of the tacking strips unless you intend to replace them.

Probably the most common problem with the top frame is severe rusting of the header. If yours has small rust holes, they can be repaired by welding or with fiberglass resin. Larger holes will necessitate fabricating patch panels out of metal or using fiberglass cloth and resin. Just make

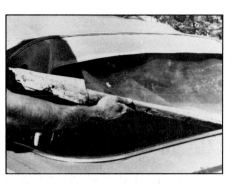

After removing bolts from trim sticks, simply lift top and trim sticks away from the body.

sure you do a good job as any sizeable imperfection in the top frame will show through the top or may even damage it.

Often, some of the tacking-strip retaining tabs will be rusted off. If this is the case, you'll need to weld new tabs on. The tabs are spaced equally along the center of the header. However, along the last 6 in. of the header, the number of retaining tabs double. And that holds true for both outboard edges. See illustration nearby.

If your header looks like a piece of swiss cheese, you're probably better off replacing it than patching it. Currently, only used headers are available, so be sure to examine it before you lay down your hard-earned dollars.

To replace a header, remove the large hex-head screw and nut, just rearward of the top latch, from both sides of the header. Pull the old header from the top frame. Install the replacement header to the top frame with the old screws and nuts, but leave them loose enough to slide the header back and forth.

Latch the header to the windshield and check the alignment between the front and center side rails. They should be even with each other, with no bowing. If they do bow, slide the header on the frame until they're even.

Tacking Strips— If you need to replace the old tacking strips, new tacking-strip material is available from **Kanter Auto Products**, among other vendors. This material comes in a roll and is sold in two different sizes: a smaller one for the front of the top and a larger one for the rear bow. Simply cut the material to length with side cutters.

To replace the old tacking strip on the underside of the front header, carefully bend the retaining tabs back and lift the old strip away. Then lay the new tacking strip in place and cut it with side cutters. Finish the job by bending the retaining tabs firmly against the strip.

The tacking strips on the top of the head-

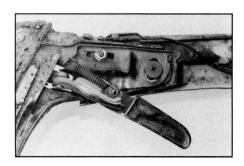

This severely rusted top header must be replaced with a used header from a donor car. Regardless, top header can be adjusted fore and aft by loosening the nut above the straight slot. Top latch can be adjusted side to side by loosening the large-head Phillips screw.

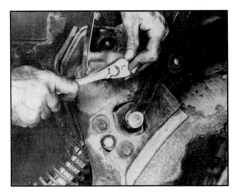

Before loosening hinge linkage, scribe around bolt washers for reference when reinstalling top frame. Note assist spring for manual-type top.

er are also secured with tabs, but only one on its outboard edge. The rest of the strip is stapled to the header with heavy-duty staples. You can replace these strips by bending the tabs away, then prying the strips off with a screwdriver. Then lay the new strip material in its place and cut it.

Because there isn't any easy way to fasten the strips to the header with staples, the new tacking strips should be glued and screwed into position. Use a waterproof adhesive, such as weatherstrip cement and four sheet-metal screws. Make sure the screws are evenly spaced and the heads of the screws are countersunk into the tacking strips. Also, mark the locations of the screws on the header with a china marker. This will help prevent hitting the heads of the screws with staples when you install the top pads.

If the tacking strip in the rear top bow requires replacement, it must be pried out with a screwdriver and the replacement strip glued and screwed into position, just like the tacking strips on the top of the header bow.

Now that you've got the top frame ready to go back in the car, you just put it in, put the new top on and you're done right? Well, yes and no. You're on your way, but there are still some checks and adjustments to make. And these checks and adjustments wouldn't have been necessary if you didn't remove the frame. But this is the price you pay for perfection.

Weatherstripping—After mounting the top to the car, you'll need to install the weatherstripping. Start at the front of the top where it meets the windshield header and work rearward. This will allow you to adjust the spacing between the side roof-rail weatherstripping, if you're using the factory pieces. The nice thing about the factory weatherstripping is that it has slots that allow it to move forward and back. This makes it easy to adjust the gap between weatherstrips so that they almost touch each other when the top is up and latched.

Unfortunately, some reproduction weatherstripping doesn't have slots; they use holes instead. So if you need to move the weatherstripping forward or back, you'll have to elongate the holes with a sharp drill bit.

Check Alignment—Once the weatherstripping is in position, raise the top and check its alignment to the windshield header. With the latches engaged, the front edge of the top should be parallel to the windshield header. If it's farther rearward on one side than the other, loosen all four of the top header retaining screws and nuts, then slide one side of the header forward or back until it's even with the windshield header.

Now check the sides of the top header where it meets the windshield header. The distance between the side of the top header to the edge of the windshield header should be equal on both the driver's and passenger's side.

The final check on the top frame consists of taking a careful look at the angle between the frontmost side rail of the frame and the side rail immediately behind it. The bottom edge of these rails must align with each other so that they look like they're one piece, as close to straight as possible. If they're not, you're not going to be able to adjust the side glass to get a good, leak-proof seal. This adjustment is handled by turning the larger set screw at the pivot point between the two rails until you have the proper alignment.

With the top in its proper position, you can adjust the door and quarter window glass so each properly contacts the weatherstripping. If you skip this step and install the top first, you may not be able to get the glass to properly seal against the weatherstrip.

After you're satisfied with the fit of the glass to the weatherstripping, remove all the weatherstripping so you can glue the new top to the side rails where necessary.

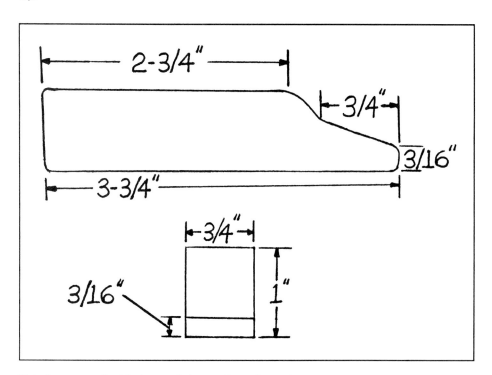

Make four top wedge blocks out of pine as shown. Insert these between windshield header and top header to stretch top as needed.

TOP INSTALLATION

Before installing the new top, there's one important factor you should be aware of—temperature. Never attempt to install a top when the ambient air temperature is below 65 degrees F. If the temperature is lower, you won't be able to properly stretch the top material, and the top will be too loose.

With that in mind, you'll be working in three distinct phases: pad installation, rear window and top material installation.

Pad Installation—The spacing between the rear and center bows is critical to correct installation of the pads and top. The correct distance between the rear edge of the center bow and the front edge of the rear bow is 14-1/2 in. If this distance varies more than about 1/4 in., you'll have trouble getting the rear window or top to fit properly.

To keep the bows properly spaced, you can simply cut two 14-1/2-in.-long spacers from a 1x2-in. strip of wood. Insert them between the bows about 5 in. from the outboard edge, then tape the bows to keep tension on the spacer sticks.

Install the pads by positioning one of the pad covers, flaps up, in the recessed area along the outboard edge of the top frame. The ends of the cover should extend over the front edge of the header and the rear bow at this point in time. The cover should also be smoothed out evenly, with no wrinkles.

Open the cover flaps and install the screws through the bottom of the cover into the center bow. Moving to the front of the cover, once again make sure the cover has no wrinkles, and with the flaps open, staple the front of the cover to the header.

With most pad kits, several sheets of thin foam padding are provided. These sheets must be layered to provide a pad thickness of about 3/16 to 1/4-in. Then they must be trimmed to fit just inside the pad covers.

So with this in mind, trial-fit the foam padding into the entire length of the pad. It must fit flat against the bottom of the pad. Note where the pad has to be trimmed, then remove the pad and trim it. Before reinstalling the pad, spray some trim adhesive on its bottom side and the inside of the cover. This helps prevent the pad from shifting later on. Keep adding layers of padding and gluing them in position until their total thickness is 3/16 to 1/4-in.

When the pads are in position, fold the cover flaps over them. The inboard flap must be folded first, then the outboard flap over it. Don't glue these flaps down as you may need to install some extra padding if

the door glass rubs against the top. This is done by lowering the top halfway and sliding more padding into the cover.

Finally, trim the pads so they fit into the trim stick pocket in the front of the header and don't extend beyond the rear bow.

Rear Window Installation—Begin installation of the plastic rear window by folding the window exactly in half, then compare the spacing of the pre-punched holes at the bottom. If these holes don't line up within about 1/8 in. of each other, they're incorrectly punched, so you'll have to make your own. These holes have to be properly spaced to ensure the back window is properly centered on the top frame.

To make your own holes, mark the lower edge of the fold line with a china marker or pencil with the top still folded in half. When you find the center of the car, make another mark about 3-in. forward of the deck lid, aligning the window marking with the car centerline. You'll also need to find the centerline of the trim stick that is immediately below the rear window.

Align the center of the trim stick with the center of the window. With the bottom edge of the window about 1/4-in. below the bottom edge of the trim stick, staple the window to the trim stick. Work from the center out to minimize the chance of window wrinkles.

Next, run an awl through the holes in the trim stick to perforate the rear window, then screw the trim stick into position and check the fit of the window. If it's skewed to one side, you'll have the remove the staples and move the window in the opposite direction to correct it.

If you're satisfied with the fit of the window, gently pull the vinyl sides into position and lightly mark them with a pencil, just even with the top of the rear pinchweld. Then unscrew the center trim stick.

Position the side trim sticks to the underside of the window, so their top edge is just below the pencil marks you made earlier. Then staple the trim sticks in this position as you pull the material forward and slightly down. Next, install the trim sticks with the proper screws, and check the fit of the window once again.

Small wrinkles in the C-pillar area of the rear window assembly are of no real consequence. However, if you're not satisfied with the fit of the window, remove the trim stick(s) in the area you're dissatisfied with. Then remove the staples and shift the window to remove any remaining wrinkles, as covered earlier.

Finish rear-window installation by pulling the top of the rear window forward and

New tacking strips are installed in the top and bottom of the header.

Plastic rain gutter channels rain from top to over wheelhouses, where it flows into and out of rocker panels. With age and heat, gutter cracks and allows water to leak onto floor. Two-piece aftermarket replacement gutter is available from Camaro aftermarket suppliers.

When installing convertible top, rearmost top bow should be about 14-1/2 in. away from next bow. Use a 1x3-in. wooden board to keep this dimension constant while top is installed.

Top pads must be tacked to three of the top bows. Make sure the tack heads don't protrude; otherwise they may cut the top.

tacking it to the rear bow. Space the tacks about 1-1/2-in. apart. Then check the fit of the window again and adjust it if necessary by restapling it to the rear trim stick(s).

Top Material Installation—When the pads are on and the back window is in, you can begin installing the top. To help make sure the top is even from side to side, fold the top in half lengthwise, then mark the fold line at the front and back of the top. When the top is unfolded and laid over the top framework, these marks should line up with the center of the windshield and the center of the deck lid.

Begin working at the rear of top, where it meets the body. Unscrew the trim sticks

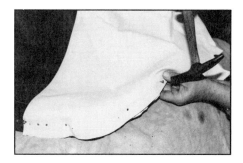

After drawing top material downward, tack it to rear trim sticks. If top C-pillar is wrinkled when you screw tacking strips into position, you'll have to reposition the top on the strips.

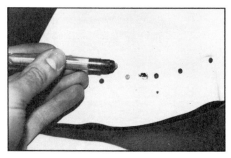

Mark and punch position of trim stick screw holes before attaching rear trim sticks.

Wooden wedge blocks are required for tacking front edge of top. The top should be drawn fairly tight to the front without wrinkles, then tacked to the wedge blocks. This allows you to mark the top front for proper fitting. Use drawing on page 160 as a guide.

from the body but leave the rear window attached to it. Then drape the new top over the top frame, in approximately the position that it'll be installed in.

Next, take one of the top C-pillars and drape it over the trim stick in the position you think it should be in with the top up. The bottom edge of the C-pillar should be about 1/4-in. or so below the bottom edge of the trim stick. This will give you a "fudge factor" to work with if the C-pillar isn't tight enough when the top is stretched to the front. Tack the material in this position. Holding a small block of metal against the area of the trim stick you're driving the tacks into will damp the movement of the trim stick, making it easier to drive the tacks in completely.

If you think you'll cheat a little the first time around and tack the C-pillar a little lower on the trim stick, think again. If your guess makes the C-pillar too tight, when you move the trim stick down to where it should be, the tack holes from the first try may show. And that means it's time for another new top. So start from the loose side and work the top until it's tight.

Tack the area of the pillar closest to the window and work your way forward. A little trick you can use to minimize any wrinkles in the C-pillar is to hold the top material tightly, just below the bottom of the trim stick and near the top of the C-pillar. This will give you a preview of any areas that will look drawn when the top is up. Slide the material down and forward as necessary and proceed with the tacking.

When you've tacked both sides, run a scratch awl through the bolt holes in the trim sticks and through the top material. This will make it easier to install the screws. Then secure the trim sticks with all the screws. This is necessary to properly evaluate the fit of the top.

With the windshield header covered with masking tape, gently tap the wedge blocks you made between the top header and the windshield header. A block should be positioned at each outboard edge of the top and in line with the top seams that run front to back.

Now stretch the top so it's taut and tack it to the wedge blocks, using two or three tacks per block. Be sure to drive the tacks through the seams to prevent any tearing, and only drive the tacks in about halfway as you'll need to remove them later.

With the top in position, locate the top seam that runs side to side above the back window. It should be directly above the rearmost top bow. This allows you to install the welt to the rear bow and hide the seam at the same time. If the top seam isn't

properly positioned, you'll have to remove the top from the rear trim sticks and move it as necessary.

When the rear seam is directly above the rearmost bow, begin tacking it to the bow. Drive three or four tacks, spaced about 1-1/2-in. apart, into both the driver's and passenger's side of the seam. Then take the head of the tack hammer and run it across the seam, making an indentation of sorts in the top. This will guide you in installing the rest of the tacks. Running your thumb across this area will provide the same type of guide.

Tack the entire seam, spacing the tacks about 1-1/2-in. apart. Don't position the tacks much closer together than this because you'll need untacked areas to install the rear welt.

Install the rear welt by draping it across the rear bow so that it hangs equally on both sides. The round portion of the welt must be toward the front of the car and the flat part of it directly above the bow. Mark the ends of the welt directly above the bow but don't cut it at this time. Now tack the welt to the bow, keeping the tacks about 1-1/2-in. apart.

When you've finished tacking, roll the front of the welt back over the tacks and lightly tap it into place with a rubber or rawhide mallet. You can even use the palm of your hand. The idea is to get the welt to cover all of the tacks, sealing them from view and the weather.

Use your hand to feel through the top for the lower edge of the rear-bow tacking strip. Each end of the welt should be cut about 1/2-in. shorter than this to allow installation of the metal welt cap screws into the tacking strip.

When you're ready to install the metal caps, place the cap in position. Stick the point of the scratch awl through the screw hole in the cap and into the top. Then remove the cap and drive the awl into the top about 3/8-in. This will reduce the effort required when you install the screw, as it takes a considerable amount of force to get the screw started. And for this same reason, as you're installing the screw, be sure to hold the tip of the screwdriver with your fingers. This helps prevent a slip of the screwdriver through your new top.

Now turn your attention to the front of the top. Find the center of the top and header and mark them with a pencil. Since the top is still pulled tight due to the wedge blocks, feel for the front edge of the header and mark the top at this point, along its entire length.

With the top marked, remove the tacks from the wedge blocks and lower the top halfway. Align the center marks of the top and header and pull the top material so the penciled outline of the header lies just under the front edge of the header. Secure the top to the header tacking strip with about ten staples, then carefully raise and latch the top to check the fit of the vinyl.

It shouldn't have any dips and should be fairly tight. Don't worry about any small wrinkles as they'll usually disappear when the top is out in the hot sun for awhile. If the top seems too loose or too tight, remove the tacks and adjust the top accordingly. When the top has a good fit, tack it the header along its entire length, starting at the center of the top. Space the tacks about 1-1/2 in. apart.

Begin installation of the front welt by folding it in half. Then cut a small notch in the excess welt material at the halfway mark to give you a reference point from which to work. Be careful not to cut through the stitching. Now align the notch with the center of the top and tack it all along the header except for the very ends.

Cut the welt about an inch beyond the edges of the top. Fold this extra material underneath the header to produce a good-looking corner, then tack it to the header.

Now turn your attention to the top cables. As you recall, each cable fits into a pocket on the outboard edge of the top. To help you install the cables, lower the top about halfway then slide a 2-ft. length of straight coat hanger into each pocket. What you're trying to do is open the pocket so the relatively soft cable will slide through without getting caught. Remember the end of the cable with the spring is positioned at the front, and a screw holds both ends of the cable.

The next step involves gluing the flaps of the top material to the side rails just above each front window and just behind the rear quarter windows. Use black weatherstrip adhesive and pull the flaps just enough to eliminate any wrinkles in the top before gluing them. Use an awl to locate the screw holes for the weatherstripping.

Well Liner Installation—When you're happy with the fit of the top, proceed with fastening the front edge of the boot well to the rear seat bulkhead. Begin by installing the rear seat backrest. Then push the leading edge of the well liner just so it touches the back of the seat. Also, make sure that the liner isn't cocked by sliding it side-to-side while watching for wrinkles. The liner is centered when all wrinkles are gone.

Starting at the center of the liner, punch a hole through the liner and through one of the holes in the top of the seat bulkhead

simultaneously with a scratch awl. Then push a retaining clip through the hole. If you don't care about authenticity, you can use a small sheet-metal screw. Work in this manner until the job is completed.

VINYL TOP

A vinyl-coated fabric roof covering was available on all first-generation Camaros and most second-generation cars as well. The cover assembly consists of three dielectrically bonded sections, glued to the

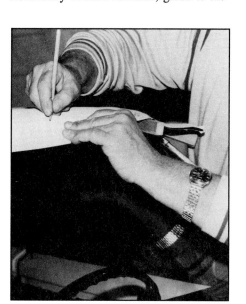

Use pencil to mark top where it passes over front of header. Then raise top halfway, pull the mark you just made so it's just under the bottom edge of the header, and tack it to the header with 25-30 tacks.

roof panel with non-staining nitrile adhesive and retained to the window openings with drive nails and trim clips. Around the doors and side windows, it's folded under the drip rails, glued and retained with the drip molding and weatherstripping. Along the bottom edge of the C-pillar and tulip panel, the vinyl is secured with more adhesive, drive nails and trim clips.

The problem facing the owners of vinyl-topped Camaros is that as the vinyl wears and cracks, the roof panel is exposed to the elements, thus giving rust a chance to gain

Rear of top is tacked to rear bow in many places. Then top welt is tacked over this bow. Finally, welt is folded back over itself, concealing the tacks.

A quality top installation has no wrinkles, is taut but not too tight to close and has straight seams. Thanks a lot, Mr. Kohl!

163

To remove old vinyl top, first remove all window reveal moldings, drip moldings and side window weatherstripping (except 1969 models). Then peel away top from roof by breaking adhesive seal with putty knife. Here, severe roof rust was caused by cracked and peeling top that admitted water.

a foothold. This is aggravated by the fact that many vinyl-roof cars were not fully painted in the roof area, and that some of the adhesives that GM used turned out to be corrosive as they decomposed. The co-author's 1973 Type LT Camaro had roof rust so bad that a new roof panel had to be welded on.

To install a new vinyl top, the windshield and backlite do not have to be removed. The old vinyl top can be scraped off after removing the windshield and backlite reveal moldings, drip moldings and side window weatherstripping (all

models except 1969), and quarter-panel and tulip-panel trim moldings. Also, cover the windshield and backlite glass with protective cloth and tape and remove the drive nails for the reveal-molding clips. A heat gun (200 degrees F maximum) applied to the vinyl can help loosen stubborn adhesive.

To install a new vinyl top, we went to renowned Camaro/Firebird specialists **Z + Z Auto in Orange, CA (714) 997-2200,** to follow the procedure with the camera. Check out the nearby how-to photo sequence.

HOW TO INSTALL A VINYL TOP

1. Joe Ureno from Z&Z Auto, Orange CA, starts vinyl top installation by marking center of roof and location of seams on masking tape strips on windshield and backlite.

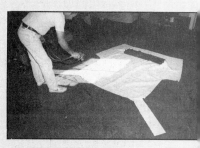

2. Joe inspects new top for defects in seams and stitching. When satisfied that no defects exist, he spreads top out upside down and applies even coating of contact adhesive.

3. After masking a spraying roof with contact adhesive, Z&Z crew carefully aligns top with roof according to marks made earlier. As long as it isn't pressed down firmly, position of top can be adjusted slightly.

4. Keep seams straight by aligning them with marks on windshield and backlite.

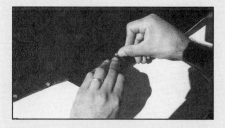

5. Once center of top is in position, stretch edges tight to eliminate folds and wrinkles.

7. After tracing window openings with chalk, trim top with a 3/4-in. margin to fold over into openings.

9. With a razor blade, cut small slots for weld studs that hold molding clips. Stretching top over these studs helps anchor edges into position and resist curling over the years.

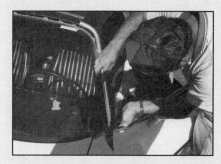

6. Around radius of sharp corners, cut slits in fabric to help prevent wrinkles.

8. Use small putty knife to press edges into window openings.

10. Once you're satisfied with the fit of the top, install the trim around the windows, using new clips whenever possible.

1. With proper care, your new vinyl top should last 10-20 years, longer if you keep it out of the sun and elements.

15

Hideaway Headlamp Systems

There's more to Camaro Rally Sport hideaway headlamps than meets the eye. Shown is 1969 vacuum-operated system.

When it comes to first-generation Camaros, what really makes a Rally Sport a Rally Sport is its hideaway headlamp system. Tucking the headlamps neatly out of sight behind the grille when they're not in use is a great styling touch. And it adds exclusivity and value; a mere 140,000 of the nearly 700,000 Camaros built between 1967 and 1969 were Rally Sports.

So much for statistics. Hideaway headlamps may be a plus when it comes to resale value, but they're definitely a minus if they're not working properly. The retractable headlamp doors are opened and closed by electric motors on 1967 models and a vacuum-servo system in 1968 and 1969. What you don't see behind those doors are numerous brackets, levers, bellcranks, arms, bushings and on 1968 & '69 models, yard upon yard of vacuum hose, a vacuum tank and large vacuum canisters called *actuators*.

If your 1967-69 Rally Sport's hideaway headlights are still working after all these years, you're one of the lucky ones. So be careful not to damage the motors or actuators by opening the headlight doors with your hands. You can do this in a pinch if they aren't working, but avoid doing so if at all possible.

Chevrolet designed around the problem of inoperative headlight doors on the 1969 model by

designing see-through headlight grilles so the headlights would shine through them in case the doors wouldn't open.

Forcing the doors open on 1967 models is probably the worst thing you can do because this system uses electric motors with plastic gears to open and close the headlights. By forcing the doors open (or closed for that matter) you run a good risk of stripping the gears in the motors. These little gems are no longer available from Chevrolet and will set you back a pretty penny for used originals or a set of new reproductions.

This wasn't a problem on 1968-69 models because they relied on a vacuum system to open and close the headlights. They didn't require gears so there's no chance of stripping them. To open the doors without turning the headlights on, push the vacuum relay valve on top of the vacuum reservoir toward the left fender. This routes vacuum to the actuators so they open the headlight doors.

If the headlights don't open automatically, pull on the outer edge of the door toward the center of the car. Do this until the doors lock in the open position.

Something to keep in mind if and when you're looking for some of the vacuum-system components for 1968-69 Rally Sports is that 1968-82 Corvettes used some of the same hideaway-headlight components. These include the vacuum

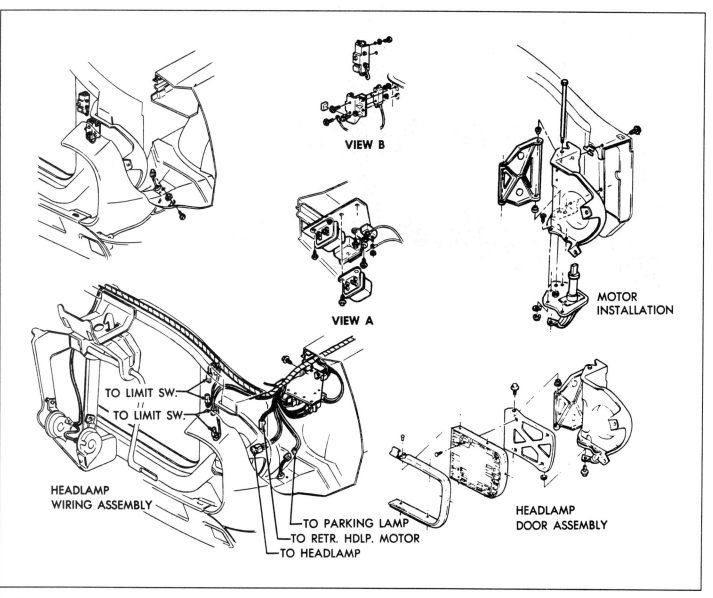

TO LIMIT SW.
TO LIMIT SW.

HEADLAMP
WIRING ASSEMBLY

TO PARKING LAMP
TO RETR. HDLP. MOTOR
TO HEADLAMP

HEADLAMP
DOOR ASSEMBLY

Details of 1967 hideaway headlamp system. Small electric motors with plastic gears are prone to stripping. Aftermarket motors with steel gears are now available. Courtesy Chevrolet.

Typical RS malady is shown on this 1967 SS/RS350—doors are stuck in open position.

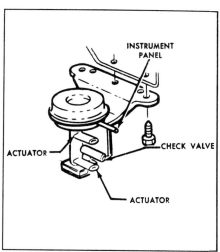

INSTRUMENT
PANEL

CHECK VALVE

ACTUATOR

ACTUATOR

Details of vacuum relay valve—1968-69 models. Courtesy Chevrolet.

relay on top of the vacuum reservoir, the system check valve and the headlight switch. Also note that 1968-69 SS350 and SS427 Impalas used a vacuum-operated hideaway headlamp system with some interchangeable parts, such as the headlamp door actuators.

The headlight switch used with 1968-69 Rally Sports can be identified by its two vacuum ports. One of these vacuum ports is linked by vacuum lines connecting the check valve and filter to the engine's intake manifold. The other port is connected to the top port on the vacuum reservoir-mounted vacuum relay.

When the headlight-switch control knob is pulled out to the headlight "on" position, vacuum is routed through the switch and control valve to the front vacuum port on the headlight actuators. This vacuum pulls on each actuator diaphragm which in turn moves the headlight door linkage and door to the open position.

When you push the headlight-switch control knob to the headlight "off" position, vacuum is routed in a similar manner to the port on the rear of the vacuum actuators which closes the headlight doors.

The actual hardware for 1968-69 Rally Sport Camaros is quite similar to the untrained eye. However, many of the pieces do not interchange.

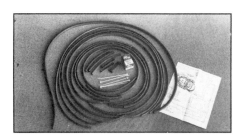

Chevy switched to a vacuum system to operate the headlamp doors on 1968-69 models. Classic Camaro repair kit contains 12 hoses in three sizes, plus filter, fender ties, bundles and instructions.

In humid areas or the Snowbelt, this pivot bolt may rust in place or break upon disassembly.

On 1968-69 models, install new vacuum filter (arrow) then connect two hoses to this vacuum fitting. One hose goes to the headlamp switch on the dash; other hose goes to a tee near the vacuum tank.

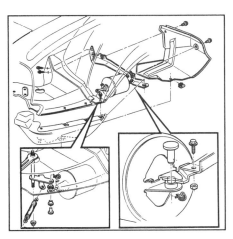

Details of vacuum actuator assembly and linkage—1968-69 models. Courtesy Chevrolet.

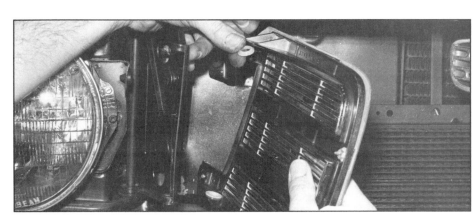

Don't forget to install new washers at pivots of headlamp doors.

Aside from vacuum leaks on 1968-69 models and burned-out electric motors and stripped gears on 1967 models, worn-out bushings, grommets and washers are biggest problem. Whenever linkage is taken apart, install a new bushing kit such as this one.

Not just any nut and bolt will work here. Special shank pivot bolt does the trick. Note see-through door on 1969 model that allows car to be driven at night if doors get stuck in closed position.

Twas the night before Christmas and all through the garage. . ... Seriously, the assembly process can be the most rewarding phase of your Camaro's restoration, because magically, methodically, a hulk and a pile of parts are transformed into a gleaming machine. A car is reborn.

It's difficult to say with any certainty exactly when a car becomes more than just a pile of pieces and metamorphosizes into a cherished member of the family. Your vision of a completely restored, gorgeous, fun-to-drive Camaro has to be a driving force. Maybe you remember what this Camaro looked like when it was new, or saw one recently on the street or at a car show that gave you inspiration, something to look forward to and strive for. Perhaps you've been dreaming about the day you first fire up your revitalized Camaro, wheel it out into the crisp morning air and embark on a great adventure.

The wonderful thing about this chapter is that you are now very close to realizing and living that dream. All that remains, really, is to assemble all of those new, used and reconditioned parts and components carefully and methodically. You've come this far; don't rush it by taking any shortcuts that you'll regret later. Through hard work and determination, you'll have presided over the rebirth of an American Classic—your Chevrolet Camaro!

Assembly is almost without exception the reverse of the disassembly process. By and large, if you remember how you took something apart, you'll remember how it goes back together. Take stock of the nuts, bolts, screws, grommets, rubber bumpers, gaskets, clips and other fasteners you have on hand.

If some of the screws have mushroomed-over slots or any of the plastic clips are missing tabs or teeth, it's wise to pop for one of the fastener kits offered by numerous Camaro aftermarket sources. Then, you know you'll be able to finish the job, even if it's at 2 A.M. or on a weekend.

UNDERCARRIAGE

Undercoating—The factory didn't coat the floorpan and wheelwells of first- and second-generation Camaros with asphaltium-based undercoating. But many rustbelt-area Chevy dealers did as part of the pre-delivery process. If your Camaro originally came with the undercoating applied or if you live in a Northern clime where road salt is used and you plan to drive the car a lot, now's the time to spray on some fresh undercoating, before the exhaust system and other underbody components are installed. These present an overspray problem of the first magnitude. Get some asphalt undercoating on an exhaust downpipe or catalytic converter and your Camaro will smell like a tar baby every time the engine gets hot.

Exhaust System—If you purchased a new exhaust system, line it up with the old one just to make sure you got the right pieces and you know how it all fits together. Do a mockup assembly in your driveway

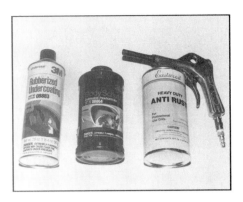

As far as the rust bug is concerned, the tender parts down low on your Camaro are tasty morsels. If you live in the Rustbelt, applying asphaltium-based undercoating to the floorpan and wheel wells is a wise move.

Don't throw that old exhaust system away until you've matched it up with a new one. Assemble the pipes and test-fit the system before tightening clamps to specifications.

Getting rid of that grungy old trunk mat and installing a new one can really wake things up under the deck lid. A fresh coat of trunk spatter paint finishes things up nicely.

and with a magic marker, mark the correct orientation of the pipes. On mild-steel exhaust systems installed on Camaros hailing from rustbelt states, it's a good idea to sandblast the pipes and brush on some *Eastwood Stainless Steel Coating*. Not only does it look good, but it protects the mild steel from rapid corrosion buildup. On the other hand, if the system you purchased is made from aluminized or stainless steel, no coating is necessary.

Whereas some of the original factory hangers can be reused, never reuse exhaust-system clamps or the asbestos exhaust-manifold-to-down-pipe gasket(s). You may need a floor jack to support the ends or middle of the pipes as you are installing them. Raise the system into position, but don't cinch down the clamp nuts until you've checked all pipes for interference with the body, rear axle, crossmember or other components. It's also a good idea to give the heat riser a good cleaning and lubricating with special high-temperature solvent. Only when you're sure the system is hung just right should you tighten the clamp nuts to about 15-20 ft-lbs. Later, when the engine is up and running, check the system for exhaust leaks.

TRUNK

Spatter Paint—Most Camaros were finished with gray and white or gray and blue spatter paint on the trunk floor and sides of the rear quarter panels and rear panel. Aerosol spray cans of spatter paint are available at most auto body supply and paint stores. If you do respray the trunk area, be careful to mask off items such as the rear wiring harness.

Trunk Mat—What a difference a new trunk mat can make in your Camaro's appearance. Before mounting the spare tire and jack, get out your old trunk mat and use it as a template to cut a new mat to size. Then install it in the car and trim as necessary around the fuel-tank filler pipe and bumper braces. Cut a hole in the mat for the spare tire mount J-hook.

Molded mats that fit the contours of the trunk are available fron various suppliers for those of you who want that "original" appearance.

ENGINE COMPARTMENT

Rubber Bumpers—Install new rubber bumpers on the hood, cowl, fenders and radiator support, replacing any hardened, cracked or missing bumpers. If any of the round bumpers prove difficult to slide over the head of the adjustable shaft, spray them with silicone. There are two adjust-

able bumpers on first-generation Camaros and five on second-generation cars. Check the height of the hood relative to the header panel, fenders and cowl and adjust the height of each threaded shaft accordingly. To adjust, you'll need to temporarily remove the rubber bumper, loosen the locknut at the base of the shaft and rotate the shaft with a large, flat-blade screwdriver.

Hood Pad—Wear a long-sleeve shirt for this and try not to rub your eyes after handling the hood pad. It's made out of a fiberglass material and skin or eye contact can cause itching and a rash to develop. If you are buying a new pad from GM, you'll have to cut it to size using your old pad as a template. Aftermarket pads come pre-cut. It helps to have a friend to help support the pad as you install the large-head plastic clips because the pad is floppy and will hang down in your face.

Battery & Tray—If you're installing a new battery tray or have derusted and refinished the old tray, it's a good idea to give the tray several coats of *Eastwood Battery Tray Coating*. This is a satin black plastic coating that helps prevent battery acid from attacking the steel tray. Install the tray and battery. Don't hook up the battery cables just yet, though.

INTERIOR

Carpet Underlayment—If you tore out the old carpet underlayment because it was cracked and coming apart or the floorpan was rusty, now is the time to install new underlayment, available from **Ssnake-Oyl Products in Dallas, TX.** The underlayment is made of a modified bitumen membrane with a polyester core and a polyethylene film on the top side. This material is easier to work with when it's warm, so if it's cold outside, warm up the underlayment in the house before taking it outside. Alternately, you can lay it in the sun or heat it with a hair dryer.

Lay the pieces in position on the floor as the label on each piece dictates. Make sure that the shiny side with the polyethylene film faces away from the floor. Start with the front pieces and work toward the rear. If you have any trouble with the underlayment not laying properly, simply heat it with a hair dryer until it does what you want it to do. The underlayment will tend to keep its shape as it cools. With the underlayment in position, you can begin laying the new carpet.

Carpet Installation—If you're installing a two-piece carpet, start with the rear half first. Position the center of the carpet over the transmission tunnel, then work out any wrinkles toward the sides of the car with your hands. Make sure the carpet is centered in the car. Push the carpeting down into the rear footwells, leaving plenty of extra on either side to fit under the sill plates. This material can be trimmed off later when you're ready to install the sill plates. Then install the front carpet in the same way.

On floor-shift or console-equipped Camaros, trace the outline of the console mounting brackets or shifter boot with a piece of chalk. Trim the carpet to fit.

Use an ice-pick or awl and X-Acto knife or scissors to find and then enlarge the holes for the seat belt bolts and front seat bolts. Poke through the carpeting from underneath with the awl, then enlarge the holes to the diameter of the bolt shanks. Otherwise, the bolts will be next to impossible to thread into the floorpan and when they do, you'll have carpet fibers wound around the bolt shanks, preventing you from torquing the bolts to specifications.

If you didn't remove the kick panels, use a blunt-edged stick to fold the carpet under them. On the driver's side footwell, fit the carpet around the dimmer switch, using a new grommet if necessary. And if you have any trouble keeping the carpet in place, use some spray adhesive.

Finally, install the carpet firewall guard at the base of the steering column and any remaining floor-area trim.

Seat Belts—If any seatbelts are inoperable due to broken or stuck latches, seized retractors or if any exhibit frayed webbing, get new, used or rebuilt replacements. Don't drive the car without functioning seat belts and don't carry any passengers you don't have belts for. Never substitute any seat belt anchor hardware for the original Grade-8 bolts. Install the belts and tighten all anchor bolts to 30-45 ft-lbs.

Rear Package Shelf—If your Camaro had a rear speaker (or two) and/or a blower-type rear defroster, cut the holes for these items in the new rear package shelf, using the old one as a template. Replacement package shelfs come in black only, so if you need to dye it another color, do so now while it's out of the car.

To install the package shelf, spray contact adhesive on the underside of its vinyl front lip and more adhesive on the upper forward edge of the rear-seat bulkhead. Slide the shelf back until it butts against the base of the backlite opening and press the shelf lip onto the bulkhead. Then install the rear speaker(s) and other items through the trunk opening.

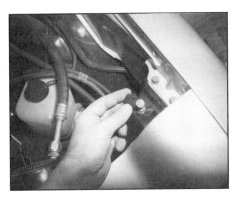
Forward corners of hood can be adjusted for height with these adjustable bumpers. Then slip new rubber caps over bumper heads.

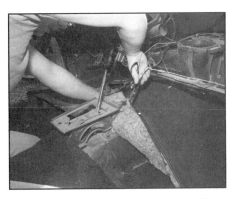
With the front section, trace outline of shifter and/or console brackets (if so equipped) and cut carpet to fit.

Most battery trays in old Camaros get eaten alive like tray at left. Bolt in a new tray and protect it from ravages of battery acid with plastic-based Eastwood Battery Tray Coating or equivalent.

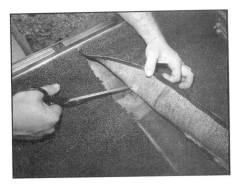
Install sill plates to keep carpet in place. Locate seat mounting-bolt holes with an ice pick and enlarge the holes to bolt-shank size with scissors.

Carpet underlayment from Ssnake-Oyl Products, Dallas TX, helps provide a quality feel and sound to the car. To flatten underlayment and help it follow body contours, heat it with a hair dryer.

Install a new grommet for the dimmer switch.

If installing a 2-piece carpet, start with the rear section. Center it on the driveshaft tunnel and make sure it overlaps onto door sills.

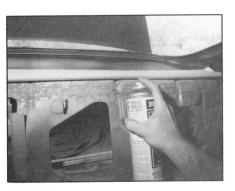
Slide new package shelf into position, flip up front lip and apply contact adhesive to lip and rear bulkhead.

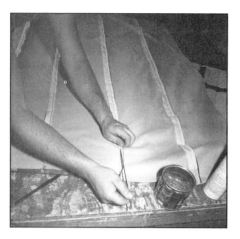

1. To install a cloth headliner, first transfer listing wires from old headliner, center in sleeves and snip off excess.

2. Spray trim adhesive around perimeter of inner roof panels and headliner.

3. Snap listings into plastic clips in roof. This is not as easy as it looks. Listings are a very tight fit in clips.

4. Stretch headliner into corners and work edges into their grooves with a curved putty knife. Cutting slits in cloth (arrow) helps eliminate wrinkles on curved sections.

HEADLINERS

Headliner (67-73 Cut & Sew Type)—
There's a bit of art involved with installing a cut-and-sew cloth headliner. It was the last of the skilled-labor procedures on the Chevrolet assembly line. Some finesse is required to "hang" a headliner that's tight and wrinkle-free. And it's not easy working with contact adhesive and fighting gravity with a piece of cloth that's hanging down in your face as you work it into position. You can do a successful job if you take your time and work the edges tight. Required tools are an awl, large scissors, a curved-blade putty-knife tool and lots of contact adhesive.

Before you begin, install all of the screws for the sun visors, rear-view mirror support, dome light, coat hooks and shoulder-belt anchors without the parts they retain. Reason? You can use the Braille method to find the locations for the fastener holes in the headliner after it's installed by feeling for the appropriate "bulges." Works every time! And make sure all the screws are in as deep as possible to minimize the chance of wrinkles when you remove them later.

Begin headliner installation by transferring the listing wires from the old headliner to the new one. On first-generation cars, the wires are different lengths, so make sure you reinstall them in the headliner according to the numbering system you used at disassembly. On second-generation cars only, measure the length of the listing-wire sleeves, center the wires in them and snip off any excess from the ends of the sleeves. Next, spray the entire perimeter of the headliner's underside with contact cement. Do the same to the perimeter of the roof's underside, taking care to avoid getting cement overspray on the windows or any trim by using a cardboard masking plate as you spray.

Moving from front to back or back to front (it doesn't matter), take the headliner inside the car and being careful to avoid getting contact cement all over, snap the listing wires into their plastic clips. This is not as easy as it sounds because the headliner sleeves and listing wires are a very tight fit in the clips. Do whatever works to get them in; spraying with silicone or sliding them in from the sides worked for us.

On first-generation cars only, the listing wires hook into the sides of the roof panel. You'll notice that there's three holes for each wire; this is to accommodate adjustment so you can change wire position to keep the headliner tight and free of wrinkles and sags.

At this point, you'll have a cloth headliner with glue on it hanging down from four or five wires. Don't panic. Mentally bisect the headliner into it north/south and east/west planes and begin stretching the headliner to the front center, sticking the headliner to the header above the windshield and spreading it outward tight and wrinkle-free as you go. Tuck the headliner over its flange lips and trim off any excess fabric. But leave an inch or so of fabric around the perimeter until you're sure the headliner is straight and tight. If you cut it off too short, you'll have to buy a new headliner and start all over again. Still with us? Repeat the stretch, tuck and trim number for the sides and the rear of the headliner. On curving sections, you can sometimes eliminate wrinkles by cutting slits on the edges of the fabric.

Finding the screw holes for the sun visors, rear-view mirror support, coat hooks, shoulder belts and dome light is next. Slowly and methodically, feel for these pieces and cut a small relief hole for the screw heads. Then remove the screws and install the trim pieces. The shoulder belt bolts should be torqued to 15-18 ft-lbs. For the dome light, find the wire leads, cut a second hole for them and pull them through. Connect the leads to the light, screw the light in position, install the bulb and cover.

Then slide on the headliner garnish windlace trim and screw the A-pillar trim pieces in position. On 1967-69 Camaros, the sail panels should be recovered with the same material used for the headliner; Stretch the material in place, glue it with contact adhesive and snap it into place inside the C-pillar with its three metal clips. On 1970-73 Camaros, the garnish moldings over the side windows snap into position.

Headliner (1973-81 Hardboard Type)—
Installing the hardboard headliner on late-model Camaros is a simple matter of bowing it slightly and feeding it through a door opening or windshield cavity, raising it up and securing it with screw-on garnish trim moldings, front, sides and rear. The side moldings also snap into position with Christmas-tree type plastic clips. Install the dome light as described above.

If your 1978-81 Camaro is equipped with a T-top, the replacement headliner will have to be cut to fit around the roof-panel moldings. Use your old headliner as a template

Other Interior Trim—The remaining interior trim can be installed now in the reverse order of removal. However, if you haven't adjusted the side windows, rear quarter windows (first-generation only), installed the door handles, lock cylinders and window felts, or installed the driver's side remote mirror adjustment, leave the trim panels off for now.

Radio installation is next to impossible after you've installed the console. By the way, never turn on the radio with the speaker wires disconnected; the radio's output transistor may blow out.

The window cranks that needed a special lock-clip tool to remove can be snapped back into position with the special horseshoe clip in place. To do this, partially slip the clip back onto the crank so that its open ends rest on the metal portion of the clip groove. Then slide the crank over the splined shaft on the regulator. Press the crank on firmly, then push the clip on with the lock-clip tool.

To give you working room inside the car, the front seats should be the last items you install. Remember that the rear seat cushion(s) are hooked into their locking tabs in much the same manner they were removed—by pressing down while simultaneously pushing back.

On second-generation Camaros, if one cushion sits higher than the other, remove it and bend the bottom seat-frame rung that hooks into the floorpan up slightly so the seat will sit lower. Sure, it's blacksmithery, but it works. And if one seat sits lower than the other, do the opposite to the bottom rung. Whatever you do, make sure both seat cushions are secured in position, not just floating around back there.

WEATHERSTRIPPING

Depending on the application, you'll need *3M Weatherstrip Adhesive* (Gorilla snot as we're fond of calling it) or equivalent, new plastic weatherstripping fasteners or short-shank screws in addition to the new weatherstripping you're installing.

Hood-to-Cowl Seal—No adhesive is necessary, but we recommend new plastic clips for this seal for 1970-81 models. On second-generation cars, the seal comes in three pieces: a long one that wraps around the cowl vent opening at the rear of the underside of the hood, and two small pieces that go at opposite ends of the cowl. Regardless of application, orient all hood-to-cowl seals with the open channel facing the rear of the car.

Door Weatherstrips—These seals come with nylon "T" type clips already installed in them and wrap around the front, back and bottom of each door. The molded ends can be attached with large-head plastic rivets or cross-head screws. A little weath-

5. Trim off any excess, but not too close. Leave 1/2-3/4-in. overlap at edges.

erstrip adhesive under each end won't hurt either, but don't go overboard with the gooey stuff.

Inner & Outer Side Window Felts— Replacement of these critical seals is covered on pages 116, 120 and 137. Without them, water can accumulate inside the doors and the windows will get scratched. On second-generation cars, short-shank screws can be used to attach the outer felts rather than the stock plastic rivets, but you may have to loosen the windows in their tracks to gain access to the screw/rivet holes or use an offset screwdriver. Do not use weatherstrip adhesive here.

Roof Rail & Quarter Window Seals— Both of these seals fit into steel channels and are glued in place with weatherstrip adhesive. However, if your Camaro has a bright steel drip molding, this must be installed before the roof rail weatherstripping is. After applying a bead of adhesive along the full length of the channel and rubber, work the weatherstripping into its groove with a putty knife or equivalent. The ends are fastened with large, cross-head screws or large-head plastic rivets.

Trunk Weatherstrip—This one-piece seal is glued in place with weatherstrip adhesive and hooks into a sheet-metal channel around the perimeter of the trunk opening on the rear quarter panels and rear panel. Butt the ends of the weatherstrip together centered above the lock cylinder. A common mistake is to cut the seal too short, causing the butted ends to separate as the seal shrinks with age. Avoid this by cutting the seal about 1/2 in. oversize which will force the butted ends together. Then put a dab of weatherstrip adhesive between the ends to keep them from separating.

6. Finally, install garnish moldings and windlace trim. Then fit dome light, visors, coat hooks, shoulder belts, sail panels and rear quarter trim panels.

Once the headliner is squared away, go ahead and bolt in the seats and console.

Apply a bead of weatherstrip adhesive into the roof rail channel, then work weatherstrip into position with a putty knife. Ends are secured with plastic rivets or large-head Phillips screws.

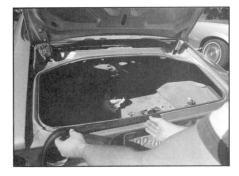

Common mistake when installing trunk weatherstrip is to cut it too short. Apply weatherstrip adhesive, work strip into channel, then cement butted ends together.

EXTERIOR

If you forgot to detail any exterior trim piece beforehand, now is a good time to do so. Trim with paint overspray can be cleaned up with a slightly dulled single-edge razor blade. Stainless steel and chrome can be cleaned to new brilliance with 000 grade super-fine steel wool. If the bumper brackets or areas behind the grille and parking lamps are weathered or have body-color overspray on them, now's a good time to spray them with a coat or two of semi-flat black paint.

Lock Cylinders—If your Camaro seems to have a different key for every lock, now is a good time to get a matching set of exterior locks—doors and trunk—that use the same key.

Deck Lid—Installation of the deck lid is accomplished by positioning the deck lid onto the hinges. Then install the lid bolts finger tight. When all four bolts are in place, shift the lid until the hinges line up with the outlines you drew. Then carefully close the lid until it's not quite latched. Check its alignment with the tulip and quarter panels. If the gaps appear to be even at all points and it opens and latches as before, the lid is aligned. If the gap spacing is uneven, refer to *Deck Lid,* page 149, for more information on aligning the

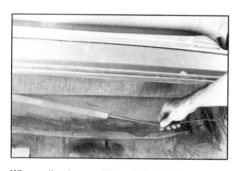

When adjusting position of deck lid torsion rods, keep rubber sleeve on them. It prevents rods from rubbing and creaking against each other.

If installing a rear spoiler on a car not originally equipped with one, go slowly, make your measurements then drill through trunk and rear quarter panels (1971-81 only) with 1/8-in. drill.

lid. For information on installing rear spoilers, refer to *Spoilers,* elsewhere on this page.

Torque Rods—When you open the deck lid on any first- or second-generation Camaro, you're assisted by two torque rods. These rods are made of spring steel and are held in such a way that they're twisted. As you raise the deck lid, they untwist, helping to lift the lid. After a period of time, the spring steel fatigues, reducing the degree of assist. This is more prevalent on second-generation Camaros with rear spoilers that add to the weight of the deck lid. In severe cases, the rods may actually break. Torque rods that are slightly fatigued can be re-tensioned using the following adjusting procedure.

Removal—To remove the torque rods, disengage them from the hooks under the tulip panel. Do this methodically and slowly. These rods possess a lot of latent energy. If you attempt to just pluck them out of the car without doing a little thinking beforehand, you might find yourself doing a lot of thinking; in a hospital bed. So here's the lowdown on removing them with a minimum of danger.

On first-generation cars, there's three hooks on either side of the trunk. Second-generation Camaros have just two hooks on each side. The idea here is to get the ends of the torque rods down to the lowest hooks. This reduces the tension in the rods. Then you remove them.

Begin removal with a piece of 1/4-in. pipe at least a foot long. Slide the pipe over the end of either torque rod. Pull up on the pipe to disengage the rod from the hook. When the rod is free, guide it into the next lower slot. Ditto for the other torque rod. Keep working like this until you get the torque rods into the lowest hooks.

Now slide the pipe over the end of either torque rod once again. Pull up on the pipe and gradually let it swing down. Do this for the other side as well.

Installation—Installation of the torque rods is practically the reverse of removal. However, as you raise the rods, the tension on them will increase instead of decrease. So be careful.

Adjustment—Adjust the tension on the torque rods so that the deck lid raises slightly when you unlock it with the key. It shouldn't fly open when the trunk is unlocked. This is accomplished by raising or lowering the torque rods in their hooks until a happy medium is reached. You can also compensate for fatigued torque rods or the added weight of a spoiler in this manner.

SPOILERS

Installation—Installation of the spoilers is essentially the reverse of removal. However, when installing the rear spoiler, use strip caulk to seal the bolts where they protrude through the sheet metal. This will prevent water leakage into the trunk and subsequent rusting.

If you're installing a rear spoiler on a car not previously equipped with one, or you've replaced the deck lid or rear quarter panels with new ones or used ones from a non-spoiler car, here's how. Begin by wrapping the ends of the spoiler mounting studs with masking tape. Do a good masking job so when you position the spoiler on your car to trace the bolt pattern, you won't scratch the paint. Place the spoiler on the deck lid of the car so that it's centered. If it's a three-piece 1971-81 rear spoiler you're installing, do the same with the end caps on the rear quarter panels. Check your work with a ruler.

When the spoiler is centered, trace the outline of the spoiler mounting studs onto the deck lid (and rear quarters) with a sharp china marker. Then remove the spoiler and recheck your measurements. If you're off, wipe the china marker off the paint and try again. When you're sure the spoiler is centered, use a prick punch to mark the center of each stud outline on the sheet metal. Don't hammer the punch too hard or you may put a sizeable dent in the sheet metal.

With the holes marked, drill through them using a 1/8- in. drill bit. Now check your work one last time by positioning the spoiler onto the deck lid (and rear quarters). If the holes line up, drill them out using a drill bit that's 1/16-in. larger than the diameter of the spoiler studs. This will give you a little room to move the spoiler for alignment purposes if you need to. It's quite likely that you'll need that adjustment room to get a three-piece rear spoiler looking anywhere near decent. Also, counter-bore any stud holes in the deck-lid where the deck lid inner panel prevents you from installing nuts on the spoiler stud. Do this on the inner panel only using a 5/8-in. drill bit.

Now slide the spoiler into the holes in the sheet metal. Then seal the studs with strip caulk and install the attaching nuts. And remember not to overtorque them, or you might create a crack around one of the studs in the fiberglass.

WINDSHIELD & BACKLITE

The windshield and backlite are bonded in position. As you may recall, you had to cut through this adhesive bond to remove these items. Technically, you can reactivate any adhesive caulk still remaining in the window openings with caulking primer, but after decades of exposure to the elements, we doubt that it's in reusable condition. It's best to scrape the old caulking out of the window openings (leaving a thin veneer of the old stuff is OK) and start off fresh for best glass adhesion and watertightness. So, if you haven't already done so, scrape out the old caulking, taking care not to scratch the paint. Also, replace any bent or distorted window reveal-molding clips at this time. These are difficult to get to once the glass is in place.

You'll need a helper to handle the glass. And for the rear window, you should have two large suction cups to lower the glass in position so you don't pinch your fingers. The windshield and backlite are large, heavy and unwieldy and if you drop one, you're out at least a hundred bucks apiece.

We recommend making a dry run first so you can position the glass in the opening exactly where you want it. There should be about a 3/16-in. gap fully around the glass in the opening. Support under the front, lower corners of the windshield with the two metal/rubber pads and tighten the supports in position. When the glass is centered, mark the position with masking tape at the right and left roof pillars and center top of the roof and on the corresponding locations on the glass, of course. This will give you "crosshairs" to aim for when the new glass is lowered into position with fresh adhesive. Once the adhesive makes contact, it'll take Herculean strength to adjust the position even 1/16 in., so get it right the first time.

Clean the glass thoroughly. If you're reusing old glass or installing used glass, clean the outer inch or so of the glass perimeter with a single-edge razor and remove any traces of old caulking. Next apply clear adhesive primer to the outer 3/4-in. or so of what will be the contact surface for the caulking. This cleans and primes the glass to accept the strip caulk you'll be applying next. The black strip caulk specially formulated for glass installation should be applied at 70 degrees F or warmer room temperature to keep it pliable. If it's cold in the garage where you're working, put the strip caulk in the oven or microwave for a minute or so at low setting.

In the windshield opening, brush on a coat of black weatherstrip adhesive and primer. Take care not to get any on the paint or interior. A thin, consistent coat is

best; don't leave any bare spots. Once applied, you have 15 minutes to get the glass in place, so don't apply the adhesive/primer to the backlite opening until after you've installed the windshield or you'll be rushed and may regret it later.

With a helper, pick up the prepared windshield and walk over to the front of the car; you on one side and your friend on the other. The side windows should be down so your helper and you can reach around the A-pillars and support the glass from inside the car. Line up your marks and make contact, then press the glass firmly against the body to make a good adhesive seal. Check the seal by pouring water around the perimeter of the glass (don't use a hose, though, as the pressure might blow through the still pliable seal). If all is well, remove the masking tape, connect the radio antenna lead (1970-81 only) and install the windshield wipers and wiper stops.

Repeat the procedure for the backlite, using a pair of suction cups to lower the glass in position.

1. When installing a windshield or backlite, clean perimeter of glass with clear adhesive primer, then stick on double-sided butyl rubber strip caulk.

2. Next, brush primer/adhesive onto window openings.

MOLDINGS & EMBLEMS

With the windshield and backlite installed, you can now snap their reveal moldings back in place. If the moldings have overspray on them, remove it with a single-edge razor. Stainless-steel moldings on 1967-69 models can be polished to surprising brilliance with 000-grade steel wool.

The remainder of the moldings can be snapped or screwed into place, taking care not to scratch the paint. Wherever possible, replace old plastic fasteners with new ones for positive retention as the edges of the clips get worn or chipped off.

In areas of the car that are inaccessible, such as the rear edges of the front fenders, use barrel nuts instead of stud nuts or speed nuts, whenever possible. Not only will this eliminate the time wasted to partially remove the front fenders to gain access to the rear of the emblems, but it also minimizes the chance of scratching the paint and makes future emblem removal (for cleaning, waxing or whatever) much easier.

SIDE MIRRORS

Using a new gasket, screw or rivet each mirror bracket to the door. Then fit the

3. With a helper, lower glass into position and press firmly to make a good seal.

4. Check for leaks by pouring water around perimeter of seal.

mirror to its bracket and install its attaching screw. On models with a remote-control driver's side mirror, route the cable through the door and out through the opening in the trim panel. The mirror will not function unless you secure the remote-control wires on the sheet-metal clip inside the door.

LICENSE-PLATE GROMMETS

If you are reusing the front and rear license-plate brackets, paint them low-gloss black and discard the old screw grommets and rubber bumpers and install new ones.

ASSORTED LIGHTS & LAMPS

Reconditioning—You can separate the lens from the housing on all parking lamps and 1967-68 Camaro tail lamps by removing the galvanized tail-lamp mounting plate. Then insert the tip of a screwdriver between the bezel and lens and carefully pry up from the inside. Because 1969 models don't use a bezel, just insert the screwdriver between the lens and the housing to separate them. From 1970-on, some tail lamps are constructed as one piece and cannot be separated.

Lens Polishing—Minor scratches on the lens can sometimes be removed with plastic polish. If they remain after polishing, try using some car paint polishing compound on a terry cloth towel and apply a little elbow grease. Rub the polish into the lens and check it frequently to avoid damaging it. When the scratches are removed, apply a coat of plastic polish.

This process usually removes all types of repairable scratches. If it doesn't, you'll probably have to buy a new lens.

Bezel Detailing—If you're working on a 1969-77 tail lamp or 1970-73 Rally Sport parking lamp the lens will sport stainless-steel trim. This trim can be cleaned with fine steel wool. Just be sure not to scratch the lens.

The bezels on 1967-68 Camaro tail lamps are made of chrome-plated die-cast metal. The outside edge of each bezel is smooth while the inner portion is textured and covered with semi-gloss black paint. If the bezels are pitted in both areas, you might be able to have them replated if the pits are small. Check with your plater to be sure.

If only the textured area is pitted, you can save the bezel. First clean each pit using a Dremel tool equipped with a small pointed stone. Then mix some structural epoxy, like *J-B Weld*, and use it to fill each hole. When the holes are filled and dry, apply more J-B Weld to each filled hole so that it closely matches the surrounding area.

When the epoxy is dry, use the Dremel again to do the final dressing of each filled hole. Then clean the bezel with a solvent, mask off the surrounding area and repaint it using a semi-gloss black paint. *Rustoleum 7777 Satin Black* is a good choice.

Gaskets—Many old gaskets can be removed by sliding a screwdriver underneath them and gently pulling the gasket up. But if your gasket won't come up this way, try using an adhesive release agent, such as the one made by 3M. Be sure to follow the directions on the can.

If you end up tearing a gasket, you can repair it using trim cement to hold it together. Or if the old, flat-style gasket cracked, dried out, was oversprayed or otherwise worn out, make a new one from bulk stock and use the old gasket as a template.

Mounting Plate—If you're reconditioning a 1967 or '68, turn your attention to the tail-lamp mounting plate. Because this part is galvanized, it's very rust resistant. Chances are the interior of the plate is fine. However, chances are the exterior isn't. Although galvanized metal doesn't rust, it does corrode. This corrosion can be removed with a wire brush mounted in a drill or by hand.

Once the corrosion is removed, you're faced with a different dilemma. The original hot-dip galvanizing has the appearance of large silver flakes. Unfortunately, there aren't any materials on the market that can duplicate the look of this finish. And painting galvanized metal is tricky in itself because many paints will not stick to a galvanized surface. So, the best you can do is to repaint it using a material that's specifically formulated for galvanized metal.

To install front-fender emblems on pre-1975 models, you'll have to loosen the fenders a bit to install the emblem stud nuts.

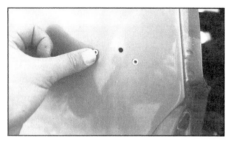

Or get smart and install new emblems with push-in nylon barrel nuts as shown here.

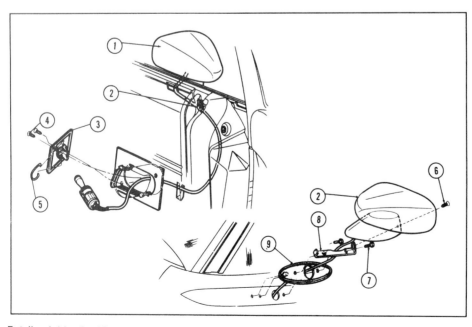

Details of driver's side sport mirror mounting and control cable routing (1973-74). Courtesy Chevrolet.

If you're redoing a 1969 model, things will be a little easier for you. Any gouges in the taillight housing can be filled with an epoxy, sanded and painted. If the reflector area inside the housing looks dull, clean it with *Prep-Sol* and repaint it using bright aluminum paint.

Installation—Make sure all bulb sockets are free of corrosion and check each bulb visually for a broken filament. Don't over-tighten the lens screws as you may crack the lenses.

GRILLE

The grilles on many first- and second-generation Camaros are plastic cross-hatch designs. And since they're plastic, they have a tendency to chip or crack when confronted with a sizeable stone. But being plastic also gives them an advantage; they can be easily repaired provided half the grille isn't missing.

Crack Repair—To repair a crack, grind a small channel on both front and back sides of it using a small grinding point. This provides more surface area for the repair adhesive, giving it a better bond. Use a name brand plastic adhesive and fill both channels with it. After the adhesive dries, use a fine file to remove any excess. Then thoroughly clean the entire grille with a wax and silicone remover like *Prep-Sol* and paint it.

Chip Repair—If you're going to repair a small chipped section and you don't have the original chipped-out pieces, get some plastic for filler material and some plastic adhesive. You can pick up the plastic at your local hobby shop. It's usually sold in sheets of varying thicknesses. Pick the size that best matches the grille.

Begin the repair by cutting out the chipped section to the nearest vertical and/or horizontal ribs. A fine-toothed hacksaw blade works quite well. Then measure and cut a replacement piece to fit. As when repairing a crack, grind a channel in both sides of the plastic at its edges. Then apply the adhesive. When the adhesive has dried, use a fine file to remove any excess. Finish the job by cleaning the grille with *Pre-Kleano* and painting it.

Aluminum Trim Repair—Now if your 1967 or '68 model grille has sustained some damage to the upper or lower aluminum trim on the grille, you have two options. If the damage is relatively minor, like a small rounded bend, you might be able to straighten it out using a pair of pliers. Just be sure to wrap some masking tape around the plier jaws to prevent damaging the metal.

Heavier dents can be taken out with a pick hammer and some patience. Work slowly and gently in a spiral pattern, working from the outside edges of the dent inward. This minimizes the effort by allowing the metal to spring back into its former shape as quickly as possible. If you push the metal too far the other way, wait until you've finished hammering the damaged area out before you begin to hammer the new damage in.

When you've brought the metal as close as possible to its original shape, use a file to remove any high spots. Work carefully to avoid going through the metal. If you do go through, use an aluminum filler rod or aluminized epoxy to fill in the hole. Unfortunately, if the hole is in an unpainted area of the grille, chances are you're not going to be able to blend the repair perfectly if it's done with epoxy.

If you can't repair the damage without damaging the grille, or if the damage is too severe, separate the aluminum trim from the plastic center section and replace the damaged piece. You can do this by drilling out the heads of the rivets that hold the plastic and aluminum together.

To secure the new pieces together, use 3/16-in. aluminum rivets with a 1/4-in. grip range. Don't use steel rivets for this procedure because the steel will cause the aluminum to corrode when it gets wet. And be sure to use backup washers on the rivets, opposite the rivet head. This will help to distribute the clamping load of each rivet over a greater area, which will minimize the chances of cracking the grille center section.

If you don't have access to a rivet gun, #10 machine screws and nuts can be used to secure the aluminum trim and center sections together. Replacement grille parts can be ordered through many Camaro aftermarket parts suppliers.

BUMPERS

Reconditioning—Depending on your desires and the condition of the bumper, you may want to recondition it. If it's a chrome-plated bumper and has only minor surface rust, it can be removed with naval jelly, extra-fine (000) steel wool or other similar substance. Bumper stickers can be shaved off with a razor blade, and the adhesive film removed with paint thinner, turpentine, *Bestine* solvent or in some cases, mayonnaise! However, if the chrome plating is pitted or if the bumper has any twists, bends or dents, you'll have to get the bumper repaired and replated or buy another one. Some Camaro specialists, such as **Z&Z Auto in Orange, CA**, sell recycled, replated bumpers on an exchange basis. Just drop off or ship your dented, rusted old bumper as a core and save at least half the cost of a new bumper.

If you opt to get it plated, make sure that you're dealing with a reputable firm. Ask friends who had plating work done recently if they're satisfied with the work and who did the work. Take your bumper to a few of these reputable platers and have them assess the situation. They should be able to tell you if the bumper can be replated and what the results will be.

However, if you decide that you're going to junk the original bumper and buy a new one, hold onto your original until you find a suitable replacement. Almost any bumper on the car is better than none.

Be sure to use new carriage-type bumper bolts upon installation. These are available from various Camaro aftermarket sources. It's also a good idea to have a helper for bumper installation to keep the bumper from slipping and damaging the paint.

License Plate Lamp—Reconditioning of the rear license-plate lamp on 1967-69

Coat of semi-flat black paint will restore faded Z/28 grille. Be careful to mask off bright surround molding.

models consists of removing the lens and then cleaning the lamp body with rust remover and painting it. Just don't get any of the rust remover on the lamp contacts inside the socket. These contacts can be cleaned with a pencil eraser if they're not very dirty or with some 400-grit sandpaper if they are. The lens can be cleaned inside and out with a plastic polish. And be sure to use new screws to retain the lens.

WIND NOISE & WATER LEAKS

WATER LEAKS

After adjusting the hood, page 170 in the beginning of this chapter, as well as the doors and side windows, pages 118 and 106 in the *Door Reconditioning* chapter, it's time to see how weathertight your Camaro really is. Maybe you've finished your car and decided to wash it for the first time. As you spray some water over it, you notice a couple drops of water gathering in the front footwell area. No big deal. You'll just wipe it up when you're done. When you're finished, you open the door and can't believe your eyes! What was once a couple of drops of water is now deep enough to launch a boat in! Such is the way of water leaks.

And to make your life difficult, the leaks almost never originate from where they appear to. Some of the most common points of leakage are the windshield, windows, cowl, joints between sheet metal, under the weatherstrip of T-tops, wheelhouses, incorrectly adjusted door latches, hoods and deck lids, and wiring-harness clips to name a few. Of all the various types of water leaks, cowl leaks are the toughest to find because they're out of sight.

Before you wrack your brain trying to find the source of the leak, you should know that weatherstripping can only seal against water up to a pressure of about 20 psi. Anything over that can cause a water leak for which there is no cure. So keep that in mind the next time you wander over to your local high-pressure car wash. If you're at a do-it-yourself car wash, avoid spraying the water directly at the weatherstripping and you'll have less water to clean up later.

Analytical Approach—When you're looking for a water leak, you'll have to think like a detective, asking yourself how severe the leak is, when does it happen and where does it show? If the leak is only a couple of drops, you might be able to put off finding it for a while. However, if the leak is anything like the previous example,

you'd better find it, and fast, before the rust does!

Many of the tools required for finding the source of water leaks can be found in your lawn shed. All you need is a garden hose, adjustable spray nozzle, a flash light and a friend. The friend will act as a "spotter" looking for leaks as you test the suspected problem areas with water from the garden hose.

Garden Hose Check—To begin, leave the spray nozzle off the end of the hose. Turn the water on so that a moderate amount of water comes out of the hose then have your spotter get in the car with the flashlight. Direct the stream of water at the lowest point of the suspected area then work your way up slowly while the spotter looks for signs of leakage. While you're checking, keep in mind that many leaks start forward of the point where the leakage is evident.

Water on the rear seat and trim can be caused by a leaky T-top. If this is the case, water is probably coming in under the weatherstrip. This is a common problem. If after testing, this proves to be the problem, seal the underside of the weatherstrip with sealer.

If you have a leak in the trunk, you should suspect a problem in the rear wheel houses as well as the deck lid seal. Also, keep in mind that trunk leaks are usually found between the seal and the pinch weld underneath the seal. To find the leak, you'll need a spotter in the trunk. Just make sure the engine is off and the deck lid is closed.

Soap & Air Check—Leaks in body seams are difficult to find because the leak is usually small. However, they can be checked using a solution of three parts water to one part liquid detergent. Apply the solution to the suspected body seam and apply compressed air to the seam from the underside. Air bubbles will indicate the source of the leak.

Caulk & Sealer—Before repairing any leaks, make sure that you use a good sealer. This includes strip caulk, brushable sealers, and so on. However, there are two types of sealers you shouldn't use; *3M Super Weatherstrip* sealer is a good product but it isn't compatible with General Motors seals. The same goes for silicone-based sealers.

Whenever you find a leak, repair it from the outside, otherwise the water will seep into the joint and cause later rusting. Strip caulk is a good product to use on the underside of the weatherstrip channels as well as the retaining screws. Be sure to apply the caulk the full length of the

channel.

Door Sealing—If you're having a problem with a bad seal at the door or window but you don't know which one is causing the problem, close the door so that it latches but isn't completely closed. Stand two feet away from the door and using the opposite hand that you usually open that door with, push against the rear edge of the door. If the door doesn't close, lower the window and try the test again. Now if the door closes, you know the problem is with the door glass. And if the door doesn't close, the problem lies with the door.

AIR LEAKS

Wind noise can take the form of a whistle, roar or rush. As a car is driven, air pressure inside the car becomes considerably higher than the air outside. It's this difference in pressure that sets the stage for some types of wind noise.

You can simulate this condition by setting the heater fan on high while checking for sources of wind noise. Smear some liquid dish soap on the areas where the glass meets the weatherstripping. If the soap begins to bubble, you've found the leakage area.

Another way of checking for wind or water leaks, either by itself or in combination with the soap, is listening for air leaks with a hose. You can do this by having a friend slowly direct an air hose or hair dryer around the seal in question while you follow his movements from inside the car with a length of heater hose held to your ear. If you think you've found the source of the problem, place a piece of masking tape over the suspect area and check it again. If the noise goes away, you've found the leak.

If the noise is caused by a portion of the weatherstripping being too far away from the glass, shim the underside of the weatherstripping to bring it against the glass. If the leakage is around the entire edge of the glass, the window probably has to be aligned.

Wind noise is a close relative to water leaks in some instances. For instance, if the side glass pulls away from the seals while you're driving, you'll hear wind roar or whistle as well as get a water leak if it's raining.

A blow-out clip is used on second-generation Camaros to help prevent the door glass from pulling away from the weatherstripping at highway speeds. This clip is located above the door glass near the top of the A-pillar. If the door glass is properly aligned to the weatherstripping, bend the outboard edge of this clip toward the glass and road test the car to see if you've solved the problem.

Another area that has a bearing on wind noise is sheet-metal alignment. Improperly aligned sheet metal can disturb the airflow around the car causing varying types of wind noise.

DETAILING

This section is not about cleaning. We're assuming that your freshly restored Camaro is already tidier than a Dutch kitchen and spotlessly clean. Keeping it that way is the subject of the following chapter, *After The Restoration*, wherein we give you the real skinny on cleaning, polishing, waxing, buffing and all those nifty things your dentist never told you to do with a toothbrush. In this section, we concentrate on the small items, details if you will, that make the really outstanding Camaros stand out from the also rans.

DECALS, APPLIQUES & STICKERS

It used to be that it was easy to spot the low-mileage original Camaros from the restored ones. All you had to do was check the engine compartment, trunk and door jamb for the original-equipment decals that got beat up as the miles piled on and exposure to the elements took its effect or got destroyed when the car was refinished as part of the restoration. The low-mileage "warehoused" cars had them; the restored Camaros didn't. After 20 years of adhesive curing, few if any decals could be safely removed without tearing, cracking or trying to take the paint with them. This presented a painting problem. If the decals were to remain in place, they'd have to be covered with masking tape or petroleum jelly to keep overspray off them. And often, there would be a definable ridge around the decal that was a dead giveaway for a repainted car.

The novice might think that he could simply buy new replacement decals from the dealer to dress up his car. But, GM never was and probably never will be in the decal business. Luckily, a few years back, aftermarket specialists stepped into the void and began reproducing these decals so restorers can attain that as-new look with all the proper decals, but without the masking and reconstructive surgery headaches.

Today, you can buy replacement reproduction decals for nearly every (and we'll get into that distinction in a minute) original application on your 1967-81

Camaro. The reproductions are not perfect. Line one up against an original and you can see discrepancies in type style and leading, but the overall effect is pleasing to the eye and captures the effect of the originals, right down to the GM part number.

Engine Compartment—You can dress up your Camaro's engine compartment with decals for cooling system/antifreeze instructions, emission- control calibrations, engine displacement/horsepower descriptions for the air cleaner and/or valve cover(s), oil filler cap, GM windshield-washer bottle instructions, Harrison A/C unit, Frigidaire A/C compressor specifications, cooling fan cautions and even a sticker imploring you to "Keep Your GM Car All GM" with AC/Delco parts. On 1970-81 Camaros, the antifreeze caution

T-tops are notorious leakers and squeakers. Make sure weatherstrips are in tip-top shape. Blow-out clip (arrow) on 1970-81 models can be bent inward to help keep glass in contact with weatherstrip.

Today, reproductions are available for just about every decal and applique that came with your Camaro for that just-off-the-showroom-floor look.

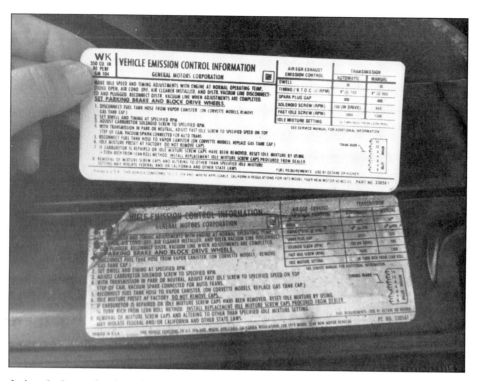

A close look reveals minor differences in typeface and leading between original and reproduction emissions-calibration sticker. But overall impact of getting rid of faded, stained and cracked stickers is impressive.

and emission-control calibration stickers mount to a sheet-metal plate that bolts to the radiator support and fan shroud. If you're worried about maintaining originality but want fresh-looking stickers, unbolt the plate and save it for authenticity documentation when you sell the car or if the Feds ever get nosy about emission-control stickers. Obtain a spare plate from another car in the junkyard, paint and detail it and install your reproduction stickers on it. It's sort of like having your cake and eating it too.

Trunk—Inside the trunk is a somewhat protected environment. If you're repainting the car the original color, perhaps you didn't have to paint the inside of the deck lid at all. Nevertheless, the decals for jacking instructions and Positraction cautions were often hastily applied at the assembly plant, with air gaps, bubbles and wrinkles already in them. Add 20 years of drying time and many of these decals are history. Not to worry. Reproductions of these decals, plus the caution decal for the inflatable Space Saver Spare are available from the aftermarket. Here's a tip: Before scraping off what's left of the old decals (or using spray-on decal remover), take a photo of their location so you can install the new ones in the same place (but without the air gaps and wrinkles!).

Body—Elsewhere on the body, the reproduction decal folks have you covered, too. Available are window stickers proclaiming factory air conditioning or Astro Ventilation for first-generation cars, fuel recommendation stickers for the glovebox door, engine-starting or catalytic converter/push-starting warning stickers for the driver's sunvisor, and for the driver's door jamb, stickers for tire pressure and GM's vaunted Mark Of Excellence. One very important door-jamb decal is no longer available for Camaros due to a GM copyright action resulting in an FBI bust of decal makers, as you'll soon read about.

Preparing for a color-change paint job (returning it from Earl Schieb dark green to its original color), one of the co-authors carefully removed the tire-pressure decal from the driver's door jamb by gently pulling after coaxing up one edge with a single-edge razor. The other sticker on the door jamb, however, which contained information on gross vehicle weight ratings, the month and year the car was built and redundant data on the VIN (car serial number), wasn't going anywhere and had to be masked. The co-author diligently masked off this sticker and painted the car. After painting, he proceeded to remove the masking tape covering the sticker, but alas, the tape was stronger than the adhesive bond of the sticker to the body and the sticker shredded itself into little slivers.

When the co-author contacted a Camaro aftermarket parts outlet, he was informed that the GVWR/VIN door jamb decal was no longer available. In fact, it is now illegal to manufacture or sell this decal. It appears that some unscrupulous types were counterfeiting cars with GVWR/VIN stickers (as well as bogus metal trim tags and serial number plates) to make pedestrian Sport Coupes into Z/28s, pace cars and other more valuable iron. The result is a loss for the majority of Camaro owners and enthusiasts in the hobby who only wanted to keep their cars correct and in top condition.

Applying Decals—First and foremost, make sure the surface is absolutely clean. If necessary, clean with an ammonia-type window cleaner, but never use a solvent that will leave an oily film and prevent good adhesion. With large decals, you may want to spray the surface with water and liquid dish soap first so you can move the decal around a bit before the water dries and the decal adhesive sets. According to Ralph Greinke of **Stencils & Stripes Unlimited, Park Ridge IL (312/692-6893),** a good choice is *Lemon Fresh Joy* because it has silicones in it that retard the adhesive from acting right away, thus giving you more time to get the decal straight. Regardless of technique, pull off the protective paper backing and hold the decal taut, so it doesn't flop down or fold over and stick to itself. Press down the center of the decal first, fanning your fingers outward from the center to stick the decal to the surface without trapping air underneath. Small air bubbles can be popped with a safety pin. With large decals, a plastic squeegee can be helpful for smoothing out wrinkles.

Removing Decals—It used to be that decals could only be removed by mechanical means—scraping, pulling, sanding or grinding. Problem was that the paint under the decal and immediately surrounding it was sometimes damaged in the process. When Camaros began using huge appliques for body graphics on mid-'70s Rally Sports and Z/28s, removing say a hood decal almost always required repainting the car, even if the paint was still in good shape. Some decals could be removed by using a heat gun or common household hair dryer to reactivate the vinyl solvents in the backing, provided the decal wasn't baked on or dried out and cracked.

Today, large decals can be safely removed with aerosol decal remover. Typically, the decal remover is a blend of methyl ethyl ketone (MEK) and xylene with isopropyl alcohol as a base. When applied at room temperature (60-90 degrees F), the solution should return the plasticizers in the vinyl into their original liquid state in 10-15 minutes. The gooey vinyl, which will now resemble wet toilet paper, should be scraped off with a firm plastic squeegee like that used to apply body filler. After the decal is removed, hose off the car with lots of water to remove all traces of decal remover and old decal fragments. The decal adhesive may remain. This can be removed with kerosene or with an aerosol decal adhesive remover containing napthol spirits, ethylene glycol (antifreeze), isopropyl alcohol and propane.

These are toxic chemicals, so a few words about safety are in order. Always use these preparations in a well-ventilated area. Don't apply in direct sunlight or in temperatures below 60 degrees F or above 90 degrees F. And don't use decal remover on a Camaro with a straight lacquer paint job, because it may lift the paint. Enamel, any acrylic or urethane is OK, however.

PAINTING STRIPES WITH STENCILS

The bold stripes that adorned 1967-73 Z/28s, SS's and Rally Sports were done in a more labor-intensive but durable manner; they were painted on. Unlike decals which will crack and fade with age, painted stripes should last as long as the paint on the rest of the car.

In the past, the challenge to the restorer of a Z/28 or other Camaro with painted-on stripes was finding a similar car to measure the location of its stripes and transfer this information to his car. While it was possible to easily measure straight sections, the compound curves were nearly impossible to get right.

But now, stencil kits are available that take all the guesswork out of where the stripes will go. You still need to be careful about masking and need sound painting techniques to make the job look good. Check out the nearby photos as John

3. After all air bubbles are squeegeed out, John carefully removes top covering from decal.

4. Voila! Decal installed. John inspects for loose corners. If necessary, these can be affixed with clear fingernail polish.

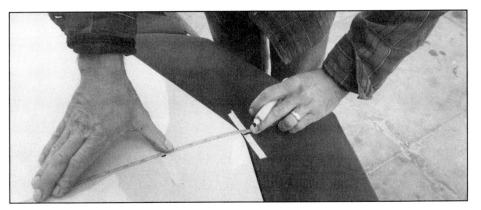

1. Late-model Z/28s used large decals for graphic impact. Here John Anderson of Z&Z Auto, Orange CA, measures for a new 1977 Z/28 hood decal from Stencils & Stripes Unlimited. Decal must be centered on hood.

2. After wetting hood with soapy water, John pulls backing off decal and positions it on hood. Decal can be moved slightly until soapy water is squeegeed out from underneath, allowing adhesive to take hold.

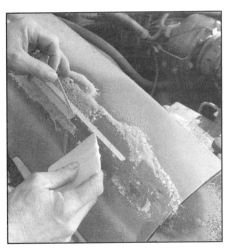

Opposite problem. Old decal has to come off, but owner wants to keep original paint. Spray on 3M Woodgrain & Stripe Remover, wait until it dries, then pull off decal using a squeegee to help break the seal.

Anderson of **Z & Z Auto** demonstrates how to stencil and paint runway stripes on a 1973 Z/28.

OTHER DETAILING TIPS

Still other items can make your Camaro look like it did when it rolled off the assembly line. In the engine compartment, original style heater hoses, radiator hoses, radiator cap and hose clamps are available from aftermarket sources. Add to that list reproduction Delco black-top batteries for top-post applications. For details on these and other items, see page 123 in the *Driveline Reconditioning* chapter.

Tires are another item that will get the judges attention. Whereas the original-style belted-bias wide-oval tires of the late 60s and early 70s wouldn't be our choice for driving these days, these can give your restored Camaro a correct appearance. Goodyear Polyglas, UniRoyal red-streak and Firestone Wide Oval tires in correct Camaro 60- and 70- series sizes are now being reproduced. See page 67 in the *Suspension & Steering* chapter for details.

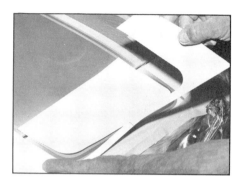

Unlike decals used on late-model Camaros, runway stripes were painted on 1967-73 Z/28s. The straight sections are easy to mask for paint application, but making the compound contours of the curved ends would be next to impossible without a stencil kit.

Mask off areas that don't need paint. Stripes can be white or black, usually selected for maximum contrast with body color.

Masked and ready for paint, dry cleaner garment bags were used to cover most of car. Note that stripe treatment on deck lid and rear spoiler varies with model year and type of rear spoiler. Follow instructions accompanying stencil kit.

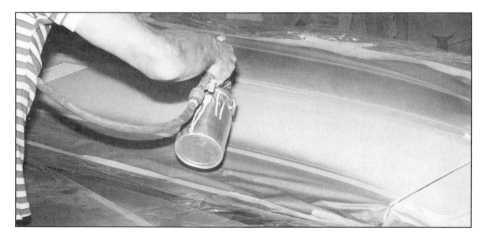

Apply paint stripes as you would paint on rest of car. After paint has dried thoroughly, remove tape and enjoy!

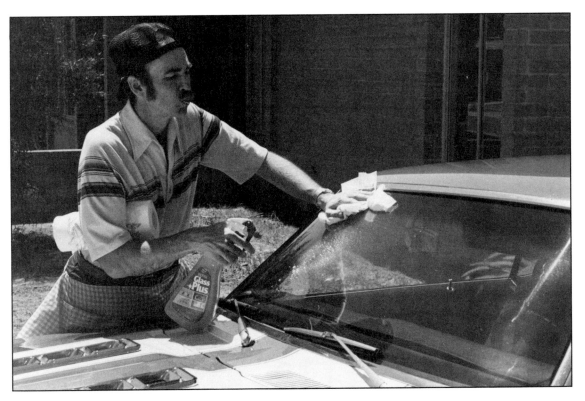

Your Camaro will never look as good as the day you finished its restoration. But keeping it clean—paint, wheels and tires, engine compartment, interior and yes, the windows—will help keep it looking good longer.

It's a jungle out there. As a gleaming tribute to General Motors' design brilliance, engineering expertise and manufacturing savvy, your freshly restored Camaro is a veritable babe in the woods. Despite your best efforts, tremendous forces of nature lurk at every turn, seeking to bio-degrade your Camaro and eventually return it to its basic elements. In the natural flow of things, seasons change and the warm sunlight of mid-day turns to the inky chill of night. Wind, rain and snow buffet your Camaro's brightwork. Relentless solar radiation ages paint, tires, weatherstripping and interior trim. Trees drip their gooey resins and insects make fatal rendezvous with your Camaro's flanks by the tens of thousands. Birds score direct hits with you know what. Small rocks, pebbles and sand get kicked up and bombard your Camaro's underpinnings.

And that's just the natural stuff. We haven't even begun to calculate the hazards of sharing highways with brain-dead drivers, parking lots with door-ding-happy dimbulbs or living downwind of industrial air pollution, let alone contemplate the scurrilous acts of larcenous souls bent on purloining part or all of your Camaro for monetary gain.

Face it. Your Camaro will never look or run better than it did the glorious moment you first fired it up and rolled it out into the light of day after its painstaking restoration. Ever notice how the Egyptian pyramids and Rome's Coliseum don't look as good as they once might have? No person or thing is immortal.

But you've got another plan. You're going to take the steps to help preserve and enjoy your Camaro for years to come. We're with you every step of the way.

CAR CARE TIPS
Car Cover—First and foremost, get a quality car cover. You wouldn't explore Antarctica in your skivvies or traverse the Sahara Desert in cut-offs, would you? It only makes sense that the longer you can keep the elements from feasting on your Camaro, the longer it can go before needing attention. Most car covers are quite effective at keeping direct sunlight, tree sap, leaves, pine needles, bird droppings, snow, ice and dirt off the sheet metal— and that's a good thing. Of less concern is rain, because any cover that won't let any moisture in also won't let it out; and that is more of a threat to paint and sheet metal. Trapped moisture between the body and car cover can create a high-humidity, greenhouse-like environment wherein rapid corrosion occurs. Even while covered, you want to make sure air can circulate over, under and around your Camaro.

As with anything else, you generally get what you pay for and car covers are no exception. We like cotton car covers because they're durable, won't scratch the paint, breathe well and are washable (car covers do get dirty and you wouldn't want to put a dirty cover on a clean Camaro). At $100-150, cotton car covers are not cheap, but then you wouldn't have gone to the trouble of restoring your Camaro if you wanted the cheap way out. Beware of canvas covers that can scratch the paint. And plastic covers deliver a double whammy; in humid climates, they trap moisture and in dry climates, they dry out and crack in no time at all. Your Camaro is probably better off with no cover at all than a bargain-basement $39.95 cheapie.

If you leave your covered Camaro outside for lengths of time, it's a good idea to get a lock for the cover. The lock utilizes a padlock and braided steel cable (like those used for bicycle locks) to tie the sides of the cover together underneath the rocker panels. So secured, the lock keeps the cover in place should a sudden windstorm pop up or another Camaro owner figure he'd like your cover for his car. It also keeps the small fingerprints and bicycle

and tricycle handlebar marks of inquiring minds off the paint should you park your Camaro where the admiring public can get their mitts on it.

Sun Shades—Obviously, most people won't bother taking a car cover everywhere they go. If you live in the Desert Southwest where the sun beats down relentlessly, you'll want to at least keep your Camaro's interior (with you in it!) from baking to death. Many folks add dark-tinted Mylar film to the side and rear windows to reduce the sun radiating into the car. On a restored car, this is not original and depending on how dark the tinting is, may be illegal in some states such as California. Plus, backing down your driveway on a moonless night can be an adventure and a test of your night vision.

So what to do? Parking under a tree has

its aforementioned bird and sap hazards. Luckily, fold-out plastic or cardboard shades or screens are available to place inside the windshield and backlite. These can keep direct sunlight off the dash pad, steering wheel and rear seatbacks, thus prolonging their serviceable life.

During the summer in desert areas, you may be forced to keep the side windows open half an inch or so just to vent the tremendous heat buildup in your car due to the greenhouse effect. Otherwise, it's best to keep side windows rolled up tight. Not only does this make it just a bit harder for a thief to slip a coat hanger down between the weatherstrip and window and pull up the door lock knob, but it also seals out dust and dirt. The less dirt that gets in, the less dirt you'll have to clean up. It's as simple as that.

Protectants—It's a fact of life that all vinyls and soft plastics harden and crack with age. This is attributable to high temperatures and environmental air pollutants such as ozone. But whatever the trigger, the reason plastics dry out is their co-solvents evaporate. You can sometimes see this evaporation as a light film inside the windshield of new cars left closed up in high heat and direct sunlight. Because many of these co-solvents are petroleum-based and known to be carcinogens—cancer-causing—the EPA limits the amount of co-solvent car makers can put in the plastic.

Protectants such as *ArmorAll* and *STP's Sun Of A Gun* are silicone-based formulations designed to retard this degradation. Some people like the wet-look sheen these impart to vinyl trim. A popular misconception is that protectants reintroduce softening agents baked out of the vinyl over time. They don't. Rather, protectants penetrate the pores and valleys in the vinyl, temporarily sealing in whatever co-solvents still are present.

Don't use protectant as a one-step cleaner; you'll only be spreading around some of the dirt and mixing it with silicone. And if you do use it—after cleaning first—apply the protectant sparingly, as it will only become a dust magnet and your Camaro's vinyl will have seemingly grown a fuzzy coating of dirt and grit in no time at all. And don't use protectant on the steering wheel, brake or clutch pedals. Silicone-lubricated controls are slippery ones, and that's a real safety hazard should your hand or shoe slip off during an emergency maneuver.

Wiper Blades—Nothing can ruin your day more than getting caught out in a sudden downpour, flicking on the wipers and

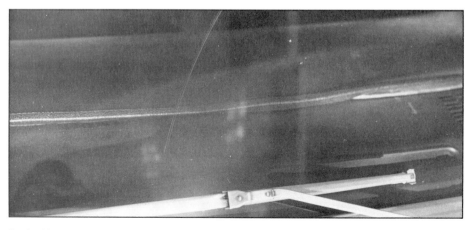

First and foremost, get a quality car cover. This attractive, washable cover from Beverly Hills Motoring Accessories, Hollywood, CA (213/657-4800) keeps the elements at bay while allowing your Camaro's finish to breathe without a buildup of moisture.

Don't skimp on wiper blades. If you were caught out in a sudden downpour and a blade cracked or flew off mid-stroke, you'd be looking through a scratched windshield like this poor fellow.

watching helplessly as the wiper arm gouges a deep scratch in the windshield right in front of your eyes. Inspect windshield wiper blades often, looking for cracks at the ends. Sitting out in the open, the rubber blades get cooked by the sun and attacked by airborne pollutants. In extreme climates, wiper-blade life can be as short as six months.

Parking Lots—Sooner or later, you'll probably be faced with that spectre of all car collectors—parking your Camaro in a parking lot. You'll likely find the most remote corner of the lot, hundreds of feet from the nearest car, and hike in to the store you need to visit. Then, you'll hike back out to your pride and joy only to find a van load of gypsies have taken up residence alongside. Your cadence turns to a gallop as you scurry up to the van side of the Camaro only to discover not one but two fresh door dings on its flanks. You confront the driver of the van with the door-ding evidence, but he retorts that they must have been there from before. Your word against his. What to do?

Aside from energizing a force field around your Camaro, there is little you can do to prevent the errant ding or dent in a dangerous world. Car covers offer minimal protection against a rapidly opening steel door, hell bent on mixing molecules with your Camaro's sheet metal. You can install body side moldings, but practical as they are, these are not original. Or you can get vinyl-covered foam-rubber strips that attach magnetically to the sides of your car. These are also a good idea if your roommate or spouse sharing the garage with your Camaro has a habit of opening doors first and looking later.

Likewise, if you're parked on a hill or are opening a door with the wind behind you, hang on to the inside handle to keep the door from getting away from you and slamming into a pole, tree or other car. It's no more pleasurable to be the dingor than it is to be the dingee.

Don't Slam It—And speaking of door dings, there's plenty of careless drivers out there looking to score a bull's eye on your car without you adding to the problem by self-inflicting some sheet-metal whacks yourself. When closing the hood, doors or decklid, apply no more force than is necessary to latch them shut. If adjusted properly and lubricated, these can be latched by guiding them into the closed position with an open hand. Panel imprinting—shallow depressions in the sheet metal around the door handles and center outboard edge of the hood and trunk—is a common sight on used-car lots and is caused by owners

actually denting the metal with the palms of their hands.

Exercise, Exercise—Like an aging prize fighter whose fakes and jabs fall short of the mark, your Camaro needs to be taken out for a run every few weeks or so just to keep its vital fluids in circulation. Drive it for at least an hour or about 20 miles to ensure that the engine, transmission and final drive get up to operating temperature. Doing so will evaporate all corrosion-causing condensation from the exhaust system and circulate coolant and oil throughout the engine. The drive will also coat the gears and other moving parts in the transmission, power-steering pump and gear and rear axle with lubricant, preventing corrosion and keeping seals wet—thus preventing them from drying out, cracking and causing a leak. The drive belts will polish the pulleys, wiping away any corrosion that may be starting, and the battery will get recharged by the alternator, helping to prevent a sulfated condition. Oh, you'll also be using up some of the gasoline in the tank before it gets gummy and varnish-like and keep the diaphragms in the fuel pump and carburetor accelerator pump bathed in gas.

The suspension gets a workout, too. By going over a few bumps, you'll be wiping corrosion off the shock-absorber piston shafts. Apply the brakes, and the linings will polish the rotors and/or drums. And the tires will have less of a tendency to develop flat spots if they don't sit in one position for months on end.

If you plan on storing your Camaro for an extended period, see the tips listed under *Long-Term Storage*, page 188, in this chapter.

Fuel For Thought—In case you haven't noticed, leaded gas is fast disappearing

from the service stations of America. High-octane leaded premium is a high-dollar ($3.00 per gallon and up) specialty item available only in bulk or from companies that sell to race tracks or high-performance marine applications. In response to an EPA crackdown on environmental lead, the few companies still offering leaded regular do so with a formulation that contains only 0.1 grams of tetraethyl lead per gallon-roughly 1/40th the percentage available in gasoline such as Sunoco 260 in the halcyon muscle-car days of the late 1960s and early 1970s.

The problem is, 1967-70 Camaros were designed to run on relatively high-octane, leaded fuel. Whereas octane can be enhanced in any number of ways (mixing premium unleaded with leaded regular, fitting water injection, adding octane boosters in a can), the lead requirement is tougher to fill. In these early Camaros, lead acts as a lubricant and anti-wear additive between the valve heads and their seats. Without this lubricant, rapid exhaust-valve seat wear can occur, leading to loss of compression and paradoxically, increased fuel consumption and a quantum leap in exhaust emissions due to incomplete combustion.

The condition is exacerbated by high temperatures, high loads and high rpm—all of which occur in high-performance Camaro engines. The fix is to mix commercially available anti-wear additives with the fuel in the tank or have the engine modified with pressed-in hard-steel exhaust valve seats. For this modification procedure, see page 122 in Chapter 11, *Driveline Reconditioning.*

Get A Tune-up—After all the hard work and long hours you put in on the restoration, you'll want the driving experience to

If your 1967-70 Camaro has the original engine, it was designed to run on leaded fuel. To prevent valve damage with today's no-lead or low-lead gas, install hard valve seat inserts or use an anti-wear additive in the tank. Gas treatments will help reduce fuel system deposits.

be as rewarding as possible. One way to ensure optimal performance is to get a complete tune-up—not just throwing in a set of points and plugs, but taking your Camaro to a shop with a scope and perhaps a chassis dynamometer. By observing the read-outs and scope patterns, a skilled technician can check for dead or weak cylinders, air (vacuum) leaks, engine-mount condition, as well as the operation of the PCV valve and, on 1973-and-later models, the EGR valve. He'll also be able to check initial and final distributor advance, and basic air/fuel and idle speed settings.

Driving a stylish, high-performance car is no fun if the engine sputters, lurches and gasps down the road. Have the engine tuned for maximum efficiency and maximum fun at the stoplight.

Easy does it when cleaning your Camaro's paintwork. Eastwood English Chamois is super soft and naturally absorbent. Courtesy Eastwood Company.

Don't forget the wheels when giving your Camaro a bath. Eastwood Spoke Brush gets into those hard-to-get spots between the spokes.

If the shop has a chassis dynamometer, the operator can simulate real-world driving conditions by placing your Camaro on the rollers. This is a good way to isolate driveability problems—sags, hesitation, lean-miss, back-firing and the like—in a controlled environment.

Washing—Aside from the fact that keeping your Camaro's finish clean will help it last longer, driving a clean car is much more enjoyable than driving a dirty one. A clean car shows pride of ownership and gets the admiring looks and thumbs-up signs. Afterall, isn't that why you restored your Camaro in the first place?

Even with your car protected by a car cover, dirt will find its way onto the finish when you take your Camaro for a drive. Particulate matter and splashed-on mud that could scratch the finish if left on and rubbed in can be easily hosed off. And most commercially available car-wash products do a good job of removing road film—a mixture of tire dust, oil droplets, asphalt and other components of air pollution. Some of these are corrosive to paint and should be removed as soon as possible. Tree sap, road tar, squashed insects and bird droppings usually require a second pass after removing the other dirt and film with a bug and tar remover. Bird droppings are acidic and can eat through paint if left on for an extended period.

Stay away from liquid dish soap as it can lead to a detergent burn of your Camaro's finish, particularly if you wash the car in direct sunlight on a hot day. And washing your car in direct sunlight is something you should avoid if at all possible. Also avoid the use of car-wash products containing silicones if you plan to do any more painting on your car. The silicones, if not removed, will cause fish-eyes in the paint finish when new paint is applied. Use only clean water and light pressure as you lather the car up. Let the water and car-wash solution foam the dirt away. Heavy pressure may cause scratches. Use only a horse-hair brush (which lifts off dirt on its follicles, a sponge or 100% cotton cloth to wipe down the car. Polyester rags can cause scratches, too. To avoid water spots, dry with a chamois or terry cloth towel. And don't forget to clean those door jambs and deck lid lips.

If your Camaro has a weathered finish, it's a good idea to use a heavy-duty car cleaner after washing. This liquid contains hydrocarbon solvents, surface-acting agents and silica, clay or other polishing agents to remove stubborn stains and worn paint. It should be followed by a thorough polishing and/or waxing.

If the water beaded up on your Camaro's finish when you washed it, chances are it does not need to be waxed. But if the water did not bead, the next step is to apply a coat of wax.

So far, we've only talked about washing with water. Car cleaning systems such as *Kozak's Dry Wash* and *Meguiar's Trigger Wash* use no water at all. These are handy for times when no water is available and for quick cleaning at car shows. With *Meguiar's Trigger Wash*, you just spray it on and wipe it off with a clean 100% cotton terry cloth towel. Special lubricants in *Trigger Wash* allow you to wipe off dirt and grit without scratching that $2000 paint job.

Cleaning—We could probably write an entire book on how to clean and detail every nook and cranny of your Camaro, but a few common-sense reminders are in order here. Frequent vacuuming of the seat upholstery and carpeting is a good way to keep dirt from being ground into the fabric, which will cause premature wear. Commercially available vinyl, cloth and carpet cleaners do a good job, but try to avoid cleaners that require getting these items soaking wet. We prefer enzyme-type cleaners that are applied without water, then are vacuumed up along with the dirt when they dry. This helps reduce water staining the carpet backing and helps prevent the onset of mold and mildew. Applying a protective coating of *Scotch-Guard* to cloth upholstery can help avoid the misery of future stains and spills.

Vinyl tops can cause big problems when they crack and allow moisture to get at the steel roof panel, promoting rust and corrosion. If your Camaro has a vinyl top, keep it clean and dressed with a protectant that helps keep the vinyl co-solvents from evaporating.

The clear vinyl rear window on convertibles can be a royal pain to keep clean and easy to see through. With age and exposure to sun, it yellows, then turns brown in blotches and cracks. An otherwise sound convertible top then makes backing up a scary adventure and ruins its appearance. You can extend the convertible-top back window's life by wiping away any dust or debris with a water-dampened cloth then polishing it with optical-grade plastic polish such as *Meguiar's Mirror Glaze #10*.

Tires and wheels can make a big difference in the way your Camaro looks. Camaros have always been designed to show a lot of wheel, with full wheel openings and plenty of room for fat, low-profile tires. Commercially available wheel cleaner does a good job of dissolving road tar, brake dust, bugs, grease and grime. Just spray it on, work it around the wheel slots with a tampico-fiber spoke brush and hose it off. The tires should also be scrubbed with a bristle brush. Whitewalls or raised-white-letter tires can be brought to brilliance with *Westley's Tire Cleaner* or *Brillo* pads. If you're going to show the car, a light coating of *ArmorAll* or other protectant is a popular touch. An old air-cleaner lid held up against the wheel is a good way to avoid getting protectant on the wheel when you spray the tires.

Chrome and stainless-steel trim can be kept shiny and bright with routine treatments of chrome cleaner, followed by a coat of wax. Minor oxidation or corrosion can be remedied by scrubbing with 000 grade extra-fine steel wool and a commercially available cleaner wax.

No matter how clean your Camaro otherwise is, peering out through grimy, smudged windows just makes a car feel dirty. Take the time to clean all windows (and the rear-view mirrors) inside and out with ammonia-based window cleaner or vinegar. Don't use towels that leave a lot of lint. Surprisingly, old newspapers work fine as cleaning towels.

Polishing, Waxing & Buffing —If water beaded up on your Camaro when you washed and cleaned it, the lustre is deep and the paint finish is free of swirl marks, scratches, orange peel, sags, runs and other blemishes, there's no need to polish or wax it now. Hang up the hose, close the garage door and take your Camaro for a ride. On the other hand, if your Camaro's finish could stand a little improving, it's time to do some polishing and buffing.

To remove heavy oxidation, acid-rain streaking, swirl marks, runs, sags, scratches and dirt in the paint, some wet-sanding is in order. Using a hard-rubber sanding block and ultra-fine (up to 2000 grit) wet/dry sandpaper, finish off the paint imperfections. Next, polish off the sanding marks with a polishing pad, then repeat the procedure to restore the original gloss. Note that different pads are required for clear-coats, urethanes and polyurethanes than for enamels, acrylic enamels, lacquers and acrylic lacquers. But remember, wet-sanding is not recommended for metallic paints.

For a normal polish job, all you'll need is some non-abrasive liquid polish and some elbow grease. But more professional results can be had if you machine-apply and machine-buff the polish. Buffers fall into two major categories: orbital and ro-

tary. Orbital buffers are no more than large vibrators and can be operated without special skills. Rotary buffers, however, require a professional technique to avoid burning through the paint. Keep in mind that the newer paints—clear-coats and urethanes—are less tolerant to heat build-up and buffer speed should be kept in the 1200-1750-rpm range. Conventional enamel and lacquer paints can handle higher heat and buffer speeds—up to 3000 rpm—without damage. Wool or foam polishing/buffing pads can be used, but always must be kept parallel to the surface.

When machine buffing or waxing, always cover the edges of character lines and adjoining panels with masking tape to prevent burning through the paint. We like natural carnauba waxes for high-gloss and long-lasting durability, but other formulations work well too. After waxing, clean out wax from around badges and trim with a plastic-bristle paint brush cut off to a 3/4-in. length. Also open the hood, deck lid and doors and wipe off any dried wax build-up along the edges of these panels. A *Q-Tip* works well cleaning out dried wax from the cowl vent louvers.

APPRAISAL & INSURANCE

Many cars ten years old or older that are used merely for transportation aren't worth carrying comprehensive fire, theft and collision insurance on. Liability insurance, yes, because it's the law in most states and you never know when a minor fender bender will land you in court with a major lawsuit brought by an opportunistic "victim" and an ambulance-chasing attorney.

But now, after a full restoration, your used car is a special-interest collector car, worth many times whatever your insurance company says is average resale value. Some insurance companies see a restoration trend here and are now offering comprehensive policies based on declared value. And why not? Most collectors take outstanding care of their cars, driving them perhaps a few dozen times a year and racking up a couple thousand miles on the odometer at the most. The cars are carefully stored, driven conservatively, mechanicals kept in safe condition—in short, these are good risks as any sharp-penciled insurance agent can see.

Beware, though, that some of these policies have numerous restrictions determining when, where and how much you can drive your collector car. Most of these are common sense—don't park the car unattended in a shopping-center, movie or fast-food parking lot, keep the miles down to a few thousand a year, don't drive at night, etc. Just be sure to read the fine print. You wouldn't want to be returning from an annual Camaro club meet, stop at a McDonalds for a bite to eat on the way home, have the car stolen and the insurance company refuse to pay off.

Still, many of these collector-car insurance policies are good, sensible deals that can save you money. Check old-car magazines such as *Hemmings Motor News* and *Cars & Parts* for insurance-company listings. There are others, but a few of the insurance companies with good reputations in this area are:

Jon K. Jacobs of the Appraisal Network, El Toro, CA (714/586-1992) checks out the co-author's 1973 Type LT Z/28. It's a good idea to open the doors, trunk and hood for the appraiser to see inside and have all necessary documentation on hand.

J.C. Taylor
320 S. 69th St.
Upper Darby, PA 19082
(215/748-0567)

American Collector's Insurance
P.O. Box 8343
Cherry Hill, NJ 08034
(609/779-7212)

Condon & Skelly
Suite 203
121 E. King's Highway
Maple Shade, NJ 08052
(609/234-3434)

Also, keep in mind that some major automobile insurance companies also offer policies for your Camaro. And some of these policies are just like those for the family car in that they don't impose any driving restrictions. So give 'em a call.

How do you determine declared value? It really depends on market conditions, car condition, and when all else fails, what a prospective buyer is willing to pay. Here, you'll need the help of a third party— someone without a vested interest in the value of the car, but with the necessary knowledge to determine what it really is worth. The insurance company will tell you to get a written appraisal. Collector-car appraisers aren't exactly jumping out of the Yellow Pages. But the insurance company might suggest one, you could peruse *Hemmings* or *Cars & Parts* classifieds for one, or you might contact one of the following organizations for a referral:

American Society of Appraisers
P.O. Box 17265
Washington, D.C. 20041
(703/478-2228)

International Society of Appraisers
P.O. Box 726
Hoffman Estates, IL 60195
(312/882-0706)

Upon selecting an appraiser, you'll need to have on hand the current registration, repair and service records, any documentation that might verify options and equipment such as a Monronie (original window) sticker, Protect-0-Plate (1967-72 only), broadcast or build sheet and so on. If this is a rare, numbers-match collector Camaro, make sure the appraiser has access to and can see these numbers on the engine, transmission and other parts of the car. Open the hood, doors and trunk— appraisers don't like to take chances on banging or breaking anything. Most appraisers won't take your Camaro for a ride, but they will ask you to start the engine. The car's serial number and odometer reading will be recorded.

For the $75-$150 fee (depending on how much research is required), you'll receive a 10-15 page report in about a week with Polaroids of the front, sides, back, interior, trunk and engine compartment to submit to your insurance company. An appraiser's estimate of your car's value is just that—an estimate of one person. But it's an opinion that your insurance company will likely accept.

LONG TERM STORAGE

We're of the opinion that Camaros are to be driven, not salted away in some warehouse. But the reality is that in Northern climes, it's best to retire your Camaro for the winter so that it will still be around to drive and enjoy in many springs, summers and autumns to come. As mentioned earlier, cars don't like to sit for long periods of time. But when necessary to do so, follow these tips to prevent this period of inactivity from doing any damage.

First and foremost, clean your Camaro from stem to stern and give it a good coat of wax. Clean and vacuum the interior and clean the windows. You'll also want to keep the windows rolled up tight and a car cover draped over its flanks, but first a little preventive medicine.

Support—To prevent spring sag, you'll want to jack your Camaro off the ground and support it on jack stands. But don't let the wheels hang freely as this can damage the springs and promote rusting of the exposed shock- absorber piston shafts. Place a second set of jack stands under the rear axle and lower control arms to raise the wheels to the normal suspension attitude. Then remove the wheels and tires, stacking them on their sides separated by large paper bags or craft paper. Remove the spare tire from the trunk and add it to the stack.

Gasoline—Gas will get "stale" and varnish-like if left in the tank for a year or more. If your Camaro will be tied up for years, drain the fuel tank. For shorter periods, draining isn't necessary, but it's a good idea to top off the tank so there's little air space inside to promote condensation and resulting corrosion. Conversely, carburetors don't like to sit for even a few months with old gas in them. Plug the fuel line leading from the tank to the fuel pump, start the engine and run it until the float bowl runs dry.

Fluids—Other vital fluids, namely the engine coolant and brake master cylinder should be topped off. If you haven't considered this already, substituting DOT 5 silicone-based brake fluid is a good bet for cars that sit in areas with high humidity for extended periods. Conventional glycol-based DOT 4 brake fluid is *hydroscopic*— meaning it absorbs moisture in the air, leading to brake system corrosion. To make the switch, all glycol brake fluid will have to be pumped out and the system flushed with alcohol before the silicone fluid is added because the two are incompatible.

Other Tips—Another way to combat the effects of humidity is to place cookie sheets filled with charcoal briquettes or commercially available dessicant inside the interior, trunk and around the car. Also necessary to combat are vermin such as packrats and mice that might want to take up residence in your immobilized Camaro for the winter. Packrats and mice can ruin wiring harnesses, sound insulation and upholstery in no time at all. Place pans of mothballs in the interior and trunk to keep these little fellas from settling in.

Take out the battery, place it on a wooden block and put it on a trickle charger every couple of weeks for a few hours. This will keep its charge up and help prevent sulfation of the plates.

In humid areas, even the cylinder bores can rust. Remove the spark plugs and with an oil can, squirt a half-teaspoon full of oil or ATF into each cylinder. Then reinstall the spark plugs.

Finally, tuck your Camaro in for a winter's hibernation and drape it with the car cover. But don't cinch the cover down too tightly. Leave enough space so air can circulate underneath.

HAVE FUN!

Here's where the authors of this book give you the "go forth and enjoy" benediction. As corny as it may sound, Camaros were mass-produced cars with sporty looks and fun-to-drive personalities that deserve to be driven—maybe not every day, but once in a while for sure. It really pains us to see some of the best examples of perhaps General Motors, most brilliant design warehoused in some industrial park by some non-enthusiast investor. True, 1967 Z/28s, 1969 ZL-ls and a host of other Camaros don't exactly grow on trees. But these are pricey exceptions to what has always been a relatively inexpensive ponycar that made a working-class performance statement, unhampered by the mechanical complexity and snobbish orthodoxy of many European makes.

So go ahead and take your freshly restored Camaro for a Sunday spin. Join a car club and show off your handiwork. Take part in car shows, picnics, Saturday-night drag racing at the track, cruisin', rallies and how-to tech sessions.

If current trends continue, values for all first-generation and most second-generation Camaros will continue to rise. So as long as you keep your Camaro in good condition, you should have every expectation of getting every dollar and bit of sweat equity out that you put in—and then some.

The neat thing about having a 20-year-old car in pristine shape is the fun you have sharing that enthusiasm with others. So go ahead, indulge yourself and share in the Camaro experience.

CAMARO EXTERIOR COLORS

1967

CODE	EXTERIOR COLOR	DITZLER	DUPONT	R-M
AA	Tuxedo Black	9300	88	A-946
CC	Ermine White	8259	4024L	A-1199
DD	Nantucket Blue	13349	4815L	A-1899
EE	Deepwater Blue	13346	4817L	A-1900
FF	Marina Blue	13364	4850L	A-1920
GG	Granada Gold	22818	4825L	A-1919
HH	Mountain Green	43651	4816L	A-1901
KK	Emerald Turquoise	43661	4818L	A-1903
LL	Tahoe Turquoise	43659	4824L	A-1904G
MM	Royal Plum	50717	4832L	A-1905
NN	Madeira Maroon	50700	4624LH	A-1711M
RR	Bolero Red	71583	4822LH	A-1907R
SS	Sierra Fawn	22813	4826L	A-1908
TT	Capri Cream	81578	4819L	A-1909
YY	Butternut Yellow	81500	4620L	A-1715

1968

CODE	EXTERIOR COLOR	DITZLER	DUPONT	R-M
AA	Tuxedo Black	9300	88	A-946
CC	Ermine White	8259	4024L	A-1199
DD	Grotto Blue	13512	4892L	A-1985
EE	Fathom Blue	13513	4899L	A-1992
FF	Island Teal	13514	4901L	A-1994
GG	Ash Gold	22942	4896L	A-1988
HH	Grecian Green	43775	4902L	A-1995
JJ	Rallye Green	43898	5070L	A-2005D
KK	Tripoli Turquoise	13516	4900L	A-1993
LL	Teal Blue	13516	4893L	A-1986
NN	Cordovan Maroon	50775	4915LH	A-1999M
00	Bronze	43775	4910	A-2010F
PP	Seafrost Green	43774	4897L	A-1989
RR	Matador Red	71634	4948LH	A-1997R
TT	Palomino Ivory	81617	4895L	A-1987
UU	LeMans Blue	2083	4908	A-2007
VV	Sequoia Green	43773	4898L	A-1990
YY	Butternut Yellow	81500	4620L	A-1715
ZZ	British Green	43895	4949	A-2011

1969

CODE	EXTERIOR COLOR	DITZLER	DUPONT	R-M
10	Tuxedo Black	9300	88L	A-946
40	Butternut Yellow	81500	5036L	A-1715
50	Dover White	2058	5033L	A-2080
51	Dusk Blue	2075	5016L	A-2098
52	Garnet Red	2076	5099LH	A-2099R
53	Glacier Blue	2077	5015L	A-2100
55	Azure Turquoise	2078	5014L	A-2101
57	Fathom Green	2079	5013L	A-2102
59	Frost Green	2080	5012L	A-2103
61	Burnished Brown	2081	5011L	A-2104
63	Champagne	22813	5064L	A-2105
65	Olympic Gold	2082	5010L	A-2106D
67	Burgundy	50700	5063LH	A-2107M
69	Cortez Silver	2059	5032L	A-2108
71	LeMans Blue	2083	5030L	A-2109
72	Hugger Orange	2084	5021LM	A-2111R
76	Daytona Yellow	2094	5026LH	A-2119
79	Rallye Green	43898	5070L	A-2005D

1970

CODE	EXTERIOR COLOR	DITZLER	DUPONT	R-M
10	Classic White	8631	5040L	A-1802
14	Cortez Silver	2059	5032L	A-2108
17	Shadow Gray	32604	5113L	A-1910
25	Astro Blue	2165	5123L	A-2261
26	Mulsanne Blue	2213	5190L	A-2262
43	Citrus Green	2334	5126L	A-2266
45	Green Mist	2171	5122L	A-2268
48	Forest Green	2173	5116L	A-2269
51	Daytona Yellow	2094	5026LH	A-2119
53	Camaro Gold	2174	5073L	A-2091F
58	Autumn Gold	2179	5117L	A-2272
63	Desert Sand	2183	5125L	A-2275
65	Hugger Orange	2084	5021LM	A-2111R
67	Classic Copper	23215	5076L	A-2276G
75	Cranberry Red	2189	5118LH	A-2278F

1971

CODE	EXTERIOR COLOR	DITZLER	DUPONT	R-M
11	Antique White	2058	5338L	A-2080
13	Nevada Silver	2327	5276L	A-2438
19	Tuxedo Black	9300	99L	A-946
24	Ascot Blue	2328	5270L	A-2439
26	Mulsanne Blue	2213	5327L	A-2482
42	Cottonwood Green	2333	5274L	A-2444G
43	Lime Green	2334	5322L	A-2445D
49	Antique Green	2337	5273L	A-2448
52	Sunflower Yellow	2338	5283LH	A-2422
53	Placer Gold	2339	5280LH	A-2449F
61	Sandalwood	2181	5325L	A-2273
62	Burnt Orange	2340	528IL	A-2451G
67	Classic Copper	23215	5323L	A-2276G
75	Cranberry Red	2189	5339LH	A-2278F
78	Rosewood Metallic	2350	5275L	A-2461

1972

CODE	EXTERIOR COLOR	DITZLER	DUPONT	R-M
11	Antique White	2058	5338L	A-2080
14	Pewter Silver	2429	5426L	A-2541
24	Ascot Blue	2328	5270L	A-2439
26	Mulsanne Blue	2213	5327L	A-2482
36	Spring Green	2433	5436L	A-2546D
43	Gulf Green	2435	5428L	A-2548
48	Sequoia Green	2439	5429L	A-2552
50	Covert Tan	2441	5431L	A-2554
53	Placer Gold	2339	5280L	A-2449F
56	Cream Yellow	2444	5443L	A-2558G
57	Golden Brown	2445	5439L	A-2559
63	Mohave Gold	2448	5434L	A-2562D
65	Orange Flame	2450	5435L	A-2564D
68	Midnight Bronze	2451	5430L	A-2565
75	Cranberry Red	2189	5339L	A-2278F

1973

CODE	EXTERIOR COLOR	DITZLER	DUPONT	R-M
11	Antique White	2058	5338L	A-2080
24	Light Blue Metallic	2523	5473L	A-2623
26	Dark Blue Metallic	2524	5478L	A-2624
29	Midnight Blue Metallic	2526	5474L	A-2626

CODE	EXTERIOR COLOR	DITZLER	DUPONT	R-M
42	Dark Green Metallic	2528	5489L	A-2628
44	Light Green Metallic	2529	5475L	A-2629
46	Green-Gold Metallic	2530	5479L	A-2631D
48	Midnight Green	2531	5480L	A-2632
51	Light Yellow	2533	5484L	A-2634G
56	Chamois	2537	5481L	A-2638
60	Light Copper Metallic	2538	5490L	A-2639D
64	Silver Metallic	2541	5476L	A-2643
68	Dark Brown Metallic	2543	5483L	A-2647D
74	Dark Red Metallic	2545	5477L	A-2649F
75	Medium Red	2546	5485L	A-2650F
97	Med. Orange Metallic	2555	5552L	A-2659

1974

CODE	EXTERIOR COLOR	DITZLER	DUPONT	R-M
11	Antique White	2058	5338L	A-2080
26	Bright Blue Metallic	2524	5478LH	A-2624
29	Midnight Blue Metallic	2526	5474L	A-2626
36	Aqua Blue Metallic	2640	42805L	A-2702
40	Lime-Yellow	2641	42800LH	A-2703
46	Bright Green Metallic	2643	42806	A-2705
49	Dark Green Metallic	2645	42803L	A-2707
50	Cream Beige	2646	42807L	A-2708
51	Bright Yellow	2677	42809LM	A-2709G
53	Light Gold Metallic	2649	42876LM	A-2710G
55	Sandstone	2650	42808L	A-2711
59	Golden Brown Metallic	2367	42875L	A-2480G
64	Silver Metallic	2541	5476L	A-2643
66	Bronze Metallic	2653	42801LH	A-2714G
74	Medium Red Metallic	2658	42810LM	A-2718F
75	Medium Red	2546	5485LM	A-2650F

1975

CODE	EXTERIOR COLOR	DITZLER	DUPONT	R-M
11	Antique White	2058	5338L	A-2080
13	Silver Metallic	2518	43537L	A-2618
15	Light Graystone	2742	43450L	A-2793
24	Medium Blue	2745	43451L	A-2798
26	Bright Blue Metallic	2746	43452L	A-2799
29	Dark Blue Metallic	2748	43453L	A-2802
44	Medium Green	2642	42802L	A-2704
49	Dark Green Metallic	2752	43454LH	A-2805
50	Cream Beige	2646	42807L	A-2708
51	Bright Yellow	2677	42809L	A-2709G
55	Sandstone	2755	43455L	A-2808
58	Dk. Sandstone Metallic	2757	43461LH	A-2810
63	Light Saddle Metallic	2759	43457LH	A-28112D
64	Persimmon Metallic	2760	43458LM	A-2813F
74	Red Metallic	2658	42810LM	A-2718F
75	Red	2546	5485LM	A-2650F

1976

CODE	EXTERIOR COLOR	DITZLER	DUPONT	R-M
11	Antique White	2058	5338L	A-2080
13	Silver	2518	43537L	A-2618
19	Black	9300	99L	A-946
28	Light Blue Metallic	2772	44141L	A-2801
35	Dark Blue Metallic	2863	44130LH	A-2930D
36	Firethorn Metallic	2811	43953LH	A-2916F
37	Mahogany Metallic	2864	44131LM	A-2931G
45	Lime Metallic	2816	44134LH	A-2934G

CODE	EXTERIOR COLOR	DITZLER	DUPONT	R-M
49	Dark Green Metallic	2752	43454LH	A-2805
50	Cream	2867	44132L	A-2935
51	Bright Yellow	2094	44139LH	A-2936D
65	Buckskin	2829	44159L	A-2939
67	Saddle Brown Metallic	2871	44138L	A-2941
75	Light Red	2546	5485LM	A-2650F
78	Red-Orange	2084	44140LM	A-2942R

1977

CODE	EXTERIOR COLOR	DITZLER	DUPONT	R-M
11	Antique White	2058	5338L	A-2080
13	Silver Metallic	2953	44716L	A-8680
19	Black	9300	99L	A-946
22	Light Blue Metallic	2955	44171L	A-8682
29	Dark Blue Metallic	2959	44718L	A-8685
36	Firethorn Metallic	2811	43953LM	A-2916F
38	Dark Aqua Metallic	2961	44714L	A-8687
44	Med. Green Metallic	2964	44719LH	A-8690
51	Bright Yellow	2094	44139LH	A-2936D
61	Light Buckskin	2869	44713L	A-2999
63	Buckskin Metallic	2970	44722L	A-8695D
69	Brown Metallic	2972	44721LH	A-8698F
75	Light Red	2546	5485LM	A-2650F
78	Orange Metallic	2976	44715LM	A-8702F

1978

CODE	EXTERIOR COLOR	DITZLER	DUPONT	R-M
11	Antique White	2058	5338	A-2080
15	Silver	3076	45177	A-9369
19	Black	9300	99	A-946
22	Light Blue Metallic	2955	44717	A-8682
24	Bright Blue Metallic	3079	45185	A-9372D
34	Yellow-Orange	3070	45277	——
48	Dk. Blue-Green Metallic	2965	44720	A-8691
51	Bright Yellow	3084	45186	A-9377F
63	Camel Metallic	3090	45181	A-9383D
67	Saffron Metallic	3091	45183	A-9384F
69	Dark Camel Metallic	3092	45182	A-9385G
75	Light Red	3095	45187	——
77	Carmine Metallic	3096	45184	A-9389F

1979

CODE	EXTERIOR COLOR	DITZLER	DUPONT	R-M
11	Antique White	2058	5338	A-2080
15	Silver	3076	45177	A-9369
19	Black	9300	99	A-946
22	Light Blue Metallic	2955	44717	A-8682
24	Bright Blue Metallic	3120	45807	A-9801
29	Dark Blue Metallic	3121	45801	A-9802G
40	Light Green	3122	45803	A-9803
44	Med. Green Metallic	3123	45804	A-9805D
51	Bright Yellow	3084	45186	A-9377F
61	Beige	3124	45804	A-9809
63	Camel Metallic	3125	45806	A-9811D
69	Dark Brown Metallic	3126	45805	A-9813G
75	Red	3095	45187	A-9388F
7	Carmine Metallic	3096	45184	A-9389F

1980

CODE	EXTERIOR COLOR	DITZLER	DUPONT	R-M
11	White	2058	5338	A-2080
15	Silver	3076	45177	A-9369
19	Black	9300	99	A-946
24	Bright Blue Metallic	3217	B-8013	A-11402
29	Dark Blue Metallic	3207	B-8007	A-11403D
40	Lime Green Metallic	3218	B-801	A-11404
51	Bright Yellow	3219	B-8619	A-11409
57	Gold Metallic	3215	B-8014	A-11410D
67	Dark Brown Metallic	3226	B-8016	A-11413
72	Red	2973	44770	A-8699F
76	Dark Claret Metallic	3212	B-8010	A-11417F
79	Red Orange	3221	B-8002	A-11419
80	Bronze Metallic	3222	B-8015	A-11420
84	Charcoal Metallic	3223	B-8012	A-11421

1981

CODE	EXTERIOR COLOR	DITZLER	DUPONT	R-M
11	White	2058	5338	2080
16	Silver Metallic	3308	B-8140	12201
19	Black	9300	99	A946/p403
20	Bright Blue Metallic	3309	B-8110	12202F
21	Light Blue Metallic	3310	B-8141	12203
29	Dark Blue Metallic	3207	B-8007	11403D
51	Bright Yellow	3219	B-8019	11409V
54	Gold Metallic	3322	B-8116	12219D
57	Orange Metallic	3329	B-8113	12222V
67	Dark Brown Metallic	3328	B-8l42	12225G
75	Red	3332	B-8115	12229V
77	Maroon Metallic	3333	B-8109	12231V
84	Charcoal Metallic	3223	B-8012	11421

CAMARO INTERIOR COLORS

1967

CODE	INTERIOR COLOR	COMMENTS
707	Yellow	Vinyl Custom
709	Gold	Vinyl Standard
711	Gold	Vinyl Custom
717	Blue	Vinyl Standard
732	Bright Blue	Vinyl Custom
739	Blue	Vinyl Standard & Custom
741	Red	Vinyl Standard
742	Red	Vinyl Custom
756	Black	Vinyl Standard & Custom
760	Black	Vinyl Standard
765	Black	Vinyl Custom
779	Turquoise	Vinyl Custom
796	Green	Vinyl Standard & Custom
797	Parchment w/Black accents	Vinyl Custom

CODE	INTERIOR COLOR	COMMENTS
720	Orange Houndstooth	Cloth Custom
721	Medium Green	Vinyl Standard
722	Medium Green	Vinyl Custom
723	Midnight Green	Vinyl Standard
725	Midnight Green	Vinyl Custom
727	Ivory w/Black accents	Vinyl Standard
729	Ivory Houndstooth	Cloth Custom

1968

CODE	INTERIOR COLOR	COMMENTS
711	Parchment w/Black accents	Vinyl
712	Black	Vinyl
713	Black	Vinyl
714	Black	Vinyl
715	Black	Vinyl
716	Ivory Houndstooth	Cloth
717	Blue	Vinyl
718	Blue	Vinyl
719	Blue	Vinyl
720	Blue	Vinyl
721	Gold	Vinyl
722	Gold	Vinyl
723	Gold	Vinyl
724	Red	Vinyl
725	Red	Vinyl
726	Turquoise	Vinyl
727	Turquoise	Vinyl
749	Black Houndstooth	Cloth

1970

CODE	INTERIOR COLOR	COMMENTS
710	Sandlewood	Vinyl Standard
711	Black	Vinyl Standard
712	Black	Vinyl Custom
713	Black w/White accents	Cloth Custom
714	Black w/Blue accents	Cloth Custom
715	Blue	Vinyl Standard
716	Blue	Vinyl Custom
720	Black w/Dark Green accents	Cloth Custom
723	Dark Green	Vinyl Standard
724	Dark Green	Vinyl Custom
725	Black	Cloth Custom
726	Saddle	Vinyl Standard
727	Saddle	Vinyl Custom
730	Sandlewood	Vinyl Custom

1969

CODE	INTERIOR COLOR	COMMENTS
711	Black	Vinyl Standard
712	Black	Vinyl Custom
713	Black Houndstooth	Cloth Custom
714	Yellow Houndstooth	Cloth Custom
715	Blue	Vinyl Standard
716	Blue	Vinyl Custom
718	Red	Vinyl Standard
719	Red	Vinyl Custom

1971

CODE	INTERIOR COLOR	COMMENTS
775	Black	Vinyl Standard
776	Blue	Vinyl Standard
777	Sandlewood	Vinyl Standard
778	Gold	Vinyl Standard
779	Saddle	Vinyl Standard
785	Black	Cloth Custom
786	Black w/Blue accents	Cloth Custom
787	Black w/Green accents	Cloth Custom
789	Black w/White accents	Cloth Custom
792	Black w/Saddle accents	Cloth Custom

1972

CODE	INTERIOR COLOR	COMMENTS
775	Black	Vinyl Standard
776	Blue	Vinyl Standard
777	Green	Vinyl Standard
778	Tan	Vinyl Standard
779	Covert (light tan)	Vinyl Standard

780	White	Vinyl Standard
785	Black	Cloth Custom
786	Blue w/Black accents	Cloth Custom
787	Green w/Black accents	Cloth Custom
788	Covert (tan) w/Black accents	Cloth Custom

1973

CODE	INTERIOR COLOR	COMMENTS
773	Black	Vinyl Type LT
774	Black w/Blue accents	Cloth Sport Coupe
775	Black	Vinyl Sport Coupe
776	Black w/White accents	Cloth Sport Coupe
777	Green	Vinyl Sport Coupe
778	Saddle	Vinyl Sport Coupe
779	Neutral	Vinyl Sport Coupe
780	Chamois	Vinyl Sport Coupe
781	Green w/Black accents	Cloth Type LT
785	Black w/White accents	Cloth Type LT
786	Black w/Blue accents	Cloth Type LT
788	Neutral	Vinyl Type LT

1974

CODE	INTERIOR COLOR	COMMENTS
775	Black	Cloth Sport Coupe
776	Black	Cloth Type LT
777	Black	Vinyl Sport Coupe
778	Black	Vinyl Type LT
779	Black w/Red accents	Vinyl Sport Coupe
780	Neutral	Cloth Type LT
781	Neutral	Vinyl Sport Coupe
783	Taupe	Cloth Type LT
784	Taupe	Vinyl Type LT
786	Green	Cloth Sport Coupe
787	Green	Vinyl Sport Coupe
792	Red	Vinyl Sport Coupe
796	Saddle	Cloth Sport Coupe
797	Saddle	Cloth Type LT
798	Saddle	Vinyl Sport Coupe
799	Saddle	Vinyl Type LT

1975

CODE	INTERIOR COLOR	COMMENTS
16D	Graystone	Cloth Type LT
19C	Black	Cloth Sport Coupe
19V	Black	Vinyl Sport Coupe
55C	Sandstone	Cloth Sport Coupe
55D	Sandstone	Cloth Type LT
55V	Sandstone	Vinyl Sport Coupe
55W	Sandstone	Vinyl Type LT
63D	Saddle	Cloth Type LT
63V	Saddle	Vinyl Sport Coupe
63W	Saddle	Vinyl Type LT
73D	Dark Red	Cloth Type LT
91V	White w/Black accents	Vinyl Sport Coupe
632	Saddle	Leather Type LT
732	Dark Red	Leather Type LT

1976

CODE	INTERIOR COLOR	COMMENTS
02M	Blue w/White accents	Vinyl Sport Coupe
02N	Blue w/White accents	Knit Vinyl Type LT
03M	Lime w/White accents	Vinyl Sport Coupe
03N	Lime w/White accents	Knit Vinyl Type LT
07M	Firethorn w/White accents	Vinyl Sport Coupe
07N	Firethorn w/White accents	Knit Vinyl Type LT
IIM	White w/Black accents	Vinyl Sport Coupe
IIN	White w/Black accents	Knit Vinyl Type LT
19B	Black	Cloth Sport Coupe
19C	Black	Cloth Type LT
19M	Black	Vinyl Sport Coupe
19N	Black	Knit Vinyl Type LT
26C	Blue	Cloth Type LT
64M	Buckskin w/Saddle accents	Vinyl Sport Coupe
64N	Buckskin w/Saddle accents	Knit Vinyl Type LT
71B	Firethorn	Cloth Sport Coupe
71C	Firethorn	Cloth Type LT
71M	Firethorn	Vinyl Sport Coupe

1977

CODE	INTERIOR COLOR	COMMENTS
02N	Blue w/White accents	Knit Vinyl Type LT
02R	Blue w/White accents	Vinyl Sport Coupe
03N	Green w/White accents	Knit Vinyl Type LT
03R	Green w/White accents	Vinyl Sport Coupe
06N	Buckskin	Knit Vinyl Type LT
06R	Buckskin	Vinyl Sport Coupe
07N	Firethorn w/White accents	Knit Vinyl Type LT
07R	Firethorn	Vinyl Sport Coupe
IIN	White w/Black accents	Knit Vinyl Type LT
IIR	White w/Black accents	Vinyl Sport Coupe
19N	Black	Knit Vinyl Type LT
19R	Black	Vinyl Sport Coupe
24C	Blue	Cloth Type LT
24N	Blue	Knit Vinyl Type LT
62B	Buckskin w/Black accents	Cloth Sport Coupe
62C	Buckskin w/Black accents	Cloth Type LT
62N	Buckskin w/Saddle accents	Knit Vinyl Type LT
62R	Buckskin w/Black accents	Vinyl Sport Coupe
64B	Buckskin w/Saddle accents	Cloth Sport Coupe
64C	Buckskin w/Saddle accents	Cloth Type LT
64N	Buckskin w/Black accents	Knit Vinyl Type LT
64R	Buckskin w/Saddle accents	Vinyl Sport Coupe
71B	Firethorn	Cloth Sport Coupe
71C	Firethorn	Cloth Type LT
71R	Firethorn	Vinyl Sport Coupe
72B	Firethorn w/Black accents	Cloth Sport Coupe
72C	Firethorn w/Black accents	Cloth Type LT
72R	Firethorn w/Black accents	Vinyl Sport Coupe

1978

CODE	INTERIOR COLOR	COMMENTS
11	White w/ Black accents	
17,19	Black	
24	Light Blue	
44	Dark Green	
62	Camel	
74	Carmine	
B,C		Cloth
N		Knit Vinyl
R		Vinyl

1979

CODE	INTERIOR COLOR	COMMENTS
12	Oyster	
19	Black	
24	Light Blue	
44	Dark Green	
62	Camel	
74	Carmine	
B,C		Cloth
N		Knit Vinyl
R		Vinyl

1980

CODE	INTERIOR COLOR	COMMENTS
12	Oyster	
19	Black	
26	Dark Blue	
62	Camel	
74	Carmine	
B,C		Cloth
N		Knit Vinyl
R		Vinyl

1981

CODE	INTERIOR COLOR	COMMENTS
15	Silver w/Gray accents	
19	Black	
26	Dark Blue	
63	Light Beige (Sandstone)	
64	Camel (Doeskin)	
75	Red	
B,C,D,W		Cloth
N		Knit Vinyl
R,V		Vinyl

SUPPLIERS' INDEX

Arizona GM Classics
3214 E. Milber
Tucson, AZ 85714
602/746-9300
(New and used parts, restoration)

Auto Accessories of America
Box 427, Rt. 322
Boalsburg, PA 16827
800/458-3475
(Parts and accessories)

Auto Body Specialties
Rt. 66, P.O. Box 455
Middlefield, CT 06455
203/346-4989
(Sheet metal and interior parts)

Beverly Hills Motoring
Accessories
200 S. Robertson Blvd.
Beverly Hills, CA 90211
800/421-0911, 213/657-4800
(Car covers and accessories)

California Classic Chevy Parts
13545 Sycamore Ave.
San Martin, CA 95046
408/683-2438
(1967-69 parts)

Camaro Classics
815 E. Weber Ave.
Stockton, CA 95202
209/941-2112
(New parts, restoration service)

Camaro Connection
34-B Cleveland Ave.
Bay Shore, NY 11706
800/835-8301, 516/242-9440
(New and used parts and
accessories)

Camaro Country
18591 Centennial Ave.
Marshall, MI 49068
800/533-9981, 616/781-2906
(New and used parts, tech
assistance)

Camaro Junction
P.O. Box 579
Collinsville, VA 24078
800/822-6276, 703/629-6858
(New parts and accessories)

Camaro Specialties
898 E. Fillmore
E. Aurora, NY 14052
716/652-7086
(1967-69 parts and restoration)

Carolina Camaro
Rt.3, Box 24
Elon College, NC 27244
919/584-3076
(1967-69 parts)

C.A.R.S Inc.
1964 W. 11 Mile Rd.
Berkley, MI 48072
800/521-2194, 313/398-7100
(New parts and accessories)

Chevrolet Specialities
4335 S. Highland Ave.
Butler, PA 16001
412/482-2670
(1967-77 parts)

Chevyland
3667 Recycle Rd.
Suite 8
Rancho Cordova, CA 95670
800/624-6490, 916/638-3906
(Parts, accessories and restoration
service)

Chevy World Parts & Accessories
P.O. Box 132
Bryant, AL 35958
205/597-3469
(Parts and accessories)

Chicago Camaro & Firebird
900 S. 5th Ave.
Maywood, IL 60153
312/681-2187
(1967-69 parts)

Classic Camaro Parts &
Accessories
16651 Gemini Lane
Huntington Beach, CA 92647
800/854-1280, 714/894-0651
(Complete line of new parts and
accessories)

Classic Motorbooks
P.O. Box 1
Osceola, WI 54020
800/826-6600
(Auto books and tapes)

Coast Camaro
451 Quebec
Longmont, CO 80501
800/222-5282, 303/678-5636
(1967-69 parts)

Coker Tire
1317 Chestnut St.
Chattanooga, TN 37407
615/265-6368
(Reproduction tires)

Competitive Automotive Inc.
45 Swan St.
Pawtucket, RI 02860
401/723-7225
(Parts)

Dagley's GM Muscle Car Parts
2454 S. 35th Ave.
Phoenix, AZ 85009
602/278-8375
(New and used parts)

D & R Classic Automotive
2101 75th Ave.
Elmwood, IL 60635
708/456-0606
(New and used parts)

GM Muscle Inc.
309 Prospect St.
High Point, NC 27263
800/243-8977, 919/885-1000
(Parts and accessories)

Guldstrand Engineering Inc.
11924 W. Jefferson Blvd.
Culver City, CA 90230
213/391-7108
(Camaro suspension specialists)

H.C. Fastener Company
Box P
Alvarado, TX 76009
817/783-8519
(Tools, fasteners and restoration
supplies)

Hampton's
P.O. Box 234
Downer's Grove, IL 60515
708/963-8347
(Parts)

Harmon Chevrolet
Highway 27 North
Geneva, IN 46740
800/348-4448, 219/368-7221
(OEM and reproduction parts)

Hydro-E-Lectric
48 Appleton Rd.
Auburn, MA 01501
508/832-3081
(Convertible top parts)

ICP Crown Inc.
624 Valley St.
Lewistown, PA 17044
800/288-6874, 717/242-2738
(Reproduction interior kits)

J.C. Whitney
1917-19 Archer Ave.
Chicago, IL 60680
800/541-4716
(Mail order parts and accessories)

Jim Osborn Reproductions Inc.
101-G Ridgecrest Dr.
Lawrenceville, GA 30245
404/962-7556
(Reproduction decals, books,
manuals)

Just Suspension
P.O. Box 167
Towaco, NJ 07082
201/335-0547
(Suspension rebuild kits)

Kanter Auto Products
76 Monroe St.
Boonton, NJ 07005
800/526-1096, 201/334-9575
(Front-end, brake, carpet and
headliner kits)

Kelsey Tire Inc.
Box 564
Camdenton, MO 65020
800/325-0091
(Reproduction tires)

Martz Classic Chevy Parts
RD1, Box 199B
Thomasville, PA 17364
717/225-1655
(1967-69 parts)

Mike Walker's Camaros Unlimited
Rt. 4, Box 47-B
Searcy, AR 72143
501/268-4866
(New and used parts)

National Parts Depot
3101 SW 40th Blvd.
Gainesville, FL 32608
904/378-2473, 800/874-7595
(Florida store)
313/591-1956, 800/521-6104
(Michigan store)
805/654-0468, 800/342-3614
(California store)
(Complete line of new parts and
accessories)

O.B. Smith
P.O. Box 11703
Lexington, KY 40577
606/253-1957
(Parts)

Obsolete Chevrolet Parts Co.
P.O. Box 68
Nashville, GA 31639
912/686-7230
(Parts)

The Paddock
221 W. Main
Knightstown, IN 46148
800/428-4319
(Complete line of new parts and
accessories)

Paddock West
1663 Plum Lane
Redlands, CA 92375
800/854-8532
(Complete line of new parts and
accessories)

Palm Springs Obsolete Automotive
555 N. Commercial Rd. Unit 3
Palm Springs, CA 92262
619/323-6998
(Restoration parts)

Rick's Camaros
120 Commerce Blvd.
Bogart, GA 30622
800/359-7717, 404/546-9217
(New and used 1967-69 parts)

RGR Camaro Specialities
14941 Ramona Blvd.
Baldwin Park, CA 91706
818/814-4779
(New and used parts, restoration
service)

Senior Mechanical Specialists
1426 SE Rural St.
Portland, OR 97202
503/236-9371
(NOS interior fabric)

SoffSeal, Inc.
1048 May Drive
Harrison, OH 45030
513/367-0028
(Reproduction weatherstripping)

Southwestern Classic Chevrolet
1230-B Dan Gould Dr.
Arlington, TX 76017
817/473-6061
(New and used parts, restoration
service)

Ssnake-Oyl Products
15775 N. Hillcrest
Suite 508-541
Dallas, TX 75248
800/284-7777
(Seatbelt restoration, carpet
underlayment, watershields)

Stainless Steel Brakes Corp.
11470 Main Rd.
Clarence, NY 14031
800/448-7722, 716/759-8666
(New and rebuilt brake parts)

Stencils & Stripes Unlimited
1108 S. Crescent Ave.
Park Ridge, IL 60068
708/692-6893
(Tape stripe and stencil kits)

Steve's Camaros
1197 San Mateo Ave.
San Bruno, CA 94066
800/544-4451, 415/873-1890
(Parts and accessories)

The Camaro Place/South Florida
2795 NW 17 Ave.
Miami, FL 33142
305/635-9935
(New parts)

The Carb Shop
2945 Randolph Ave.
Costa Mesa, CA 92626
714/642-8286
(Carburetor
restoration/blueprinting)

The Eastwood Company
580 Lancaster Ave.
P.O. Box 296
Malvern, PA 19355
800/345-1178, 215/640-1450
(Restoration tools, books and
supplies)

The Right Stuff
1369 Community Park Drive
Columbus, OH 43229
614/899-9090
(New brake, fuel and transmission
lines)

Twin City Camaro
2057 Viburnam Trail
Eagan, MN 55122
612/452-3207
(New and used 1967-69 parts)

U.S. Muscle Inc.
150 Airport Drive
Westminster, MD 21157
800/634-3884, 301/876-7857
(New and used 1967-73 parts,
restoration)

Year One Inc.
P.O. Box 450131
Atlanta, GA 30345
404/493-6568
(Complete line of new parts and
accessories)

Z & Z Auto
233 N. Lemon
Orange, CA 92666
714/997-2200
(New and used parts, restoration
service)

HANDBOOKS

Auto Electrical Handbook: 0-89586-238-7

Auto Upholstery & Interiors: 1-55788-265-7

Brake Handbook: 0-89586-232-8

Car Builder's Handbook: 1-55788-278-9

Street Rodder's Handbook: 0-89586-369-3

Turbo Hydra-matic 350 Handbook: 0-89586-051-1

Welder's Handbook: 1-55788-264-9

BODYWORK & PAINTING

Automotive Detailing: 1-55788-288-6

Automotive Paint Handbook: 1-55788-291-6

Fiberglass & Composite Materials: 1-55788-239-8

Metal Fabricator's Handbook: 0-89586-870-9

Paint & Body Handbook: 1-55788-082-4

Sheet Metal Handbook: 0-89586-757-5

INDUCTION

Holley 4150: 0-89586-047-3

Holley Carburetors, Manifolds & Fuel Injection: 1-55788-052-2

Rochester Carburetors: 0-89586-301-4

Turbochargers: 0-89586-135-6

Weber Carburetors: 0-89586-377-4

PERFORMANCE

Aerodynamics For Racing & Performance Cars: 1-55788-267-3

Baja Bugs & Buggies: 0-89586-186-0

Big-Block Chevy Performance: 1-55788-216-9

Big Block Mopar Performance: 1-55788-302-5

Bracket Racing: 1-55788-266-5

Brake Systems: 1-55788-281-9

Camaro Performance: 1-55788-057-3

Chassis Engineering: 1-55788-055-7

Chevrolet Power: 1-55788-087-5

Ford Windsor Small-Block Performance: 1-55788-323-8

Honda/Acura Performance: 1-55788-324-6

High Performance Hardware: 1-55788-304-1

How to Build Tri-Five Chevy Trucks ('55-'57): 1-55788-285-1

How to Hot Rod Big-Block Chevys:0-912656-04-2

How to Hot Rod Small-Block Chevys:0-912656-06-9

How to Hot Rod Small-Block Mopar Engines: 0-89586-479-7

How to Hot Rod VW Engines:0-912656-03-4

How to Make Your Car Handle:0-912656-46-8

John Lingenfelter: Modifying Small-Block Chevy: 1-55788-238-X

Mustang 5.0 Projects: 1-55788-275-4

Mustang Performance ('79–'93): 1-55788-193-6

Mustang Performance 2 ('79–'93): 1-55788-202-9

1001 High Performance Tech Tips: 1-55788-199-5

Performance Ignition Systems: 1-55788-306-8

Performance Wheels & Tires: 1-55788-286-X

Race Car Engineering & Mechanics: 1-55788-064-6

Small-Block Chevy Performance: 1-55788-253-3

ENGINE REBUILDING

Engine Builder's Handbook: 1-55788-245-2

Rebuild Air-Cooled VW Engines: 0-89586-225-5

Rebuild Big-Block Chevy Engines: 0-89586-175-5

Rebuild Big-Block Ford Engines: 0-89586-070-8

Rebuild Big-Block Mopar Engines: 1-55788-190-1

Rebuild Ford V-8 Engines: 0-89586-036-8

Rebuild Small-Block Chevy Engines: 1-55788-029-8

Rebuild Small-Block Ford Engines:0-912656-89-1

Rebuild Small-Block Mopar Engines: 0-89586-128-3

RESTORATION, MAINTENANCE, REPAIR

Camaro Owner's Handbook ('67–'81): 1-55788-301-7

Camaro Restoration Handbook ('67–'81): 0-89586-375-8

Classic Car Restorer's Handbook: 1-55788-194-4

Corvette Weekend Projects ('68–'82): 1-55788-218-5

Mustang Restoration Handbook('64 1/2–'70): 0-89586-402-9

Mustang Weekend Projects ('64–'67): 1-55788-230-4

Mustang Weekend Projects 2 ('68–'70): 1-55788-256-8

Tri-Five Chevy Owner's ('55–'57): 1-55788-285-1

GENERAL REFERENCE

Auto Math:1-55788-020-4

Fabulous Funny Cars: 1-55788-069-7

Guide to GM Muscle Cars: 1-55788-003-4

Stock Cars!: 1-55788-308-4

MARINE

Big-Block Chevy Marine Performance: 1-55788-297-5

HPBOOKS ARE AVAILABLE AT BOOK AND SPECIALTY RETAILERS OR TO ORDER CALL: 1-800-788-6262, ext. 1

HPBooks

A division of Penguin Putnam Inc.

375 Hudson Street

New York, NY 10014